Introduction to the
history of
MYCOLOGY

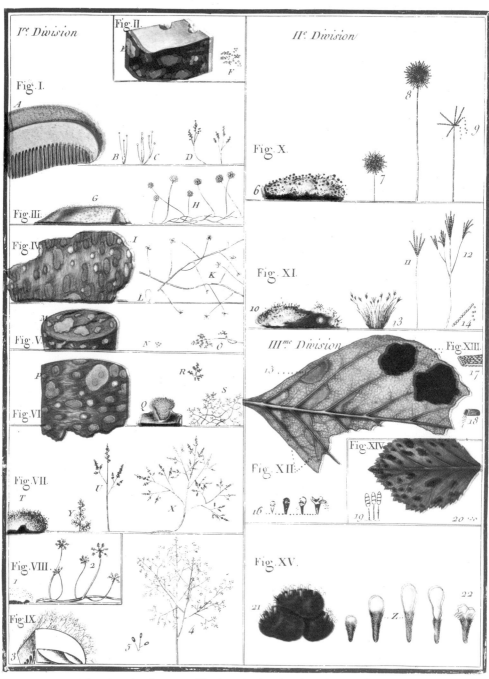

Microfungi. Pierre Bulliard, *Histoire des champignons de France*, 1791: pl. 504.

Frontispiece

Introduction to the
history of
MYCOLOGY

G. C. AINSWORTH
Formerly Director of the
Commonwealth Mycological Institute, Kew

CAMBRIDGE UNIVERSITY PRESS

CAMBRIDGE

LONDON · NEW YORK · MELBOURNE

Published by the Syndics of the Cambridge University Press
The Pitt Building, Trumpington Street, Cambridge CB2 1RP
Bentley House, 200 Euston Road, London NW1 2DB
32 East 57th Street, New York, NY 10022, USA
296 Beaconsfield Parade, Middle Park, Melbourne 3206, Australia

© Cambridge University Press 1976

First published 1976

Printed in Great Britain at the University Printing House, Cambridge
(Euan Phillips, University Printer)

Library of Congress Cataloguing in Publication Data
Ainsworth, Geoffrey Clough, 1905–
   Introduction to the history of mycology.
   Bibliography: p.
   Includes indexes.
   1. Mycology – History. I. Title.
QK603.A53   589.2   75-21036
ISBN 0 521 21013 5

To the memory of
ARTHUR HENRY REGINALD BULLER
(1874–1944)
and
JOHN RAMSBOTTOM
(1885–1974)

Contents

Preface	*page*	vii
Acknowledgements		ix
1	Introduction	1
2	The origin and status of fungi	12
3	Form and structure	35
4	Culture and nutrition	82
5	Sexuality, cytology, and genetics	114
6	Pathogenicity	140
7	Poisonous, hallucinogenic, and allergenic fungi	183
8	Uses of fungi	205
9	Distribution of fungi	225
10	Classification	241
11	Organisation for mycology	270
Epilogue		296
Notes on the text		297
Chronology and Bibliography		308
Names index		340
Subject index		351

Preface

Although much has been written during the past 200 years on the history of mycology in general and of diverse special aspects of the study of fungi this literature is widely scattered and much of it difficult of access. An attempt is here made to bring together for the first time in one volume a documented outline of the development of the main branches of mycology. In other words, an attempt is made to discharge for mycology one duty of a historian which is said to be to record 'significant change'.

With the main exception of the last few decades, which have been treated more lightly, the development of mycology as reflected in the published record has been covered as evenly as possible, and while a knowledge of the essentials of mycology is assumed technicalities have been kept to the minimum in the hope that not only mycologists but other biologists and historians of science in general will find something of interest in the history of the study of an important and fascinating group of organisms.

One approach would have been to treat the whole subject on a strictly chronological basis. The alternative adopted has been to take a number of important themes and to trace each one from early times to the present day – the topics being arranged in roughly the chronology of their appearance in mycological studies. This has resulted in most chapters being more or less self-contained. The duplication implicit in such an approach has been kept to the minimum and it is hoped that the historical structure of mycology will become apparent when the chapters are read consecutively.

The concluding chronology and bibliography of the sources on which these studies are based should emphasise the workers active, and the topics under consideration, at any one time.

G.C.A.

Treligga
Delabole, Cornwall
April 1975

Acknowledgements

I owe much of my interest in the history of mycology to the two mycologists to whose memory this book is dedicated.

Before the First World War Professor A. H. Reginald Buller wrote a history of mycology for which his presidential address to the British Mycological Society in 1914 on 'Fungus lore of the Greeks and Romans ' was intended as the opening chapter. I have had access to the greater part of Buller's unpublished typescript* which has provided many useful clues. Also, as a young mycologist I, on several occasions, submitted myself as the captive audience required by Professor Buller for the compulsive retailing, at great length, of his current mycological interests.

I first met Dr John Ramsbottom forty-three years ago, when attending my first autumn foray of the British Mycological Society; since then his numerous and wide-ranging publications on historical aspects of mycology (particularly his presidential address to the Linnean Society of London in 1939), the discussions we had together, the many kindnesses he showed me, and his interest in this book during his final illness were a constant stimulation.

I am also grateful to the late Professor Raffaele Ciferri – who as director of the Istituto Botanico of the University of Pavia lived in the house which was once Spallanzani's laboratory – for enabling me to visit his Institute and other mycological centres in Italy in 1955, under the auspices of the British Council, when I made several memorable mycological pilgrimages.

The indebtedness I have incurred during the last few years while writing this book has been considerable. The work could not have been done without access to major libraries and I am

* Of which there are copies at the Royal Botanic Gardens, Kew (together with other supporting material), Utrecht University Biohistorical Institute, and the Stevenson Library of the National Fungus Collections, United States Department of Agriculture, Beltsville, Maryland.

particularly grateful for the facilities provided by those of the British Museum, British Museum (Natural History) (especially that of the Department of Botany), Commonwealth Mycological Institute, Kew, Linnean Society of London, Royal Botanic Gardens, Kew, and the Wellcome Institute for the History of Medicine; and I acknowledge the help so freely given by all those in charge of these collections

I am also grateful to a number of antiquarian booksellers (both British and foreign), and particularly to Messrs Wheldon & Wesley Ltd, for helping me to acquire a number of historically useful mycological publications.

I have to thank the many friends – including Colin Booth, Martin B. Ellis, Geoffrey N. Greenalgh, Philip H. Gregory, Stanley H. Hughes, C. Terence Ingold, Joan B. Moore, Alan G. Morton, John A. Stevenson, Alfred S. Sussman, Keisuke Tubaki, and John Webster – who have supplied me with information or offered me advice either in answer to my requests or after reading sections of the manuscript; and also my wife who has read and re-read successive drafts and helped me with the proofs.

Mr W. G. Hoskin, Head of the Photographic Department of the University Library, Exeter, took many of the photographs for the illustrations and Mrs M. G. Samuels prepared the final typescript.

The Wellcome Trust has generously contributed to the cost of the coloured frontispiece.

For permission to use quotations I am indebted to the American Phytopathological Society, Annual Reviews Inc., the Linnean Society of London, Nature, and the Royal Society of Canada and to reproduce illustrations to the following copyright holders or owners of the originals copied:

Academic Press Inc., Fig. 62; Professor V. Amadjian, Fig. 43; British Library Board, Figs. 14, 31c; British Museum, Fig. 78; British Museum (Natural History), Figs. 3, 5, 44, 74, 83, 87; British Mycological Society, frontispiece, Figs. 36b, 102; University of Cambridge (Department of Botany), Figs. 7, 45, 46 (Scientific Periodicals Library), Fig. 50; Centraalbureau voor Schimmelcultures, Figs. 105, 106; Commonwealth Mycological Institute, Figs. 38, 88; Deutsche Gesellschaft für Pilzkunde, Fig. 95; Dr Charles Drechsler, Fig. 89; University of Exeter, Fig. 29; Professor Nils Fries, Fig. 96 Hunt Institute for Botanical

Documentation, Pittsburg, Fig. 22; Istituto Botanico dell'Universitá, Florence, Figs. 31*d*, 34; Dr F. Mariat, Fig. 50; Museo Nazionale, Naples, Fig.1; *Mycologia*, Fig. 77; *Nature*, Fig. 58; Oxford University Press, Figs. 36*a*, 37, 55; Professor G. Percebois, Fig. 90; Royal Botanic Gardens, Kew, Figs. 2, 6, 9–13, 16, 17, 19, 28, 30, 32, 41, 42, 47, 56, 75, 91–2, 94; Professor E. C. Stakman, Fig. 68; Wellcome Institute for the History of Medicine, Figs. 8, 15, 18, 97.

Finally, I must record my thanks to the members of the staff of the Cambridge University Press for the helpful critical interest and meticulous care they have given to the production of this book.

1. Introduction

Fungi, as agents of decay, play an essential role in maintaining the earth's biosphere. They are of ancient lineage and were plentiful during the Devonian period (some 400 million years ago) in association with the first land plants as both saprobes and parasites. Many and diverse fungi have from earliest times been used by man as food and, but for long unknowingly, in the preparation of fermented drinks and leavened bread. The poisonous nature of some was early recognised and among man's oldest written records are references to what can only have been fungal depredations of his crops. In addition, man has for centuries put fungi to medical uses and employed them in ritualistic practices.

Many fungal names in current use also have long ancestries but the origins of most are obscure and the senses of many have been modified by time. In 1847, Charles Badham, elucidating the etymology of 'mycology' and 'fungus', wrote:

By the word Μύκης, ητος, ου, ὁ, whereof the usually received root ‘μῦκος (*mucus*), is probably factitious, the Greeks used familiarly to designate certain, but indefinite species of funguses, which they were in the habit of employing at table. This term, in its origin at once trivial and restricted to a few varieties, has become in our day classical and generic; Mycology, its direct derivative, including in the language of modern botany, several great sections of plants, (many amongst the number of microscopic minuteness,) which have apparently as little to do with the original import of μύκης as smut, bunt, mould, or dry-rot have to do with our table mushrooms. A like indefiniteness formerly characterized the Latin word *fungus*, though it be now used in as catholic a sense as that of μύκης; this, in the classic times of Rome, seems to have been confined (without any precise limitation, however,) to certain sorts which might be eaten, and to others, which it was not safe to eat...Some melancholy etymologists...would beget fungus out of *funus* [a funeral], but Voss judiciously rejects so harsh and forced derivation.

According to *The Oxford English Dictionary* (which offers a derivation of the Latin *fungus* from the Greek *sphonggis*, a sponge; cf. the Italian *spugnola*, morel (*Morchella*)), the use of 'fungus' in English dates from the early sixteenth century but the anglicised plural 'funguses' (or 'fungusses') is now virtually obsolete and the Latin plural 'fungi'[1]* standard practice. 'Mycology' (also 'mycologist') seems to have been first used by the Rev. M. J. Berkeley in 1836 although in 1860 he did employ, if somewhat apologetically, the Latin–Greek hybrid 'fungology' in the title of his well-known book, *Outlines of British fungology.*

The Greeks and Romans had a number of special names for different sorts of fungi.[2] Among those used by Latin writers were: 'agaricum' (*Fomes officinalis*), 'boletus' (*Amanita caesarea*), 'pezica' (*Lycoperdon bovista*), 'suillus' (*Boletus edulis*) and 'tubera' (truffles; the 'misy' (*Terfezia*) and 'ceraunium' of Pliny).[3] All these names have in recent times been taken up, usually in different or restricted senses, as generic names and although none have survived in everyday English usage several have given rise in their generic senses to common names (e.g. 'agaric', 'bolete') much used in technical and semi-popular writing.

The English language is alone (if the Italian *fungo* (pl. *funghi*) be excepted) in applying the international term 'fungi' to fungi in general. In most languages one of the words originally applied to some fungi has been adopted for the whole group. For example, the German *Pilz* (a corruption of 'boletus') is now applied to any fungus although it was originally restricted to fleshy fungi in distinction to *Schwamm* (a polypore; also a sponge) and *Schimmel* (mould). Similarly the French *champignon*, of which the only surviving use in English is for *Marasmius oreades*, the fairy ring champignon. Mushroom is also of French extraction (*mousseron* (old French *moisseron*), *Clitopilus prunulus*; cf. *moisissure*, mould) while the superficial derivation of 'toadstool'[4] is as apparent as the reason for the association of toads with agarics is obscure.

The sudden development and frequently strange appearance of the fruit-bodies of larger fungi and moulds have always attracted attention. As a result there is a large non-scientific literature on fungi. Fungi are the subject of much folk-lore and they are mentioned, frequently metaphorically, in poetry, drama, and prose from the time of the classical writers of Greece and Rome, through Shakespeare, to the authors of detective novels and science fiction of today;[5] while the scenario of at least one film, Sacha Guitry's *Le Tricheur*, has a dramatic mycological opening. Recently, a ballet group, originating from Dartmouth

* See Notes on the Text, pp. 297–307.

College, New Hampshire, has been performing as the Pilobolus
Dance Theatre.

References to fungi by the poets are rarely complimentary.
William Browne (*fl.* 1591–1643) provides a typical example.

> Downe in a vallye, by a Forest side,
> Neere where the christall Thames roules on her waves,
> I saw a Mushrome stand in haughty pride,
> As if the Lillyes grew to be his slaves;
> The gentle daiseye, with her silver croune,
> Worne in the brest of many a shepheards lasse;
> The humble violett, that lowly downe,
> Salutes the gaye Nimphes as they trimly passe:
> Those, with many more, me thoughte complaind
> That Nature should those needless things produce,
> Which not alone the Sun from others gain'd,
> But turne it wholy to their proper use:
> I could not chuse but grieve, that Nature made
> So glorious flowers to live in such a shade.

Fig. 1. Fresco of fungi from Pompeii. (Museo Nazionale, Naples, 8647.)

Artists, too, have made many references to fungi from the naturalistic representations of edible fungi in Roman frescoes at Pompeii (Fig. 1; see also p. 35) and of russulas and other toadstools by the Dutch painter Otto Marseus van Schrieck in the seventeenth century to the emotive use of agarics by Paul Nash in the twentieth.[6]

Mycology, the scientific study of fungi, is of relatively recent origin because the proper elucidation of the structure and nature of fungi had to await the development of the microscope in the seventeenth century and the introduction of pure culture techniques in the second half of the nineteenth century. If, however, a date must be set for the birth of mycology it is 1729, the year of publication in Florence of Pier' Antonio Micheli's *Nova plantarum genera*, a germinal work much in advance of its time. It may also be noted that professionalism in mycology, as in other branches of biology, is characteristically a modern development and in the past the accumulation of knowledge on fungi owed much to the painstaking work of amateurs (particularly those engaged in the professions of medicine and the church) who are still able to play a useful, if now minor, role in mycological endeavour.

THE PATTERN OF MYCOLOGICAL DEVELOPMENT

In the chapters which follow an attempt is made to outline the history of the major branches of mycology, so to conclude this introduction a brief sketch of the overall pattern of the development of mycology may be helpful as a background. The main line of mycological endeavour has been in the broadest sense taxonomic, and taxonomy (Chapter 11), if less popular than it was, is still the most worked on aspect of mycology; its literature currently accounts for approximately one-quarter of the total. This trend is likely to continue because the primary census of fungi is so very far from complete. Although references to fungi go back for some three millenia, the few fungi mentioned in the first printed herbals (published towards the end of the fifteenth century) were scattered at random through the volumes and it was not until the late sixteenth century that l'Obel grouped agarics, polypores, and truffles together, an innovation accepted by subsequent authors. During the seventeenth century the number of larger fungi described rapidly increased and a climax

4

was reached in the compilations of Jean Bauhin and John Ray while in the closing years of the century the fungal nature of lichens was first suspected. Concurrently, the invention of the compound microscope during the last decade of the sixteenth century[7] enabled the seventeenth-century microscopists Hooke, Leeuwenhoek, and Malpighi to give the first descriptions of some common microfungi. From then, accounts of hitherto undescribed fungi increased exponentially until the last few decades when, in round numbers, the annual output has levelled off at approximately 100 new genera and more than 1000 new species.

Up to the middle of the nineteenth century, mycological studies were bedevilled by erroneous views on the origin of fungi, views which frequently closely resembled speculations made by the Greeks and Romans 2000 years before. Fungal spores were first observed by della Porta in 1588 but for the next 200 years, in spite of the convincing experiments on the growth of fungi from spores reported by Micheli in 1729, it was widely held that fungi originated by spontaneous generation or, if parasitic, by heterogenesis from the diseased tissues of the host. Such views were finally silenced by de Bary and his contemporaries operating in the climate of opinion created by the classical work of Spallanzani, Pasteur, and Tyndall and the acceptance of the reality of pathogenicity. Uncertainty as to the origin of fungi also affected speculation as to their status – whether they were animal, vegetable, or even mineral in nature – and although it became traditional to classify fungi as plants this convention, as a result of ideas tentatively advanced a century ago, has, during the last twenty-five years, been rejected by the increasing number of workers who would treat fungi as a kingdom distinct from both animals and plants. Another difficulty resulted from the plasticity of fungal form and the recognition that one species may exhibit several different types of spores (see p. 26). This led uncritical nineteenth-century practitioners of 'pure culture' techniques to believe they had established the conspecificity of many common moulds and even moulds and bacteria, beliefs again finally dispelled by de Bary. Sexuality, a topic first seriously investigated by Micheli, was another major problem which also took 200 years to solve, mainly because evidence was first looked for in agarics and other large basidiomycetes. A century of error preceded Ehrenberg's demonstration in 1820 of zygospore formation in *Syzygites megalocarpus*. By the mid eighteen-hundreds sexuality in ascomycetes had been established but the elucidation of the sexual mechanisms in hymenomycetes was a twentieth-century development stemming from the discovery of heterothallism by Blakeslee in 1904. The experimental study of nutrition has been another prominent theme during the past hundred years. It began in 1869 with Raulin's classic publication in which the basic nutrient requirements of *Aspergillus niger* were characterised and the need for trace elements revealed. At the turn of the century growth factor requirements were discovered and activity in this field appears to have peaked between the two world wars. Nutrition has now, as suggested to me by Professor A. S. Sussman, become largely a 'service field' for other types of mycological experimentation. Currently, most branches of mycological research are rapidly becoming more and more physiological and biochemical.

The development of mycology has been influenced by several groups of factors. Perhaps the most important group is related to the effects of fungi on man and his affairs. Presumably one of man's first interests in fungi was as food and he was early impressed that some fungi are poisonous (Chapter 7). This is reflected by the first taxonomic division of fungi into edible and poisonous categories. The proof that fungi cause disease in both animals and man (Chapter 6) led, during the first half of the nineteenth century, to a surge of medical interest in the potentialities of fungi as agents of disease but this interest was virtually lost during the rest of the century after the discovery of the dominant role played by bacteria, and later by viruses, in the aetiology of human disease. At the turn of the century, associated with the researches of Sabouraud, there was a revival of interest in ringworm and its causal fungi but medical and veterinary mycology only came into its own after the Second World War. Since the recognition, during the opening years of the nineteenth century, of the importance of fungi in the decay of structural timber, particularly of ships of the navy, and subsequently, as a result of the economic and social havoc wrought by potato blight in Ireland in the eighteen-forties, of their importance as agents of plant disease, the study of plant pathogenic fungi (Chapter 6) has made a major and continuing contribution to mycology in general. This is in part because most students of fungi have had a botanical training and in part because fungi are, both numerically and economically, the most important agents of disease in plants.

A second group of factors concerns the effect on the study of mycology of developments in other branches of science. The elucidation of sexuality in flowering plants hindered rather than helped that of sex in fungi but the announcement of the 'cell theory' in 1838–9, the general acceptance of evolution after 1859, the rediscovery of Mendelism and the birth of modern genetics at the turn of the century, and the breaking of the genetic code in 1953, all directly influenced mycological studies with outstandingly fruitful results. Equally important was the rise of organic chemistry during the nineteenth century and of its daughter-branch, biochemistry, which had a major mycological root in the discovery by Buchner, in 1896, of zymase and its role in the alcoholic fermentation of sugar by yeast. Further, the rise of immunology as a branch of medical science has found direct application in the diagnosis and therapy of mycotic disease in man and has been used to assess affinities within the fungi, as has numerical taxonomy made possible by computerisation.

The main element in the third group of environmental factors concerns the financial support for mycological studies, an aspect of the organisation for mycology considered in the final chapter (Chapter 11). As already noted, early students of fungi were mostly amateur naturalists, particularly doctors and clergymen, but a few were 'professional' (as for example Micheli who was keeper of the public gardens at Florence), but, like Robert Hooke (a paid servant of the Royal Society), not always biologists. Many more were university teachers and, since the eighteenth century, universities have been the main funding agency for mycological research. But this support was for long indirect because the many mycological contributions made by university professors and lecturers have resulted from personal interests and not from any direction imposed by conditions of employment. Only in recent years have lecturers in mycology been specifically appointed and, but more rarely, professorial chairs of mycology instituted. Chairs and departments of plant pathology have, on the other hand, been more frequent following a lead given in 1883 by the appointment by the Royal Veterinary and Agriculture College, Copenhagen, of a professor of plant pathology. State support for mycology was, like that by universities, at first indirect – for example, the members of the Paris Academy of Science, a few of whom had mycological interests, were pensioned so that they could devote their energies to scientific investigation – or via ex-gratia payments such as that given to James Sowerby for his advice on the prevention of decay in ships' timber (see p. 90). It was government departments of agriculture which first gave major support to mycology as part of the research and advisory work they sponsored on pests and diseases of economic crops. (In passing it is perhaps of interest to note that man's concern for his own health and that of his animals and plants is a descending series; private medical practice has always been lucrative, plant pathology always a public service.) Stimulated by the researches of Pasteur, the brewing industry began to employ mycologists, and the founding of the Carlsberg Laboratory in Copenhagen in 1876[8] was a major mycological and microbiological event. Subsequently, mycological research was undertaken in connection with the citric acid and other fermentation industries and after the introduction of penicillin therapy vast sums of money were expended by major pharmaceutical firms in efforts to discover and develop other fungal antibiotics. Mycology has never attracted the large financial support so characteristic of such disciplines as astronomy and atomic physics or research on cancer but in most countries some public money is now available for the support of at least the more fashionable aspects of mycology, as are grants from many national and international philanthropic funds and foundations.

Finally, attention may be drawn to some characteristics of mycologists. Up to the middle of the last century their distribution was mainly confined to Europe. The van of mycological advance then began to shift from France and Germany to the United States and the geographical scatter of mycologists is now world wide, with the greatest numbers in North America and Europe and notable concentrations in India and Japan.

The first European mycologists wrote in Latin (see p. 271) and frequently latinised their names, which can create a certain confusion today. (In this history the selection of names for the early mycological writers, though consistent for any one author, is not uniform. Sometimes the Latin name is used (e.g. Clusius not l'Éscluse), at others the vernacular (e.g. Bauhin not Bauhinus) in an attempt to follow common usage and avoid ambiguity.)

8

Until the First World War most mycologists worked and published as individuals – professors were still bench workers. Then with the growth in size of university departments and increasing numbers of research students more and more of an eminent teacher's time was absorbed in administration, travel, and in the supervision of his research school. One result of these pressures is papers published under two or more names. Concurrently the increasing complexities of research work frequently require the involvement of specialists from other fields which reinforces the development of research teams whose members share the credit for any publications. This trend is well brought out by the random sampling of standard bibliographies. From Lindau & Sydow's *Thesaurus* it appears that before 1910 the average number of authors per mycological paper was of the order of 1.05. By 1950 this figure had increased for taxonomists, the most individualistic mycologists, to 1.25 and in 1974 to 1.50; the corresponding values for plant pathologists being 1.43 (1950) and 1.9 (1974), while for medical mycology, where mycologists frequently have to collaborate with pathologists and clinicians, the 1974 figure approaches 2.5.

Because of this increased collaboration between authors and the increasing specialisation of their studies most research workers today, even the most eminent, tend to make smaller and more technical contributions to the advance of mycology. It thus becomes increasingly difficult for a contemporary reviewer to make generalisations and to allocate credit to individuals. This is one reason why the final sections of this open-ended history are treated more unevenly and lightly than are developments before 1950.

The activities of mycologists are much influenced by current fashion. For the past 250 years most of the publications of any one period contribute little to 'significant change' in mycological knowledge, they merely 'fill in', or exploit a newly introduced concept or technique and so are not retained by the sieve of history. To cite a rather extreme example, in 1958 it was discovered that a daily, oral, one-gram dose of the fungal antibiotic griseofulvin eliminated ringworm infection of the hair or skin of man and during the next few years hundreds of papers confirming this fact were published in medical and microbiological journals in all parts of the world and several national or international symposia were held – no doubt to the manufacturer's great satisfaction even if the advance in knowledge was minimal. More recently the introduction of 'small particle' griseofulvin has given rise to many additional reports that a smaller dose is equally effective. Concurrently, in the same field, the introduction of the hair-baiting technique for the isolation of dermatophytes or dermatophyte-like fungi from soil led to a large volume of publication, much of it repetitious. Attention has frequently been drawn to this aspect of scientific research which must not be decried for it is an essential feature of the scientific method that results claimed should be tested by independent experimentation, as is the workmanlike 'filling in' whereby significant advances are consolidated and additional data accumulated. Most of this work is, however, outside the scope of the present history in which emphasis is placed on the solution of major problems which have confronted students of fungi over the centuries and novel discoveries which have given new insights.

It is difficult to plan for the discovery of novelty (which can rarely be bought) but one undoubtedly favourable factor is isolation, and frequently a confirmatory test that a discovery is novel is its rejection by the majority from whose orthodoxy the discoverer has escaped. Among the classic examples of biological discoveries made in isolation is that of the concept of the origin of species by survival of the fittest which occurred to Charles Darwin from his experience as the solitary biologist on H.M.S. Beagle in the Eastern Hemisphere, and twenty years later independently to Alfred Russel Wallace when alone in the tropics of the Old World, while one of the understatements of all time is the remark of the president of the Linnean Society of London when reviewing the 1858–9 session during which the Darwin–Wallace papers were read that 'The year which has passed...has not, indeed, been marked by any of those striking discoveries which at once revolutionise, so to speak, the department of science on which they bear.' Again, at the end of the eighteenth century, Adanson, when alone in Senegal, realised the importance of the correlations between facts and, to the bewilderment of his contemporaries, propounded, as only recently appreciated, the essentials of numerical taxonomy.[9] In the mycological field the self-taught Micheli, though he became well known to the leading biologists of the time, worked outside the famous Italian universities and his now classical spore

germination experiments were received with controversy, while
his description of some nine hundred different fungi earned
him the displeasure of Linnaeus for unnecessarily distinguishing so many. Proof of the pathogenicity of fungi for plants by
Prévost in France in 1807 and for animals by Bassi in Italy shortly
afterwards were both made by isolated amateurs. Many more
examples could be cited and examples still occur. For instance,
the axenic culture of rusts was first announced not from one of
the major centres for the study of cereal rusts but by Cutter from
the Woman's College of the University of North Carolina (and
on his untimely death in 1962 the writer of the obituary notice in
Mycologia tactfully made no reference to this work). A decade
later, a 'letter' from a group of workers in Sydney, Australia,
reporting their successful axenic culture of *Puccinia graminis* was
declined for publication by the editor of *Nature*. Isolation may
affect whole countries. The Soviet Union is isolated, both
politically and by language, and the many papers and monographs on mycotoxicoses of farm animals were virtually ignored
elsewhere until the importance of these disorders was recognised in the West where their investigation is now high fashion.
Isolation can also be brought about in other ways – by illness,
disinclination to travel and to attend scientific gatherings, or by
being a specialist in some other branch of biology as is instanced
by the discovery of penicillin by a bacteriologist and of parasexuality by a geneticist who was experimenting with fungi for his
own ends.

On the other hand, it is probably true that a certain amount of
overcrowding (which encourages communication) of research
workers engaged on applied research and on post-graduate
studies in universities, and not too lavish facilities, favour
productivity although this may be in part because of the
pressures created by the attraction of workers to any centre
where stimulating work is being done under inspired direction.

2. The origin and status of fungi

The origin and status of fungi have since ancient times been subject to much speculation. Pliny (A.D. 23–79) marvelled at truffles. In his *Naturalis historia* (Book xix, sect. 11) he wrote:

Among the most wonderful of all things is the fact that anything can spring up and live without a root. These are called truffles (*tubera*); they are surrounded on all sides by earth, and are supported by no fibres or hair-like root-threads (*capellamentis*); nor does the place in which they are produced swell out into any protuberance or present any fissure; they do not adhere to the earth; they are surrounded by a bark, so that one cannot say they are altogether composed of earth, but they are of a kind of earthy concretion; they generally grow in dry sandy places which are overgrown by shrubs; in size they are often as large as quinces and weigh as much as a pound. There are two kinds: one is sandy and injures the teeth, the other without any foreign matter (*sincera*); they are distinguished by their colours being red, or black, or white within; those of Africa are most esteemed. Now whether this imperfection of the earth (*vitium terrae*) – for it cannot be said to be anything else – grows, or whether it has at once assumed its full globular size, whether it lives or not, are questions which I think cannot easily be explained. In their being liable to become rotten these things resemble wood.[1]

A few pages later (*loc. cit.*, Book xix, sect. 13) he continued:

peculiar beliefs are held for they say that they are produced during autumn rains, and thunderstorms especially, which are the main reason of their growing, and that they do not last more than a year, and are best for food in the spring. Some think that they are produced from seed, because those which grow on the shore of the Mityleneans only appear after floods, which bring down seed from Tiara where many truffles are found. They grow on the shore where there is much sand.[2]

Truffles are not infrequently associated with thunder by other classical authors. Juvenal (A.D. 60–140) (*Satires*, v, 116–19) writing in praise of truffles associates their origin with thunder[3] while Plutarch (A.D. 46–120) includes in his *Symposiacs* (Book iv, question 2) a long dissertation on 'Why truffles are thought to be produced by thunder' where it is argued that, since during thunderstorms flame comes from moist vapours and deafening noises from soft clouds, there need be no surprise that when lightning strikes the ground, truffles, which do not resemble plants, should spring into existence.[4] According to Lowy (1974) there is a tradition in the Guatemalan highlands and southern Mexico which links *Amanita muscaria* (the fly agaric) with the thunderbolt, and Wasson (1968) records that according to the Hindu Rig Veda 'Parjana, the god of thunder, was the father of Soma' which he identifies as *A. muscaria* (see p. 193).

The association of the origin of fungi with the soil can be traced back to the Greek physician, grammarian, and poet Nicander (*c.* 185 B.C.) who in his *Alexipharmaca* referred to 'the evil ferment of the earth' which 'men generally call...by the name of fungus'.[5] Finally, it may be recalled that Ovid (43 B.C.–A.D. 17) (*Metamorphosis*, vii, 392-3), on an even more speculative note, stated that in Corinth 'the ancients record that in the first age of the world mortal bodies were produced from fungi which sprang up after rains'.[6]

During the Middle Ages little was added to the knowledge of fungi and at the Renaissance one of the first uses of the printing press was to give wider currency to ideas derived from the classical authors. The German herbalist Jerome Bock (Hieronymus Tragus) (1552:942) wrote:

Fungi and Truffles are neither herbs, nor roots, nor flowers, nor seeds, but merely the superfluous moisture of earth, of trees, or rotten wood, and of other rotting things. This is plain from the fact that all fungi and truffles, especially those that are used for eating, grow most commonly in thundery and wet weather, as the poet Aquinum says: *Et facient lautas optata tonitrua coenas.*[7]

Thirty years later the Italian Andrea Cesalpino in his famous *De plantis libri xvi*, 1583 (Lib. I, cap. xiv: 28) stated:

Some plants have no seed; these are the most imperfect, and spring from decaying substances; and they therefore only have to feed themselves and grow, and are unable to produce their like; they are a sort of intermediate existences between plants and inanimate nature. In this respect fungi resemble Zoophytes which are intermediate between plants and animals, and of the same nature are the Lemnae, Lichens, and many plants which grow in the sea.[8]

The year 1588 saw a landmark in the development of mycology: the publication of the first observation of fungal spores, by Giambattista della Porta (Fig. 2) who was both precocious and versatile. In addition to voluminous publications on many aspects of natural philosophy, cryptograms and mnemonics, he composed Italian comedies, re-invented the camera obscura, and founded in Naples the Accademia Secretorum Naturae (the 'Accademia dei Oziosi') and, in the opinion of George Sarton, 'might have done better work had he

Fig. 2. Giambattista della Porta (*c.* 1538–1615), aet. 50. (G. della Porta, 1588.)

not been sidetracked by the massiveness of his erudition (much of it unsound) and by an unwieldy imagination'.[9] In his *Phytognomonica* (p. 240) he wrote:

From fungi I have succeeded in collecting seed, very small and black, lying hidden in oblong chambers or furrows extending from the stalk to the circumference, and chiefly from those which grow on stones, where, when falling, the seed is sown and sprouts with perennial fertility. Falsely therefore has Porphyrius said that fungi, since they do not arise from seed, are children of the Gods. So also in truffles, a black seed lies hidden. On this account, they come forth in woods where they have frequently been produced and have rotted away. And I have often also seen them [*presumably 'flowers of tan'*, *the myxomycete* Fuligo septica] arising where the washings of tan or the tan itself is thrown away.[10]

This fundamental observation did not, however, greatly influence the writings of either della Porta himself (see p. 82) or of his contemporaries. Certainly, J. Goedart in his *Metamorphosis et historia naturalis insectorum*, 1662, (which includes the earliest descriptions known to me of mushroom flies – Mycetophilidae) described, in the section on spiders, *Cyathus*,[11] one of the bird's nest fungi, and took the peridiola to be the seeds[12] but John Ray, in 1704, while agreeing with Goedart's interpretation added dryly '*Una Hirundo, ut aiunt, non facit Ver*' (Ray (1686–1704) **3**: 17). Gaspard Bauhin in his *Pinax* (1623; and also in the second edition of 1671), considered fungi to be 'nothing but the superfluous humidity of soil, trees, rotten wood and other decaying substances' and even the perceptive Robert Hooke held a similar view. In *Micrographia*, 1665, he recognised the affinity of mould and mushrooms:

The Blue and White and several kinds of hairy mouldy spots, which are observable upon divers kinds of *putrify'd* bodies, whether Animal substances, or Vegetable, such as the skin, raw or dress'd, flesh, bloud, humours, milk, green Cheese &c. or rotten sappy Wood, or Herbs, Leaves, Barks, Roots, &c. of Plants, are all of them nothing else but several kinds of small and variously figur'd Mushroms, which, from convenient materials in those *putrifying* bodies, are, by the concurrent heat of the Air, excited to a certain kind of vegetation [Observ. xx]

and concluded (p. 127)

First, that Mould and Mushroms require no seminal property, but the former may be produc'd at any time from any kind of *putrifying* Animal or Vegetable Substance, as Flesh, &c. kept moist and warm...

Next, that as Mushroms may be generated without seed, so it does not appear that they have any such thing as seed in any part of them; for having considered several kinds of them, I could never find anything in them that I could with any probability ghess [*sic*] to be the seed of it, so that it does not yet appear (that I know of) that Mushroms may be generated from a seed, but rather they seem to depend upon a convenient constitution of the matter out of which they are made, and a concurrence of either natural or artificial heat.

A few pages later (pp. 130–1) he restated the same view more metaphorically:

I must conclude, that as far as I have been able to look into the nature of this Primary kind of life and vegetation, I cannot find the least probable argument to perswade me there is any other concurrent course then such as is purely Mechanical, and that the effects or productions are as necessary upon the concurrence of those causes as that a Ship, when the sails are hoist up, and the Rudder set to such a position, should, when the wind blows, be mov'd in such a way or course to that or t'other place.

A decade later Malpighi, in the section of his *Anatome plantarum*, 1679, dealing with 'De plantis quae in aliis vegetant' wrote: 'either Fungi, Mucedo and Moss have their own seed, by which their species is propagated, or they sprout from the growth of fragments of themselves as happens in other plants'; but it was nearly two hundred years before these speculations became established as orthodoxy.

ORIGIN: 1700–1850

Although from the beginning of the eighteenth century there were, at every period, those who held correct views on the origin

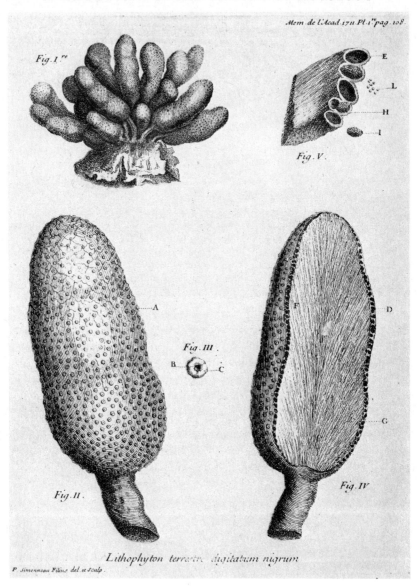

Mem. de l'Acad. 1711 Pl. 1ère pag. 108

Fig. 3. *Xylaria polymorpha*. (J. Marchant *fils* (1711): pl. 1.)

of fungi, this question, by becoming involved with that of spontaneous generation and with the development and elucidation of the concepts of contagion (see Chapter 6), putrefaction, and fermentation (see Chapter 8), gave rise to a wide diversity of opinion. The situation was further complicated by being overlaid by much, frequently wild and erroneous, speculation on the status of fungi, particularly supposed relationships with protozoa and insects. In the account which follows the separation of the various threads (for the convenience of exposition) and limiting the illustrations to representative examples (in the interests of readability) inevitably falsifies by oversimplification the prevailing climate of opinion and tends to minimise the difficulties that even a critical observer must have had in distinguishing between fact and fiction. Supplementary details relating to this period may be disentangled from the writings of Lütjeharms (1936) and Ramsbottom (1941*a*).

In 1707 J. P. de Tournefort in his monograph on mushroom culture (see p. 83) was correct in his speculation on the origin of mycelium when he wrote:

> According to appearance these white threads are none other than the developed seeds, or germs, of mushrooms, and all these germs were enclosed in the horse droppings in so small a volume, that one can only perceive them whatever care one takes, after they have grown out into little hairs.

Four years later, in 1711, Marchant *fils* observed and illustrated the spores (grains 'having some resemblance to those of vanilla, but infinitely smaller, and less glistening') of *Xylaria polymorpha* (Fig. 3) which he carefully described and concluded to be 'a true terrestrial lithophyte, and not a fungus'.

Earlier views on the origin of fungi were still, however, prevalent. Particularly that elaborated by Luigi Fernando Marsigili (who after retiring from a military career had, about 1710, founded the Institute of Bologna for the advancement of science) and set out at length in *Dissertio de generatio fungorum*, 1714. This was supplemented by a commentary by the famous epidemiologist and physician to several popes, Giovanni Maria Lancisi (to whom the dissertation is dedicated), who supported Marsigli's view that fungi are a product of corruption, mycelium being the intermediate state between corruption and the fungal fruit-body (cf. p. 65).

Similar views were held by both the English clergyman, Stephen Hales (1727) (see p. 144) and by Dillenius (1719: appendix pp. 71–2):

> A fungus is a sterile kind of plant, that is to say destitute of flower and seed, arising from putrefactive fermentation (wherefore they arise chiefly during a moist and rainy period and consist for the most part of a soft and spongy substance), yet retaining its characteristic look which it owes to a definite and specific juice of decay from which it originated.

The second major landmark in the development of mycology was the work of the Florentine Pier' Antonio Micheli (see p. 50) who not only observed spores in many groups of fungi but by cultural experiments (see p. 84) showed that under appropriate conditions spores could be induced to give rise to the same species of fungi from which they originated, studies which he summarised in *Nova plantarum genera*, 1729. Micheli thus offered a prima-facie case that fungi were autonomous organisms but this view though accepted by some was questioned by many for the next hundred years. The reasons for this were various. Those who attempted to repeat Micheli's spore germination experiments were not always equally successful; for example, Gleditsch (1740) and Mazzuoli (1743) confirmed Micheli's findings, as did Seyffert (1744); Monti (1755) did not and concluded that moulds develop on vegetable substances whether inoculated with spores or not and even after being boiled. Many hypotheses on the origin of fungi (which are listed chronologically by Blottner, 1797; Nees von Esenbeck, 1816; and Ehrenberg, 1820a), even if erroneous, gained wide currency when backed by an accepted authority. Apart from such off-beat speculations that fungi were crystals condensed from the decaying mucous of leaves, as advocated by C. Medicus (director of the botanic garden at Manheim) in 1788,[13] or the even more fantastic claim (cited by Ehrenberg, 1820a) of Frenzel who, as late as 1804, believed that fungi resulted from shooting stars (*Sternschnuppen*), the heresies were mainly associated with ideas on spontaneous generation or heterogenesis.

The idea of spontaneous generation may be traced back to Aristotle and J. B. von Helmont is notorious for still believing, at the end of the sixteenth century, that mice could arise spontaneously in heaps of corn or rags. In 1688 Francisco Redi published the results of his now classical experiments proving that maggots in decaying flesh developed from eggs deposited by flies and by the mid-eighteenth century the problem of spontaneous generation had been narrowed down to the origin of micro-organisms (or microbes). The familiar story of the solution of this problem, which inadequate aseptic experimental procedures made so intractable, by the investigations of Spallanzani in the seventeen-sixties to those of Pasteur and Tyndall a hundred years later which clinched the matter, has often been told – never better than by Bulloch (1938: 62–125) – but as bacteria were the major

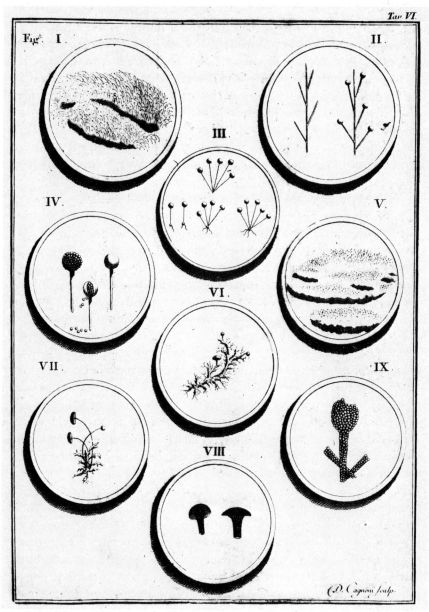

Fig. 4. *Rhizopus stolonifer*. (L. Spallanzani (1776, French transl.) **2**: pl. 6.)

problem much of the detail is outside the scope of the present review.

Influenced by the writings of 'Sir' John Hill, John Turberville Needham, a Welsh Roman Catholic priest, then living in London, became convinced by his own experimental findings of the reality of spontaneous generation for which he became an ardent advocate. Subsequently, when living in France, he collaborated with the famous Comte de Buffon, then working on his *Histoire naturelle*, who developed similar views. It was in opposition to the opinions of Needham and Buffon that Lazzaro Spallanzani, from 1769 professor of natural history at the University of Pavia, published two outstanding contributions to the controversy. The first of these, *Saggio di osservazione micro-scopiche concernenti il systema della generazione dei Signori di Needham e Buffon* appeared in 1765 and in 1769 in a French translation by the Abbé Regley together with extended critical comments by Needham. Spallanzani replied to Needham in his *Opuscoli di fisica animale e vegetabile*, 1776, which in the following year was issued as a two-volume French translation by J. Senebier together with letters from the philosopher and naturalist Charles Bonnet of Geneva in support of Spallanzani's views. It is this last publication that is of particular interest to mycologists because in the final section Spallanzani describes his observations and experiments on the origin of moulds.

Spallanzani worked principally with what was clearly *Rhizopus stolonifer*. He commented on the basal rhizoids although he misinterpreted the appearance of discharged sporangia as true but minute mushrooms and illustrated them as such (see Fig. 4). Like Micheli, Spallanzani made cultural experiments on damp bread and pieces of various fruits but in addition to the development of the mould from spores he showed that the growth was influenced by temperature and was most profuse in high summer and was more luxuriant when the culture was enclosed in a container, because of the higher humidity, than when kept in the open. Further he showed experimentally that *Rhizopus* differed from flowering plants in showing no tropic response to either gravity or light. He tried to eliminate the possibility that the spores merely functioned as a fertiliser which made possible mould growth from the substrata and to test this he inoculated inert substances such as glass, metal, blotting paper, writing paper, cotton, sponges, etc. under humid conditions to favour the growth of moulds but with negative results except for a few filaments which developed on the sponge. Experimental evidence was also adduced to show that the spores would resist a degree of heat that killed many seeds and Spallanzani suggested that this was the reason for Monti's mistaken view that moulds arose spontaneously on boiled vegetables. Spallanzani satisfied himself that moulds originate from spores.

Others too were convinced. In 1788 James Bolton was writing:

The plants which now compose the Order *Fungi*, were formerly supposed to be of equivocal generation, the sport of Nature, the effect of Putrefaction, or the brood of Chance; but that they owe their original to the seeds of a parent plant, is now well known [Bolton (1788–91) **1**:xiv].

Bulliard (1791–1812) held a similar view and Ehrenberg (1820*a*), who at the end of August 1819 had observed development of *Rhizopus nigricans* (= *R. stolonifer*) from spores, knew of no instance in which spontaneous generation need be invoked to explain the appearance of fungi. Additional evidence for the origin of fungi from spores was provided by the more critical experiment of Joseph Schilling (1827) who described how from 9 o'clock on the morning of 20 March 1826 to 8 o'clock in the evening of 22 March he had watched the germination of the spores of *Aspergillus glaucus* under the microscope, and the development of spores from the resulting mycelium (Fig. 5).

In spite of the findings of Spallanzani and subsequent workers, the view that some fungi at least may originate by spontaneous generation, or heterogenesis, continued to be held even by some leading mycologists including Persoon, Nees von Esenbeck, and E. M. Fries; and also by Unger (1833) and Naegeli (1842). Such views were particularly prevalent among those investigating pathogenic fungi and the question, discussed at greater length in Chapter 6, as to whether these fungi were the result rather than the cause of disease was not finally resolved until the middle of the nineteenth century.

STATUS OF FUNGI

As already mentioned, the development of views on the origin of fungi was paralleled by the development of views on their status, that is, their affinity to other organisms, the relationship of one fungus to another, and the taxonomic position of the fungi as a group.

Archiv f. d. ges. Naturl. B.X. Taf. IV.

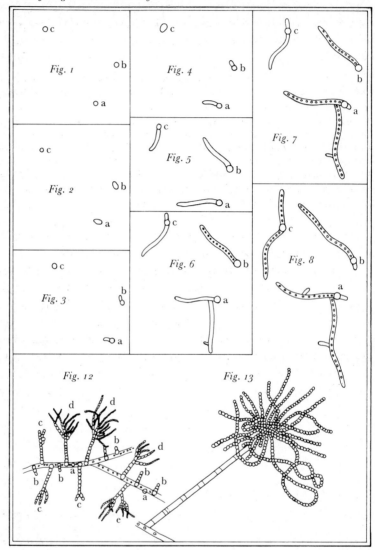

Fig. 5. First continuous observation of the growth of a fungus (*Aspergillus glaucus*) from spore to spore. (J. Schilling (1827): pl. 4 (redrawn).)

There was much speculation on the affinity of fungi to animals, particularly to the members of the Animal Kingdom variously known as 'animalcules', 'infusoria', and 'polyps'. In 1766, Baron Otto von Münchausen wrote of experiments he had made:

Fungi, when they become old, and especially Lycoperda, and all moulds, scatter a blackish dust; if this is observed under a good magnifying glass we find semi-transparent not dissimilar spheres filled with black points inside, of the substance of a polyp. I have kept such dust in water and at a moderate temperature, when the spheres swelled up and changed into oval, mobile, animal-like balls. These little animals (at any rate I will call them so because of their resemblance) move about in the water; and when one observes them further the next day they form clumps of hard weft and from these arise moulds or fungi. Where fungi grow, at first one sees white veins which one regards as roots but which in fact are nothing else than tubes in which polyps move back and forth and which form a large structure.[14]

These observations briefly published in an obscure publication would probably have escaped notice if von Münchausen had not been a correspondent of Linnaeus who set 'Mundus invisibilus' as the thesis subject for one of his students. As a result, von Münchausen's findings were considered to have been vindicated and similar microscopic organisms were seen associated with germinating smut spores. On the strength of this evidence Linnaeus introduced a genus *Chaos*, under Vermes in the twelfth edition of his *Systema naturae*, 1767 (p. 1356), two of the six species being *C. ustilago* for the organisms from various smuts and *C. fungorum* for those developed from the spores of *Lycoperdon*, *Agaricus*, *Boletus*, *Mucor*, and other fungi. Concurrently Linnaeus was in correspondence with the English zoologist John Ellis, best remembered for his *Essay towards a natural history of Corallines*, 1755. In a letter dated 1 January 1767 Linnaeus recounted his observations to Ellis:

With regard to *Fungi*, you may pick up, in most barns or stacks of corn, spikes of wheat or barley, full of black powder, which we call *ustilago* or smut. Shake out some of this powder, and put it into tepid water, about the warmth of a pond in summer, for three or four days. This water, though pellucid, when examined in a concave glass under your own microscope, will be observed to contain thousands of little worms. These ought first to be observed, to prevent ocular deception. In mould, *Mucor*, you will find the same, but not so easily as in the larger *Fungi*.[15]

and in October of the same year, after hearing that Ellis, from experiments with the 'seeds' of *Agaricus campestris*, concluded that the microscopic organisms were *animalcula infusoria* unrelated to the fungus, reiterated his viewpoint:

With respect to the *animalcula infusoria* themselves, unless I am totally mistaken, I think I have seen these to be the living seeds of Mould, *Mucor*. But before I venture to put forth such an opinion, I beg of you to lend me your lynx-like eyes...Everybody wonders at the *animalcula infusoria* being produced by an infusion of pepper and such substances; whereas the difficulty vanishes if they belong to *Mucor*; for pepper if long kept moist, is as liable to grow mouldy as anything else.

Having once discovered the little worms in *Ustilago*, by the help of the microscope, I can now see them with my naked eyes, though less distinctly.

To this Ellis made a forthright reply:

I have received your obliging letter about the seeds of *Fungi* being animated. By your letter, you seem to think that the seeds of the *Fungi* are animated, or have animal life, and move about; my experiments convince me of the contrary. I must first let you know, that I am convinced that in almost all standing, or even river, water there are the eggs, and often the perfect animals, of those you call *animalcula infusoria*. As soon as these meet their proper *pabulum*, they grow and increase in numbers.

From subsequent correspondence Linnaeus appears to have received this conclusion with a certain obtuseness. Ellis continued to crusade against the view that the seeds of fungi become transformed into vermicular animals. In a polemical open letter to Dr Matthew Maty, Secretary to the Royal Society, published in the *Gentleman's Magazine* for 1773 (p. 316), he wrote:

When men of eminence introduce a new discovery as the result of experiment, the generality of mankind are too apt to receive it implicitly for a fact, without giving themselves the trouble of *repeating* the same *experiments* to confirm the truth of their assertions. It is this want of attention that has occasioned some ridiculous absurdities to pass upon the world for surprising discoveries on the properties of animal and vegetable bodies, which nobody ever thought of before.

The illustrious Baron *Münckhausen's* new doctrine of *vegetables changing into animals* and then into *vegetables* again, is of this kind.

This credulity of the 'generality of mankind' – and also of practising scientists – is still prevalent.

At the same time other observers were making similar claims. As recounted by Lütjeharms (1936: 178–203) and Ramsbottom (1941a: 301–10) the spores of toadstools and mushrooms were considered to be the eggs of insects by D. S. A. Büttner in 1756 and as animalcules by Georg Wilk in 1768 while in the same year the Danish naturalist O. F. Müller was puzzled by seeing slender white worms swimming in the developing sporangiophores of *Pilobolus* from horse dung.[16] These views were refuted by Jonas Dryander in his thesis entitled *Fungus regno vegetabile vindicans*, 1776, but revived a few years later by R. Villemet (1784).

Entomogenous fungi, especially species of *Cordyceps*, were another source of much fantastic speculation which is also summarised by Lütjeharms (1936: 203–15) and Ramsbottom (1941a: 318–25). It was R. A. F. de Réaumur (1726) who first drew attention to *Cordyceps sinensis* by a report on specimens from Peking of what the Chinese called *tung-chhung- hsia-tshao* ('in winter an insect, in summer a herb') but he misinterpreted the upward growing fungus fruit-body attached to the caterpillar as a root. Another exotic example was the 'vegetable fly' first described by J. Torrubia, a Spanish Franciscan friar, in 1756.[17] According to the description quoted by William Watson (1764) in a communication to the Royal Society:

The *vegetable fly* is found in the island of Dominica, and (excepting it has no wings) resembles the drone both in size and colour more than any other English insect. In the month of May it buries itself in the earth, and begins to vegetate. By the latter end of July the tree is arrived at it's [sic] full growth, and resembles a coral branch; and is about three inches high, and bears little pods, which dropping off become worms, and from thence flies, like the English caterpillar.

Having been informed that Dr John Hill had examined examples of the vegetable fly Watson wrote to him and Hill replied:

There is in Martinique a fungus of the Clavaria kind, different in species from those hitherto known. It produces soboles from its sides. I have called it therefore Clavaria Sobolifera. It grows on putrid animal bodies, as our fungus ex pede equino from the dead horses hoof.

The Cicada is common in Martinique, and in it's nympha state, in which the old authors call it Tettigometra, it buries itself under dead leaves to wait it's [sic] change; and when the season is unfavourable, many perish. The seeds of the Clavaria find a proper bed on this dead insect, and grow.

The Tettigometra is among the Cicadae in the British Museum: the Clavaria is just now known.

This you may be assured is the fact, and all the fact; though the untaught inhabitants suppose a fly to vegetate; and though there exists a Spanish drawing of the plant's growing into a trifoliate tree; and it has been figured with the creature flying with this tree upon its back.

So wild are the imaginations of Man; so chaste and uniform is Nature!

But as so often the truth made little headway.

Theodor Holmskiold (or Holmskjold) in 1790 gave a good, well-illustrated account (in Latin and Danish) of the familiar European species, *Cordyceps militaris*, and recognised the fungus as being entomogenous. De Bary (1867–9; and summarised in his textbook, 1887) fully elucidated the details of infection of caterpillars by ascospores of *C. militaris* and the subsequent life history of the parasite.

PLEOMORPHISM

By the middle of the nineteenth century it had become generally accepted that fungi were a distinct group and that, even if aspects of their sexuality were still very obscure, they resembled higher plants in having their origin from 'seed' and not by spontaneous generation. There was, however, one major problem still unresolved. What was the relationship of one fungus with another? How plastic was fungus form? In particular, did one species exhibit more than one type of spore?

The attitude of Elias Fries to this last question had been ambivalent. In 1830 he expressed the view that fungi, 'and above all the lower ones', are propagated 'not only by sporidia alone but also by granules analogous to the gonidia of Lichens'. And in the *Systema mycologicum* (1821–32, **3**: 263) he wrote of hyphomycetes that those 'devoid of spores must be entirely deleted as autonomous plants! but, at the same time, it must be carefully noticed that not all the granules interspersed among the flocci are sporidia; those that are separated from the flocci are sporidia; those that are separated from the flocci as *bare* and *quite simple* bodies are conidia'. Later (*Systema mycologicum*, **3**: 363) he denies that two kinds of sporidia occur on the same plant which, as the Tulasnes suggested, is 'as if he had heard, sounding in his ears the loud voice of Linnaeus crying "It would be a remarkable doctrine – that there could exist races differing in fructification, but possessing one and the same nature and power; that one and the same race could have different fructifications; for the basis of fructification, which is the basis of all botanical science, would be destroyed and the natural classes of plants broken up"'.[18]

After further quotations from the writings of Fries and discussion the Tulasnes concluded there was no doubt

that the intimate bond of connection that exists between some ascophorous fungi and others that are ectosporous did not escape the notice of the sagacious mycologist of Uppsala. But he interpreted the connection merely as the congruence of types that were analogous but of unequal dignity, or the bond of a normal form with an abnormal one, never, if we mistake not, as the specific identity of one and the same fungus under diverse forms that were equally normal and equally typical.[19]

It was the Tulasne brothers who convincingly demonstrated the phenomenon of pleomorphism in ascomycetes.

Louis-René Tulasne was born in 1815 and after a legal training worked as a notary until the death of his father in 1839 when his

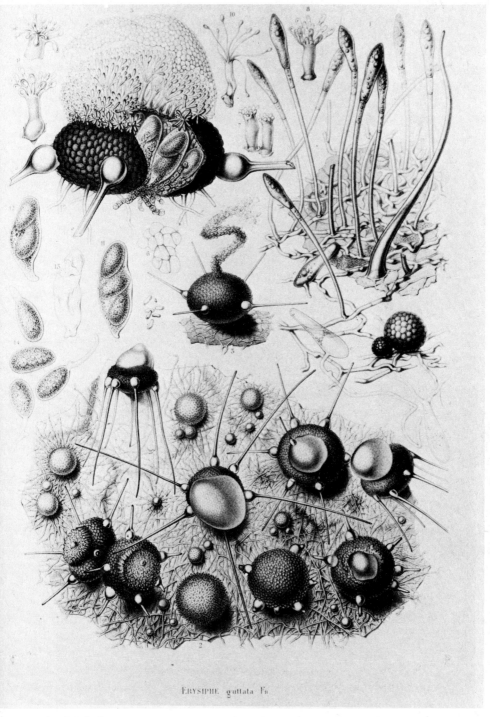

ERYSIPHE guttata Fu

Fig. 6. *Phyllactinia guttata.* (L.-R. & C. Tulasne, *Selecta fungorum carpologia*, **1**, pl. 1 (1861).)

inheritance enabled him to join his younger brother Charles (born 1816) who was studying medicine in Paris. Subsequently Louis became assistant naturalist to Adolphe Théodore Brongniart at the Muséum d'Histoire Naturelle, Paris, and Charles abandoned his practice of medicine to collaborate with his brother in a most notable series of botanical and, particularly, mycological works to the illustration of which the artistic genius of Charles contributed so much (see Fig. 6). In 1864, the ill health of Louis compelled them to leave Paris for Hyères on the Mediterranean where they lived quietly for the rest of their lives. Both were devout Catholics. Charles died in 1884, Louis the next year.

As already mentioned, the Tulasnes' major contribution to mycology was the establishment of pleomorphism among fungi by a wealth of accurate and detailed observation, first in a series of papers dealing with pyrenomycetes discomycetes, and basidiomycetes published between 1851 and 1860 and summarised in the sixth chapter, entitled 'The manifold nature of the seeds of the same species of fungus', of their magnificent three-volume *Selecta fungorum carpologia* (1861–5), 'exhibiting especially those facts and illustrations which go to prove that various kinds of fruits and seeds are produced, either simultaneously or in succession, by the same fungus', to quote from the title page of the English translation. The earlier observations on pleomorphism were extended in the *Carpologia* to cover Erysiphei (vol. **1**) (see Fig. 6), Xylariei, Valsei, Sphaeriei (vol. **2**) and Nectriei, Phacidiei, Pezizei (vol. **3**).

Similar conclusions had been reached by Anton de Bary who in 1854 had demonstrated the connection between *Aspergillus glaucus* and *Eurotium herbariorum* (Fig. 7).

These findings, and the elucidation of the relationships of the different spore forms in rusts which is reviewed in Chapter 6, were received with critical interest but they initiated much error even from the hands of skilful observers.

Schilling, as noted above, observed the development of *A. glaucus* from spore to spore but as the Rev. Berkeley observed: 'Unfortunately, experiments on the evolution of Fungi from single spores, require nice manipulation and complete leisure. A few hours' avocation is sometimes fatal to such observations'.[20] They also require something more, pure culture techniques and it was lack of these that had led F. T. Kützing in 1837 to believe

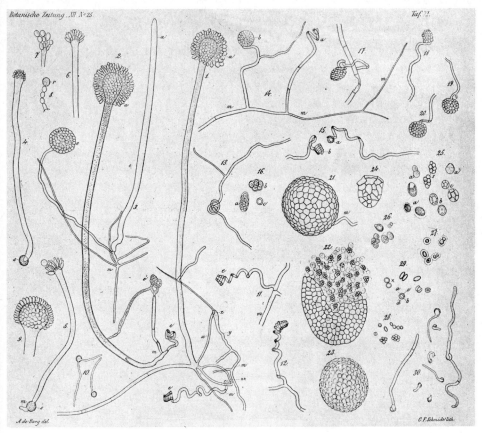

Fig. 7. *Eurotium herbariorum* and the imperfect state, *Aspergillus glaucus.* (A. de Bary (1854): pl. 11.)

that yeast cells developed into *Sporotrichum* and *Mucor* and had misled Berkeley twenty years later when he wrote:

It has been proved by myself and Mr Hoffman, by following up the development of individual yeast globules in fluid surrounded in a closed cell with a ring of air, that the proper fruit is that of a *Penicillium*, and as this *Penicillium* has, on more than one occasion, been observed to grow on fermenting matter, it is quite clear that yeast is merely an abnormal state of a Fungus, very different in habit, and forced into a peculiar mode of development by its submerged position. I believe equally that *Saprolegnia* and *Achlya...*, with their active zoospores, are merely submerged states of species of *Mucor*.[21]

These conclusions and speculations seem reasonable compared with those of less critical workers. C. A. E. T. Bail, who was the first to record the fact that mucor develops by budding when grown submerged in sugar solutions, in a series of papers beginning with his doctoral thesis *De faece cerevisiae*, of 1857, considered yeast to be transformed into *Mucor, Penicillium,*

Empusa, Isaria, etc. and also claimed that *Empusa musae,* which he identified with *Mucor mucedo,* became transformed into *Achlya prolifera.* In 1863 the English dermatologist W. Tilbury Fox in *Skin diseases of parasitic origin* presented evidence that there was but one essential fungus – *Torula* sensu Turpin [yeasts such as *Torulopsis* and *Saccharomyces*] – at least as far as the mucedinous fungi were concerned for he was 'most desirous not to ride a hobby'. The limit of absurdity was reached by Ernst Hallier, professor of botany at Jena and a prolific writer, who designed special and complicated apparatus of two kinds: the 'Cultur-apparat', essentially a dish, in which the experimental culture was made and its development and transformation observed, sealed under a bell jar standing in water and the 'Isolir-apparat' designed to isolate the experimental culture from external contamination and used to confirm observations made on the 'Cultur-apparat'.[22] He convinced himself that many pathogenic bacteria were merely stages of common saprobic moulds. For example, he related the cocci of vaccinia to *Eurotium herbariorum,* the micrococci of enteric fever to *Rhizopus stolonifer,* and believed the micrococci of gonorrhoea to be stages of an unknown *Coniothecium.* Such sensational results were given wide publicity both by Hallier himself and in the scientific and popular press until they were refuted, mainly by the criticisms of de Bary who in the second edition of his textbook (1887: 127), after dismissing these 'pleomorphic extravagances' in a few words, was able to conclude that 'Such of them as, like Hallier's especially, only belong to the scientific *chronique scandaleuse* will not be further noticed'. Hallier, unrepentant but disheartened, devoted his later years to the study of philosophy and aesthetics rather than botany.

Dimorphism

A type of pleomorphism which has attracted much attention during recent years is that distinguished as 'dimorphism', in which one fungus, which may or may not show sexuality, exists in nature in two distinct morphological forms. Dimorphism has been applied particularly to fungi pathogenic for man and higher animals when the appearance of the pathogen in host tissue differs in morphology from that of the saprobic state in culture. This alternation is typically yeast-mycelial. The term has also been applied to the conversion of mycelial mucors into true yeasts by manipulation of the cultural conditions.

Among mycoses of man (see p. 177) sporotrichosis (*Sporothrix schenckii*), blastomycosis (*Blastomyces dermatitidis*), paracoccidioidomycosis (*Paracoccidioides brasiliensis*) and histoplasmosis (*Histoplasma capsulatum*) (and also epizootic lymphangitis of horses (*H. farciminosum*)) are all caused by pathogens which are yeast-like in the parasitic phase, mycelial in laboratory culture, while in coccidioidomycosis (*Coccidioides immitis*) the parasitic phase takes the form of spherical, sporangium-like bodies (spherules). Among plant pathogenic fungi the reverse is exhibited by ascomycetes, such as *Taphrina*, and smuts (Ustilaginales) which are mycelial when parasitic, yeast-like in in-vitro culture.

There have been many studies on the conditions which determine mycelial/yeast conversions *in vitro*. In *Sporothrix schenckii* (first by Lutz & Splendore, 1908), *Blastomyces dermatitidis*, and *Paracoccidioides brasiliensis* the transformation is temperature-controlled. The morphology is mycelial at 25 °C, yeast-like at 37 °C. For *Histoplasma capsulatum* the yeast form is only produced at 37 °C on cystein-containing media (such as blood agar) and transformation to the yeast state by *H. farciminosum* (Bullen, 1949) and *Mucor* (Barnicki-Garcia & Nickerson, 1962) requires increased carbon dioxide tension, as does spherule production *in vitro* by *Coccidioides immitis* (Henry & O'Hearn, 1957).

TAXONOMIC POSITION

The categorisation by Linnaeus, on the first page of his *Philosophia botanica*, 1751, of the three kingdoms of nature – 'LAPIDES crescunt. VEGETABILIA crescunt & vivunt. ANIMALIA crescunt, vivunt & sentiunt.' – epitomised an age-long tradition. To which kingdom do fungi belong? In Chapter 10 aspects of the taxonomic treatment with the group Fungi are reviewed. Here attention is drawn to views advanced on the taxonomic relationships of fungi to other living things in attempts to answer the question posed.

From the time of the herbalists, fungi, even if confused with corals and other organisms, have been associated with plants and it was Antoine de Jussieu in 1728 in a paper, read to the Paris

Academy, entitled 'De la nécessité d'établir dans la Methode nouvelle des Plantes, une Class particulière pour les FUNGUS, à laquelle doivent se raporter non seulement les Champignons, les Agarics, mais encore les LICHEN', who proposed with supporting arguments that lichens and fungi should be combined as two sections of one class of plants (*Fungus*); a more advanced treatment than that of Linnaeus in the *Species plantarum*, 1753, where the lichens are classified as algae. By 1789 Antoine Laurent de Jussieu (nephew of Antoine de Jussieu) in his *Genera plantarum* was writing:

Fungi, which are analogous in part to animal zoophytes, begin the vegetable series, being as it were intermediate between the two; they agree only with certain Algae, but are quite different from other plants in structure, flowering, and habit.[23]

Four years later Bulliard was crusading for the view that fungi were plants –

If those who obstinately refuse to grant to Fungi a place among the productions of the vegetable kingdom, if those who pretend that all fungi are engendered only by corruption, that they have no seeds at all, no constant characters by which they can be distinguished, etc., had taken the trouble to study their organisation, to follow them in their growth, to analyse its details, and to compare them, they would doubtless blush for their error:[24]

and Persoon (1793) was of the same opinion.

The curious fungus stone (*Polyporus tuberaster*) and relationships to infusoria, etc., claimed for fungi led some to alternative views. As already noted, Marchant (1711) described *Xylaria polymorpha* as a 'terrestrial lithophyte' and in 1783 Villemet concluded from the astonishing variety of fungi that they might best find a place in the mineral kingdom as 'pseudo-zoo-lithophytes' (Villemet, 1784). In the same year Necker (1783) considered that as fungi must be excluded from the sexual system of Linnaeus, as he had been maintaining for the previous eight years, and are of such different constructions to animals, plants, and minerals they should be treated as 'intermediate organisms' (*mésymaux*) in a distinct kingdom – *Regnum mesymale*. V. Picco (1788) thought the animal nature of fungi more than sufficiently proved and in 1803 Lichtenstein designated fungi 'aerial zoophytes'. By then, however, that fungi were plants was well established as the orthodox opinion and this is still today the most widely held view, but with increasing reservations by many.

While the division of living things into plants and animals seems only common sense at the everyday level – the distinction between a dog and an oak tree is so patent – increasing knowledge has made this dichotomy more and more difficult to maintain as every biologist is uncomfortably aware and for more than a century biologists who have taken a broad look at the diversity of life on our planet have felt the need to propose additional kingdoms. The most important early innovator was the nineteenth-century German zoologist Ernst Heinrich Haeckel (for more than forty years professor at Jena), who over a period of some thirty years developed a four-kingdom system to which the four- or five-kingdom schemes of H. F. Copeland, F. A. Barkley, and R. H. Whittaker all owe a debt.

Table 1. *Biological kingdoms accepted by some recent authors and their dispositions of the fungi*

Haeckel ([1866*]1894)	Copeland ([1938*]1956)	Barkley ([1939*]1970)	Whittaker ([1947*]1969)
		VIRA	
PROTOPHYTA	MYCHOTA	MONERA	MONERA
PROTOZOA	PROTOCTISTA	PROTISTA	PROTISTA
METAPHYTA			
Phylum:			
Thallophyta (Algae, Mycetes [Fungi], Lichenes)	Phylum: Inophyta [Fungi]	Phylum: Mycophyta [Fungi]	FUNGI
Phyla: Diaphyta, Anthophyta	PLANTAE	PLANTAE	PLANTAE
METAZOA	ANIMALIA	ANIMALIA	ANIMALIA

Names of kingdoms (in small capitals) at one level are synonymous or approximately so.
* Date at which the system was first introduced.

These four systems are contrasted in Table 1 where the status accorded to fungi by each author is also indicated. Haeckel's disposition of fungi was conventional – in the plant kingdom in the phylum Thallophyta along with the algae and lichens. The three living authors all remove the fungi from the plant kingdom, Barkley and Copeland as a phylum of the Protista[25] (from which it is now widely believed that most fungi had a polyphyletic origin) while Whittaker gives fungi the status of a kingdom distinct from the Protista in a scheme in which the more advanced eukaryotic organisms are distinguished on the basis of nutrition which is photosynthetic in plants, ingestive in animals, and absorptive in fungi.

The arguments for these dispositions will not be rehearsed. The reviews of Copeland (1956) and Whittaker (1969) may be consulted for details and for further information. Neither will the substance of the two excellent complementary reviews by G. W. Martin (1955, 1968) on the status of fungi be repeated, while aspects of the alleged relationship of fungi to algae and other phylogenetic speculations are touched on in the account of taxonomic developments within the fungi given in Chapter 10. The status of fungi is still a current problem but when looked at in historical perspective it does seem probable that the continued acceptance of fungi as plants is yet another example of the inertia of orthodoxy and conservatism delaying the acceptance of advancing knowledge.

3. Form and structure

Fungi were at first mainly characterised by negative features. Theophrastus (*c.* 300 B.C.), the famous pupil of Aristotle and the first Greek author to mention fungi, writing of truffles in his *Historia plantarum* (I, 1, § 11) described them as having neither root, stem, branch, bud, leaf, flower, nor fruit; neither bark, pith, fibres, nor veins; and these words echoed down the ages. Some two thousand years later, in the opening paragraph of the section on fungi of the second (1671:369) edition of Gaspard Bauhin's *Pinax theatri botanici* a similar view is reiterated – 'Nam Fungi sicut et Tubera, neque plantae, neque radices, neque flores, neque semina sunt'.

EXTERNAL MORPHOLOGY OF LARGER FUNGI

The earliest descriptions of fungi are imprecise and in the absence of illustrations even the external features can rarely be deduced with confidence. As indicated in the Introduction (p. 2), Greek and Roman writers certainly distinguished between agarics, polypores, and truffles (including species of both *Tuber* and *Terfezia*). The Romans must also be credited with the oldest known naturalistic illustration (Fig. 1)[1] of a fungus (presumably edible and which has been variously identified) in a wall fresco at Pompeii preserved for posterity by the notorious eruption of Vesuvius in A.D. 79, the year of Pliny's death.[2] The herbalists (usually doctors interested in the medical uses of plants), who paid little attention to fungi, maintained the classical distinctions and from the end of the fifteenth century, which marked the advent of the printed herbal, they provided increasingly accurate illustrations, incorporating first-hand observation, of larger fungi. This did much to elucidate external morphology and facilitate identification.

The earliest wood-block illustrations of fungi, such as those of six agarics (Fig. 8) and two truffles (which resemble boletes)[3] in the *Ortus sanitatis*, 1491, are crudely executed, as is that in *The*

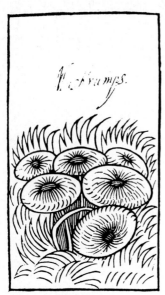

Ca.cciij.

Fungus. Nidorus. Fungi dicunt
eo qʒ aridi ignem acceptū ꝑcipiāt
ſi os em ignis eſt. vñ et eſca fulgo
dr̄ eo qʒ lit foones ignis ʒ nutrimentū. Alij
dicunt fungos vocatos eo qʒ ſunt ex eorū
genere quidam interemptoꝛi. vñ ʒ deſun
cti. Galienus. Cōplexio coꝝ frigida ʒ hu
mida multū. ꝓpter hoc ſunt prope medi
cinas moꝛtiferas. ʒ ꝓpe quoꝝ ſubſtātie
admiſcetur aliqua putrefactio.

Operationes.

A Et dixit Galin li. de dieta Boni ex
eis ſunt non nociui ʒ ſunt frigidi. ʒ ſi quis
vtitur eis nimis generant hūoꝛes malos
Ex eis ſunt ſpecies male moꝛtifere. ʒ iam
vidi bōiem qui paſſus eſt ab eis anxieta
tem anhelitus. ʒ coartationem ʒ ſincopi
ʒ ſudoꝛeʒ frigidū. ʒ euaſit poſtqʒ potauit
vinū in q̄ decoctū fuerat pulegiū. ʒ proie
ctum eſup ipͫ baurach ʒ ſic euomuit fū
gos illos ex quibus generati fuerunt hūo
res groſſi in ſtomacho.   B Unde q̄

Fig. 8. The first wood-block
illustration of a fungus.
(*Ortus sanitatis*, 1491, cap. cciii.)

De fungis. Muſſherons: Ca.C.lxxi.

Fungi ben muſſherons. They be
colde and moyſt in the thyꝛde degree and that is ſhewed by theyꝛ
bꝛolent moyſture. There be two maners
of them/one maner is deedly ʒ ſleeth them
that eateth of them and be called todeſto-
les/and the other dooth not. They that be
not deedly haue a groſſe gleymy moyſture
that is dyſobedyent to nature and dygeſty
on/and be perillous and dꝛedfull to eate
ʒ therfore it is good to eſchew them/ſuche
as eate them and feare not to fall inconue
nience ſethe them in water and medle them
with gynger / peper/ catup/ calament/oꝛ
oꝛygan and ſuche other/ and than dꝛynke
olde wyne/ pure and ſtronge. And they ſ
be of colde cōplexyon/ after them take gꝛe

Fig. 9. 'Muscherons' from *The grete
herball*, 1526.

36

PETRI ANDREAE
MATTHIOLI MEDICI
Senenſis Commentarii,
IN LIB. TERTIVM PEDACII DIOSCORIDIS
ANAZARBEI, DE MEDICA MATERIA.

 VPERIORIBVS, chariſſime Aree, commentarijs, tradidimus de aromatibus, vnguentis, oleis, arboribus, earum fructibus, & lacrymis: item de animalibus, cerealibus, oleraceis, & herbis acrimonia præditis. In hoc autem tertio radices, ſuccos, ſemina, herbas, & quæ vernacula, & inter ſe cognata cenſentur, quæque pluribus ſcatent remedijs, proſequemur.

Agaricum.   Cap. I.

AGARICVM radix fertur laſerpitij ſimilis, ſed facie ſumma ſolutior; & rariore, fungoſoque tota contextu. Duo eius genera: fœmina, quæ præfertur, rectis intus uenarum diſcurſibus conſtat: mas rotundus eſt, & vndique compactior. Vtrique guſtus in initio dulcis, mox ex diſtributione in amaritudinem tranſit. Gignitur in Sarmatiæ regione, quæ Agaria dicitur. Sunt qui radicem eſſe plantæ affirment: alij ut fungos naſci in arborum caudicibus, quadam putredine. Gignitur in Galatia Aſiæ, & Cilicia in cedris, ſed friabile, & infirmum. Adſtringendi, calfaciendíque naturam habet. efficax eſt contra tormina, cruditates, & rupta: prodeſt itidem contuſis, & ab alto deuolutis. Datur duobus obolis in uino mulſo febrim non ſentientibus: febriculoſis ueró in aqua mulſa: arquatis quoque, ſuſpirioſis, iocineris, renúmque vitio laborantibus, dyſentericis. Item ſi urina ægrè reddatur, ſi uuluæ ſtrangulatus urgeat, ſi pallor membra decoloret, drachma una dari ſolet: ſi tabes infeſtat, ex paſſo: ſi lien negotium exhibet, ex aceto mulſo: ſi diſſolutio ſtomachi eſt, ita ut cibi tenax non ſit, manditur, & ſine ulla humoris ſorbitione deuoratur: ſimili modo acida ructantibus propinatur. ſanguinis reiectiones ſiſtit, tribus obolis ex aqua ſumptum. Facit ad coxendicum, & articulorum dolores, & comitiales morbos, ſi cum aceto mulſo, pari pondere aſſumatur: menſes ciet: fœminis, quas inflatio vuluæ uexat, utiliter æquali modo datur: horrores ſoluit, datum ante febrium acceſſiones: aluum purgat, drachma vna, aut altera, ſi cum aqua mulſa bibatur: venenorum antidotum eſt denarij inſtar, cum diluta potione ſumptum: ſi ſerpentes ictum uibrarunt, morſúmque ſubfixerunt, tribus obolis ex uino potum, mirè auxiliatur. In ſumma, internis omnibus uitijs conuenit, pro uiribus, & ætate datum, nunc ex aqua, nunc ex uino: ijs ex aceto mulſo, alijs ex aqua mulſa.

AGARICVM *fungus eſt in arboribus naſcens. Plura de eo diximus libro primo in laricis arboris hiſtoria.* Agarici conſideratio. *Præſtantiſſimum habetur in Tridentinis montibus, ubi ſæpiſſimè proprijs manibus illud diuulſimus, adacta ſecuricula.*

F 3   *la.*

Fig. 10. 'Agaricum' (*Fomes officinalis*). (P. A. Mattioli (1560): 341.)

grete herball, 1526, where the ring of 'muscherons' surrounding an oak (*Quercus*) tree (Fig. 9) presumably symbolises Pliny's view that the best kind of 'boletus' (*Amanita*) grew around oak roots. The Italian Pierandrea Mattioli in his herbal of 1560 (basically an exposition of Dioscorides) has three illustrations of higher quality. Those of black and white truffles and of agaricum (*Fomes officinalis*) on larch (*Larix*) (Fig. 10) are the first book illustrations that can be identified with confidence. The third (Fig. 11) is an attractive arrangement of a number of agarics, one group associated with a tree stump, the remainder terricolous; and the incorporation of two snakes in the design is a reminder that according to Dioscorides, as elaborated by Pliny, poisonous fungi owe their deleterious nature to the breath of serpents, thus fungi growing 'neere to serpents dens' (in the words of Gerarde, 1633) should be avoided.

Fig. 11. A group of agarics. (P. A. Mattioli (1560): 545).

The dozen illustrations in the *Kruydtboeck* of Mathias de l'Obel (Lobelius) (1581) show a great advance. The gills of agarics are clearly shown, there are recognisable figures of morels (*Morchella*) (Fig. 12), the Jew's ear (*Hirneola auricula-judae*) (Fig. 12), the chanterelle (*Cantharellus cibarius*) (Fig. 13) and puff balls (*Bovista*), while truffles (*Tubera*) are included among fungi for the first time. In addition there is a figure of the stink-horn (both 'egg' and fruit-body) which in 1562 had been described and illustrated (see Fig. 14) by the Dutch physician Adriaen Jonghe (H. Junius) who gave the name 'Phallus' to it in the first independently published mycological monograph.[4]

Boom-Campernoellé Judas-ooren gelyckende.
In Latijn/ Arborum fungi auriculæ Iudæ
facie.

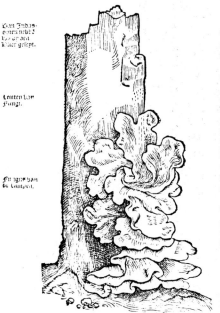

Van Judas-
oore niet/
toe op den
Baer gelept.

Lonten van
Fungi.

Fungus van
de campen.

Vt de strncken van oude boomen: sommighe swarte sommighe andere soorten, van de welcke euelijcke sijn wit oft geelachtich, ende gheremept als Judas-ooren/taey als leder.

De andere sijn spongieachtich voos/ van Aenwas substantie ende ghedaente den Agaricus niet zeer enghelijck/ van verwe insgelijcks als een spongie oft bruyn grau-geel. Dese worden in leeghen oft met aschen gheseeten en hersoden en ghedroocht om lemetten te maecken/ want sy vincken soo wel/ datse in sommighe plaetsen toer daer mede hendt/ Jck hebbe sien ghebruycken ende mede ghebruyckt inden erijch voos lonter. Men pleeghse d'antwerpé aen de beerse te vercooyen met vierslachen om vier te vatten. Dese soude wel moeghen wesen de lampe-fungus van Cornarius, het welcke berijst Matthiolus, seggaehende dat fungi vande lampen sijn de kellekens/ die aen de lemetten vande lampen/ vergaderen/ taey verduyst en de. seggende alsmen die inde lampen siet/ dat teecken van reghen is, voort benaehende de veersskens van Virgilius Libro primo Georg. ende d'authoriteyt van Plinius.

Nec nocturna quidem carpentes pensa puellæ
Nesciuere hyemem: testa cum ardéte viderét
Scintillare oleum,& putres cocrescere fungos,
Nec minus ex imbri soles,& aperta serena
Prospicere, &c.

Campernoellen de honich-raten ghelijckende. In Latijn/ Fungi suaginosi,siue fungi rugosi fauis mellis similes.

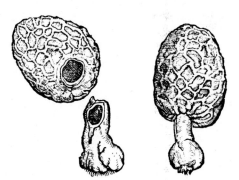

De voorleden iaeren sijn dese Campernoellen van mijn Heer vander Aluy gheschoncken / die de selue wt Hongaryen vercreghen hadde waermer veel af hendtdoor gulsichendt vande menschen/ berendende cokende ende sioutende de selue als de andere . Mijn-Heer van Renneutre saligher ghedachteniffe heeft de selue hier vooytijts noer t'leuen doen conterferten wt versche planten die hier te lande beuonden waeren maer op wat plaetse, is my ende den schildere dese ghemaeckt heeft noch onbekent.

Fig. 12. Jew's ear (*Hirneola auricula-judae*) on tree stump and two morels (*Morchella*). (M. de l'Obel (1581): 308.)

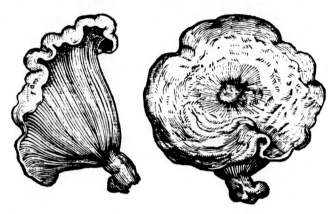

Fig. 13. Wood-block illustration of the chanterelle (*Cantharellus cibarius*). (M. de l'Obel (1581): 307.) (Compare with Fig. 21.)

L'Obel was born in Lille in 1538 and after practising medicine in England and on the Continent he returned to England where he became Botanist to King James I. His *Kruydtboeck* (and the earlier issue, as *Plantae seu stirpium historia*, 1576) were both published by the famous Plantin Press at Antwerp and the blocks for the illustrations were much used in later publications including the herbals (also published by the Plantin Press) of his contemporaries in the Low Countries, Rembert Dodoens, in 1583, and Clusius, in 1601; and in Gerarde's *Herball* (1633) and John Parkinson's *Theatrum botanicum; the theater of plants*, 1640, published in England. Finally, a selection (from new blocks) appeared in the complete edition of Jean Bauhin's *Historia plantarum universalis*, 1651.

After l'Obel the next important mycological contribution was made by Jules-Charles l'Éscluse (or l'Écluse; Latinised as Clusius) in his *Rariorum plantarum historia*, 1601. Clusius (Fig. 15) was born at Arras in 1526; studied at the University of Montpellier, where he was a pupil of the botanist and physician Guillaume Rondelet, but he never qualified in medicine. Subsequently, as a free-lance botanist, man of letters, and humanist, his whole life, hampered by ill health, was a struggle against poverty. He was an accomplished linguist and among his many translations are those made for Christophe Plantin of the Flemish herbal of Dodoens into French and the Portuguese herbal of Garcia de Orta into Latin. In 1573 he was invited to Vienna by the Emperor Maximillian II but later lost favour at court because of his religious opinions and after more travel lived in Frankfurt. He was finally awarded a professorship at Leiden where he died in 1609.

It was during the fourteen years in Vienna that Clusius laid the basis of his major contribution to mycology. Clusius, in company with István de Beythe, the chaplain of his patron Boldizsár de Batthyány, collected fungi of the region which formerly comprised the Roman province of Pannonia, now parts of Austria, Hungary, and Yugoslavia, and a watercolourist was engaged to make drawings from nature of the fruit-bodies,

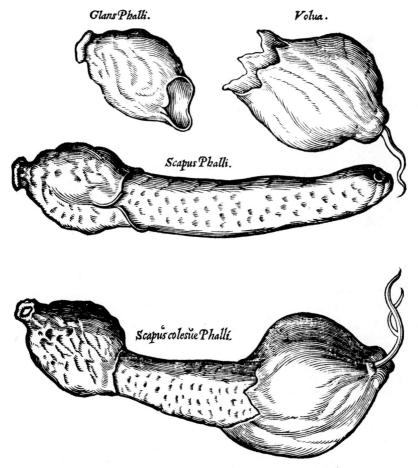

Fig. 14. Stink-horn (*Phallus*). (A. Jonghe 1562; from the 1601 issue.)

the work being supervised by Batthyány himself. The result was the unique *Clusius Codex* ('Le Code de l'Éscluse') a collection of eighty-seven sheets of illustrations, eighty-six in watercolour and one (added later) in oils (Fig. 98). In his last years at the University of Leiden Clusius brought together re-edited versions of earlier works as his *magnum opus*, the fine folio *Rariorum plantarum historia*, 1601, the concluding section of which was devoted to his hitherto unpublished observations on fungi – the title page of the part reading:

FVNGORVM

IN PANNONIIS OBSERVATORVM

BREVIS HISTORIA,

A

CAROLO CLVSIO ATREBATE

CONSCRIPTA.

The *Code* should have provided the basis for the illustrations but it was apparently mislaid by the publishers and inferior woodcuts were prepared. Some seventy-five years later the *Code* came into the possession of van Sterbeeck (see p. 48) who made much use of it to illustrate his *Theatrum fungorum*, 1675. The *Code* was lost sight of for the next two hundred years when it turned up in the library of Leiden University and in 1900 the Hungarian Istvánffi published a high quality reproduction together with a facsimile of the fungus section of the *Rariorum plantarum historia*, an exhaustive treatment of the relationship of the *Code* to the publications of both Clusius and van Sterbeeck, and other supporting data.

In spite of the loss of the *Code* Clusius's 1601 monograph was a notable publication. One hundred and five species arranged in

Fig. 15. Carolus Clusius (Jules-Charles l'Éscluse) (1526–1609).

forty-seven numbered genera are recognised and of the thirty-three illustrations (see Figs. 16, 17) all but two (*Morchella esculenta* and *Bovista nigrescens*), which are attributed to l'Obel, were original. The remaining illustrations from l'Obel (1581) were, at the request of the publisher, reproduced (with Latin translations of the original Flemish descriptions) as a supplement which was preceded by a reprint of 'De Fungis' from the second edition of *Phytognomonica* by G. della Porta (see p. 114).

The illustrations showed new details such as the decurrent gills of *Pleurotus ostreatus* (Fig. 16) and the annulus of *Lepiota procera* (Fig. 17). Both upper and lower aspects of the fruit-bodies are frequently shown (see Fig. 16) so that agarics and boletes are readily distinguished, while among species illustrated for the first time are *Ramaria botrytis* (Fig. 17) and (in an appendix) *Clathrus cancellatus*.

43

GENVS VI,

VI. Genus esculentorum Fungorum.

Fig. 16. *Pleurotus ostreatus.* (C. Clusius (1601): 216.)

XVIII. Genus esculent. Fungorum·   XIX. Generis esculent. Fung. 1. species.

Fig. 17. *Lepiota procera (left)* and *Ramaria botrytis (right).* (C. Clusius (1601): 274.)

IOHANNES BAVHINVS MED
ET CHIRVRGIAE DOCTOR.

Gallus eram, Chriſtum tandem confeſſus, & Anglis
Et Medicus Belgis, & Baſilea tuus.
VAL. THILO L.

Fig. 18. Jean Bauhin (1541–1613).

The two Bauhin (Bauhinus) brothers, Jean (Johannes) (Fig. 18) and Gaspard (Casparus) are easily confused for their lives and work show much similarity, a similarity which Linnaeus no doubt had in mind when he named a leguminous genus characterised by leaves consisting of twin leaflets or two nearly equal lobes *Bauhinia* in their joint honour. Both brothers were born in Switzerland of French parents living in exile because of their religious opinions, both became physicians like their father, travelled widely, and will always be remembered for their contributions to systematic botany. Both also left much unpublished material. Gaspard is particularly noted for his ΠΙΝΑΞ [*Pinax* = chart or register] *theatri botanici*, 1623 (second edition, 1671), a comprehensive concordance of plant names published a year before his death. Jean's *magnum opus*, the *Histoire universelle des plantes*, in which he aimed at the compilation of descriptions of all known plants, appeared posthumously. In 1619 his son-in-law, J. H. Cherler, brought out an abridgement while the *editio princeps* edited by D. Chabrey, entitled *Historia plantarum univer-salis*, was not published until 1650–1. This major work, in three large folio volumes, includes descriptions of five thousand plants and Liber xl of volume **3** (1651) is devoted to fungi, classified as 'Excrementa terrae', and illustrated by forty-three woodcuts. Most of these are taken from l'Obel and Clusius (but appear as mirror images from new blocks recut on a smaller scale, the text being printed in two columns) although a dozen are new and of higher quality. Among the latter those of St George's mushroom (*Tricholoma georgii*) and the field mushroom (*Agaricus campestris*)

Fig. 19. 'Fungi verni, mouceron dićti, odori & esculenti' (*Tricholoma georgii*) (*left*) and 'Fungus campestris, albus superne, infernè rubens' (*Agaricus campestris*) (*right*) from Jean Bauhin (1651): 814.

show different stages of development (Fig. 19), the figure of *Cantharellus cibarius* the thickened gills, and the volva of *Amanita caesarea* is clearly displayed while *Lycoperdon bovista* and the pores of *Boletus scaber* are illustrated for the first time. The accompanying Latin descriptions, though largely compilations, are more detailed than those by earlier authors and this account may be considered as the climax to the treatment of fungi by the herbalists.

The next advance in mycological illustration was a technical one, the use of copper plates on which the design was engraved by hand and then transferred from the inked plate to paper by pressure applied by passing plate and paper between two metal rollers. Later the hand engraving was replaced by drawing with a

Fig. 20. Title-page of F. van Sterbeeck's *Theatrum fungorum*, 1675, which includes portrait of the author, aet. 43.

stylus on the waxed surface of the plate which was then immersed in acid which etched the exposed surface. The disadvantage of this method – that the illustrations could not be printed, as with wood-blocks, at the same time as the text but had to be printed independently as 'plates' – was far outweighed by the possibility of illustrations of much higher quality and finer detail. Copper-plate illustrations were used in Hooke's *Micrographia*, 1665 (see Fig. 29), but their first use to illustrate larger fungi was in van Sterbeeck's *Theatrum fungorum*, 1675 (Fig. 20).

Frans van Sterbeeck, a Flemish priest of noble extraction, was born and died in Antwerp where he lived for the greater part of his life. During the eight years following his ordination in 1655, while suffering from a chronic illness, he turned his attention to botany, with particular reference to fungi and soon became a recognised expert. He was on friendly terms with other Flemish botanists and in May 1663 was visited by John Ray who admired the rare plants in van Sterbeeck's garden. It was in 1672 that Adriaan David, an Antwerp pharmacist and amateur botanist

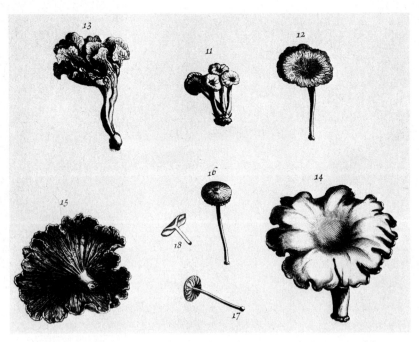

Fig. 21. Copper-plate of *Cantharellus cibarius* (14, 15) (compare with Fig. 13, a wood-block of the same species) and *Craterellus sinuosus* (11–13) by Claude Aubriet from Vaillant (1727: pl. 11).

PETRVS ANTONIVS MICHELIVS
VIXIT ANNOS LVII DIES XXII IN TENVIRE
BEATVS OMNIS HISTORIAE NATVRALIS
PERITISSIMVS MAGNORVM ETRVRIAE
DVCVM HERBARIVS INVENTIS ET SCRIPTIS
VBIQVE NOTVS AC PROPTER SAPIENTIAM
SVAVITATEM PVDOREM OPTIMIS
QVIBVSQVE AETATIS SVAE EGREGIE CARVS
OBIIT IV NONAS IANVARIAS MDCCXXXVII
AMICI AERE CONLATO TITVLVM POSVERE

Fig. 22. Pier' Antonio Micheli (1679–1737).

Fig. 23. Memorial to P. A. Micheli in the church of Santa Croce, Florence.

(he contributed a sonnet to the collection of adulatory tributes to the author which preface the *Theatrum fungorum*), took the celebrated *Code de l'Éscluse* (then owned by Dr Syen, professor of botany at Leiden University) to van Sterbeeck who made considerable, if somewhat surreptitious, use of it as a source of illustrations for the *Theatrum fungorum* in which he claimed to have based all but a few of the figures on direct observation from nature. However, according to Istvánffi (1900), of the 135 illustrations of hymenomycetes no less than seventy are taken from the *Code*, seven from Clusius's *Historia*, and a further fourteen from l'Obel and J. Bauhin. Many of the plates (see Fig. 97) are of good quality, if some are less satisfactory, and the text (which is in Flemish, not the usual Latin) includes new material.

Among the illustrated books published during the next fifty years that deserve mention are: J. P. de Tournefort, *Élémens de botanique*, 1694, in which six plates are devoted to larger fungi (including *Geastrum* and *Tulostoma*; see Fig. 98); P. Boccone,

Museo di piante rare, 1697, which includes the first illustration of the fungus-stone (*Polyporus tuberaster*);[5] Sébastien Vaillant, *Botanicon Parisiense, ou dénoument des plantes qui se trouvent aux environs de Paris*, 1727,[6] with its beautiful plates, by Claude Aubriet, which include seventy-eight figures of fungi (see Fig. 21); and finally, in 1729, *the Nova plantarum genera* by P. A. Micheli, an outstanding work, far ahead of its time, and generally recognised as marking the birth of mycology.

Pier' Antonio Micheli[7] (see Fig. 22), born in Florence in 1679, was of humble origin and his education scanty. His parents, it is said, wished him to become a bookseller but he taught himself Latin and with no immediate prospects devoted himself to the study of plants. Due to the intervention of the marquis Cosmo de Castiglione he found favour at court and in 1706 was appointed botanist to Cosmo III, the penultimate member of the Medici family to be Grand Duke of Tuscany, with responsibility for the public gardens of Florence. This enabled him to continue his botanical studies by which he achieved an international reputation. He made a number of collecting trips and in the autumn of 1736 while on an excursion to the north of Italy contracted pleurisy from which he died in Florence at the New Year, unmarried. He was buried in the church of Santa Croce where his epitaph (Fig. 23) (in the English translation by Buller (1915b: 11) reads:

Pier' Antonio Micheli lived 57 years and 22 days, happy although in moderate circumstances, an expert in natural history, a leading botanist of Tuscany, well-known everywhere for his researches and writings, and much beloved by all the worthy men of his age on account of his wisdom, sweetness of disposition, and modesty. He died on the second of January, 1737. His friends gathered contributions and erected this tablet.

Subsequently a statue of Micheli was included in the Uffizi Colonade (between Redi and Galen) and streets in Florence and Rome were named in his honour.

The Grand Duke made a gift to the young Micheli of de Tournefort's three-volume *Institutiones rei herbariae*, 1700, soon after its publication in Paris, and this work influenced Micheli throughout his life as is evidenced by the title and the concise style of the book by which he will always be remembered, *Nova plantarum genera juxta Tournefortii methodum disposita*, 1729 (Fig. 24). This work covers all groups of plants but strongly reflects Micheli's special interests – the genus *Carex* and mosses and

N O V A
PLANTARVM GENERA

I V X T A
TOVRNEFORTII METHODVM DISPOSITA

Quibus Plantæ MDCCCC recenfentur, fcilicet fere MCCCC nondum obfervatæ, re-
liquæ fuis fedibus reftitutæ; quarum vero figuram exhibere vifum fuit, eæ
ad DL æneis Tabulis CVIII. graphice expreffæ funt; Adnotationibus, atque
Obfervationibus, præcipue Fungorum, Mucorum, affiniumque Plantarum fationem, ortum, & incrementum fpeſtantibus, interdum adieſtis.

REGIAE CELSITVDINI
IOANNIS GASTONIS
MAGNI ETRVRIAE DVCIS.

. A V C T O R E
PETRO ANTONIO MICHELIO FLOR.
EIVSDEM R. C. BOTANICO.

FLORENTIÆ. MDCCXXVIIII.
Typis BERNARDI PAPERINII, Typographi R. C. MAGNÆ PRINCIPIS
VIDUÆ AB ETRURIA.

Propè Ecclefiam Sancti Apollinaris, fub Signo Palladis, & Herculis.
SUPERIORUM PERMISSU.

Fig. 24. Title-page of P. A. Micheli's *Nova plantarum genera*, 1729.

liverworts among green plants. It is, however, his treatment of fungi, especially the 'additional notes and observations on the planting, origin and growth of fungi, mucors and allied plants' specified on the title-page, that makes the book a mycological classic. Of the 1900 plants (1400 observed for the first time) enumerated in the *Nova plantarum genera*, 900 were fungi and no less than seventy-three of the 108 copper plates deal with lichens and fungi, the forty illustrating larger fungi being, in general, accurate representations. Among the new taxa he described, illustrated, and named were such familiar genera as *Aspergillus*, *Botrytis*, *Polyporus*, *Clathrus* (Fig. 25), and *Geaster*. He introduced the name *Puccinia* (for *Gymnosporangium*) after Tomaso Puccini, professor of anatomy at Florence, and he provided the first figures of *Sphaerobolus* (as *Carpobolus*) and indicated the method by which the glebal mass is discharged. He observed seeds (spores) in all groups of fungi, noted the quaternary arrangement of basidiospores on the gill surface of various dark-spored agarics, and he was the first to describe asci (see p. 66) and to culture fungi from spores (see p. 184). He illustrated the mycelial cords attached to the fruit-bodies of agarics (Micheli (1729): tab. 75) and was the first to describe hairs on the gill edge (which he interpreted as apetalous flowers) and cystidia (his 'corpori diaphani'; in some species conical and in others pyramidal). Micheli recognised that the function of cystidia projecting from the gill surface in *Coprinus* was to keep the gills apart and facilitate spore dispersal. He wrote: 'They are made by the wise device of nature so that one lamella does not touch another, to the end that seeds produced between them should not be hindered in their development or that they should not fall except when they ought to fall; the same bodies fall to the ground with the mature fallen seed'.[8]

Micheli had difficulty in financing his publications and the *Nova plantarum genera* is prefaced by a list of nearly 200 patrons including those who defrayed the cost of the plates (each of which cost Micheli forty-two giulios, the equivalent of one guinea) and had plates dedicated in their honour. Among the latter were William Sherard (the founder of the Sherardian chair of botany at the University of Oxford) and his younger brother James (an apothecary and owner of the botanic garden at Eltham, near London, for which Dillenius (see p. 246) compiled a catalogue). Both corresponded with Micheli who had

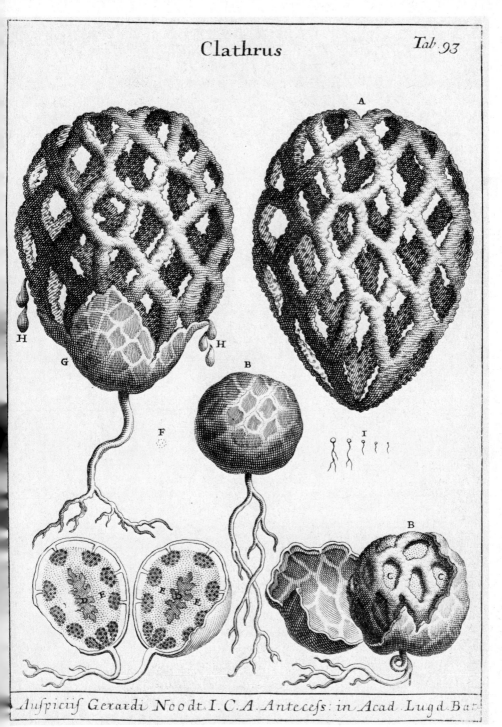

Fig. 25. *Clathrus cancellatus.* (P. A. Micheli (1729): pl. 93.)

HISTORIA
FUNGORUM
Circa
HALIFAX
Sponte
NASCENTIUM.
TOM.I

A.*Agaricus.* E.*Clathrus.* I.*Lycoperdon.*
B.*Boletus.* F.*Halvella.* K.*Sphaeria.*
C.*Hydnum.* G.*Peziza.* L.*Mucor.*
D.*Phallus.* H.*Clavaria.*

J.Bolton del: et S 7:

Fig. 26. Title-page of James Bolton's *An history of fungusses growing about Halifax,* 1788.

much impressed William Sherard when on a visit to Florence in
1717. Another patron was Cipriano Antonio Targioni (later
Targioni-Tozzetti) who, after Micheli's death, together with his
son Giovanni Targioni-Tozzetti purchased for 1381 scudi
Micheli's library, herbarium, fossils, etc., including sixty copper
plates of marine plants intended for the next part (of four
projected, according to G. Targioni-Tozzetti) of the *Nova
plantarum genera.*[9] C. Targioni-Tozzetti made little use of the
collection but he did, in 1748, bring out Micheli's catalogue of
the Florence public gardens, *Catalogus plantarum horti caesarei*

Florentini. The material was subsequently inherited by Giovanni's son Ottavanio and grandson Antonio and the manuscripts (in seventy-four volumes) are now in the University Herbarium at Florence. In 1863 the British Museum bought for £15, from Antonio's widow, Micheli's own copy of the *Nova plantarum genera,* the plates of which are supplemented by many of the original drawings.[10]

By the middle of the eighteenth century the morphological range of the larger fungi was on pictorial record. The publication of illustrations of these forms continued and the last half of the eighteenth century and the early years of the nineteenth saw

This Agaric grew in Northowerum in a little wood called Trough of Bolland. Sq. 1787. Drawn and Etched from Nature by J. Bolton at Stannary near Halifax.

Fig. 27. *Marasmius peronatus.* (J. Bolton (1788–91): 1, pl. 58.)

the production of a number of regional volumes, with artistic hand-coloured plates,[11] now much sought by book collectors. Among the more notable of these works are: *Fungorum agri Ariminensis historia*, 1755, by the Italian Giovanni Antonio Battarra with its forty excellent, but uncoloured, plates; the four volumes of *Fungorum qui in Bavaria et Palatinata circa Ratisbonam nascuntur icones*, 1762–74, by Jakob Christian Schaeffer, an evangelical clergyman of Regensburg; *An history of fungusses growing about Halifax*, 1788–91 (Fig. 26), by James Bolton (believed to have been an art teacher) with its very sensitive drawings, many made directly on to the copper plate from nature (Fig. 27), and the first and only English mycological book to be translated into German (with the figures redrawn and rearranged); and the *Coloured figures of English fungi or mushrooms*, 1797–1815, by James Sowerby, who was the founder of a dynasty of notable botanical illustrators. The comprehensive *Histoire des champignons de la France*, 1791–1812, by Pierre Bulliard with 383 plates is the first mycological book with printed colour plates (see Frontispiece); the process devised by the author.[12] This tradition continued with the introduction of colour printing by chromolithography and later processes and such twentieth-century works as the *Icones mycologicae*, 1905–10, of Émile Boudier, which also includes smaller discomycetes and certain microfungi, and the collotypes of the *Flora agaricina Danica*, 1935–41, in five volumes, by Jakob Lange, are both scientifically invaluable and superb examples of modern printing.

MICROFUNGI

Knowledge of the morphology of microfungi and the detailed structure of macrofungi was impossible without the microscope,[13] an invention fundamental for the development of both mycology and microbiology (Fig. 28). One of the most familiar names in the early history of microscopy is that of the versatile Robert Hooke, described by John Aubrey as 'but of midling stature, something crooked, pale faced, and his face but a little belowe, but his head is lardge; his eie full and popping, and not quick; a grey eie. He haz a delicate head of haire, browne, and of an excellent moist curle'.[14] Hooke was for many years Secretary of the Royal Society and he recorded his examination of a wide range of objects, with a compound

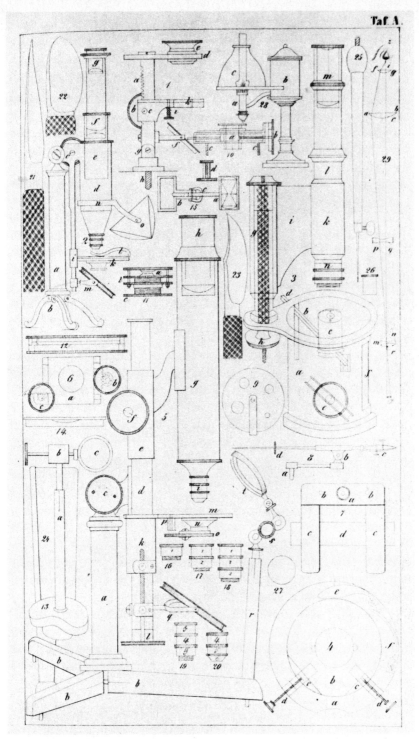

Fig. 28. The microscopist's equipment, from A. J. C. Corda, *Anleitung zum Studium der Mykologie*, 1842: pl. A, including a simple microscope by Plössl (Vienna) (fig. 1) and compound microscopes supplied by Chevalier (Paris) (fig. 2), Plössl (fig. 3) and Plössl and Schiek (Berlin) (fig. 5).

microscope of his own design and construction, in *Micrographia*, 1665 – 'the most ingenious book I ever read in my life', according to Samuel Pepys; and one that kept the diarist out of bed until two in the morning. One plate (Fig. 29) gives the first illustrations of microfungi – the sporangia of *Mucor* and the teliospores of rose rust (*Phragmidium mucronatum*) – and it should be noted that the scale of each of the two figures is carefully indicated. Another well-known pioneer microscopist was Antony van Leeuwenhoek, a draper and haberdasher who later became Chamberlain of the Council Chamber of the Worshipful Sheriffs of Delft, his native town where he lived throughout his life and died in 1723 at the age of ninety. Van Leeuwenhoek, like Robert Hooke, made his own lenses and microscopes which were of a simple design; a single lens or a doublet mounted in a plate of metal (frequently silver and occasionally gold) with a device for screwing the object into focus, and capable of magnifications of *c.* ×40 to ×200 or even more. At his death he left no less than 247 complete microscopes (each generally provided with an object) together with 172 additional mounted lenses. Van Leeuwenhoek made many original observations on micro-organisms and although he never divulged his 'particular manner of observing very small creatures' and kept his best microscopes 'for himself alone' (see the scholarly monograph by Dobell, 1932) he communicated the results of his observations in a long series of letters to the Royal Society of London into which he was elected as a Foreign Member. Van Leeuwenhoek is most renowned for his discovery of bacteria, from 'pepper water' and his own teeth, as described in his eighteenth (9 October 1676) and thirty-ninth (17 September 1683) letters to the Royal Society. He also examined some fungi. In the first letter (1673) he described his observations on 'mould' (possibly a *Mucor* species), the mouth parts and eye of the bee, and the louse, but his most significant mycological observation was that of *Saccharomyces cerevisiae* in fermented beer which having been poured into two small clean glasses was turbid due to the presence of a great many small particles of which he noted:

Some of these seemed to me to be quite round, others were irregular, and some exceeded the others in size and seemed to consist of two, three or four of the aforesaid particles joined together. Others again consisted of six globules and these last formed a complete globule of yeast. [Letter 32 (14 June 1680); see Chapman, 1931.]

Fig. 29. The first illustrations of microfungi. *Mucor* and *Phragmidium mucronatum* (rose rust). (R. Hooke, *Micrographia*, 1665, pl. 12.)

'The first Figure...is nothing else but the appearance of a small white spot of hairy mould, multitudes of which I found to bespeck & whiten over the red covers of a small book, which, it seems, were of Sheeps-skin'.

'The whole Oval oooo in the second *Figure*...represents a small part of a Rose leaf, about the bigness of the little Oval in the hillock, c, marked with the Figure X'. (R. Hooke (1665): 125, 122.)

It is thus clear that van Leeuwenhoek must be credited with the first observation of yeast cells, both single and in groups resulting from budding.

The previous year, Marcello Malpighi, professor of medicine at the University of Bologna, devoted a plate of the second part of his folio *Anatome plantarum* (published by the Royal Society in 1679) to fungi, all but one (an agaric which he designated *Fungus*) of common moulds (*Mucedo*) from inferior cheese, putrescent melon pericarp, lemons, oranges, wood, and bread, including *Rhizopus stolonifer*, other mucoraceous fungi, and, though with less certainty, *Penicillium* and *Botrytis*. Also figured, are portions of mycelium, some apparently septate. Malpighi's figures are inferior to those published by Micheli in 1729 who, as already mentioned, coined the generic names *Aspergillus*, *Botrytis*, and *Mucor* in the senses they are used today. Micheli also illustrated myxomycetes belonging to the genera *Lycogala*, *Mucilago* (two more names he introduced), *Arcyria* (as *Clathroides*), and *Stemonitis* (as *Clathroidastrum*); also *Reticularia* and *Fuligo*. Further, Miss Lister (1912) identified Micheli's *Puccinia ramosa* as *Ceratiomyxa fruticulosa* and noted that his characterisation of one species of *Mucilago* as 'alba, ramosa, radices arborum fibrosas simulans' (Micheli, 1729:216) is the first description of a plasmodium. The only previous detailed description of a myxomycete was that given two years before by the Frenchman J. Marchant (after whose father the liverwort *Marchantia* was named) of 'flowers of tan' (*Fuligo septica*) (see p. 14) which he classified as a 'sponge' (Marchant, 1727).

During the next hundred years many more microfungi were described and named and though spore shape and septation was indicated with increasing frequency the descriptions remained largely qualitative, only for larger fruit-bodies was size quantified. Even during the first quarter of the nineteenth century, Persoon, Fries, and the other founders of modern mycological systematics paid scant attention to microscopic details. The main reason for this was that although the compound microscope was invented towards the end of the sixteenth century it remained little changed for some two hundred years. The major defects of early microscopes were the spherical and chromatic aberrations of the objectives and it was not until 1830 that objectives of higher quality were designed, first by J. L. V. and C.-L. Chevalier in France. In 1840 Giovanni Battista Amici

(1784–1860) introduced a water-immersion objective but the modern microscope did not originate until 1878 when the Zeiss factory at Jena began to manufacture microscopes to the designs of F.. Abbé (1840–1908) incorporating oil immersion objectives, the substage condenser (first designed by Abbé for the bacteriologist Robert Koch) and finally, in 1886, the apochromatic objective.

Few authors followed the example of Hooke in indicating the scale of their illustrations or attempted, as did van Leeuwenhoek, to calculate the size of the objects under examination. A. C. J. Corda was the first mycologist consistently to give spore sizes, usually in terms of the 'Paris inch' (= 27.9 mm), using the abbreviations 'p.p.' or 'p.p.p.' for 'parts per Paris inch'. Another common unit of measurement of length for small objects was the 'line' (2.1167 mm = $\frac{1}{12}$ inch), the 'Paris line' (Parisier Linie or P.L.) being slightly larger (2.2558 mm). Subsequently the *micron* (a thousandth of a millimetre) became standard usage and it was W. F. R. Suringar who, in 1857,[15] introduced into biological writing the symbol 'μ' for this unit (now μm according to the Système International d'Unités).

It would be inappropriate to detail the many publications by many authors in which microfungi are treated in greater or less detail. It must suffice to mention a few of the outstanding mycologists and some of their principal publications in this field, to most of which reference is made elsewhere for details relevant to other topics. At the turn of the eighteenth to nineteenth century H. F. Link, Persoon (see p. 255), Fries (see p. 259), Montagne, J. H. Léveillé (see p. 263), and Nees von Esenbeck all made notable contributions to the knowledge of microfungi but it was A. C. J. Corda who raised the study of these forms to a new level, especially in the five folio volumes of his *Icones fungorum hucusque cognitorum*, published in Prague between 1837 and 1840, with its hundreds of small illustrations of microfungi in the author's very characteristic style (see Fig. 30). H. F. Bonorden's *Handbuch*, 1851, is still referred to by students of microfungi as are the many outstanding publications of the Tulasne brothers, for example their memoirs on the rusts and smuts (Tulasne, 1847, 1854) and ergot (Tulasne, 1853) (see Fig. 76), and their *Selecta fungorum carpologia*, 1861–5, the illustrations of which have never been surpassed for artistry (Fig. 6). Later there were the *Beiträge zur Morphologie und Physiologie der Pilze* of Anton de

Bary and M. Woronin (in five parts, 1864–81) and the fourteen parts of the *Untersuchungen aus dem gesammtgebiete der Mykologie*, 1872–1912, by Oscar Brefeld in which so much original observation is so beautifully recorded (Fig. 81).

As already noted (p. 14), della Porta was the first to observe fungal spores and Micheli recorded spores for all the groups of fungi which he studied. Both these authors, like their successors, called the spores 'seeds' and it was not until 1788 that Hedwig introduced the term *spore* (*spora*) but he merely differentiated spores from seeds by the development of the former in conceptacles which he called 'sporangia' and used *spora* and *semen* interchangeably when describing *Agaricus*. The term spore was, however, widely taken up for cryptogams and L. C. M. Richard in 1808 clearly recognised the essential difference between spores and seeds when he wrote 'Les *Sporules* diffèrent des Graines, non seulement par leur mode de formation, mais encore et surtout par leur défaut d'Embyron'; as did his son Achille Richard in *Nouveaux élémens de botanique*, 1819. In spite of this, even as late as the eighteen-sixties the Tulasnes were still inclined to the view that spore was a superfluous term because, they claimed, the whole fungus spore 'may be regarded as an embryo. Made up of only a simple cell, or of several united together, it presents a faithful picture of its purely parenchymatous parent.' This view did not impede the general acceptance of the term spore by mycologists and in the course of the next hundred years more than a hundred names (many of them still-born) were coined to distinguish spores of different shapes, modes of production, and functions.[16]

The first observation of motile spores (zoospores) in fungi was that by Bénédict Prévost (1807) when he clearly described the development of zoospores from the sporangia of a species of *Albugo* (presumably *A. portulacae*) from *Portulaca oleracea* in water when:

An hour or two after immersion, sometimes only 40 or 45 minutes, doubtless according to the temperature, which, while I have observed them, has varied from 12 to 16° R., the larger and more convex extremity opens so that the whole resembles a bottle whence a good part of the neck has been removed.

Very soon one globule appears on the outside, immediately followed by three, four, five, or six others, which come together on an instant in a ball, and move together for some time...Ordinarily, the globules then separate...They move like balls, but with much more agility.[17]

The diversity of non-motile spores proved greatest among the

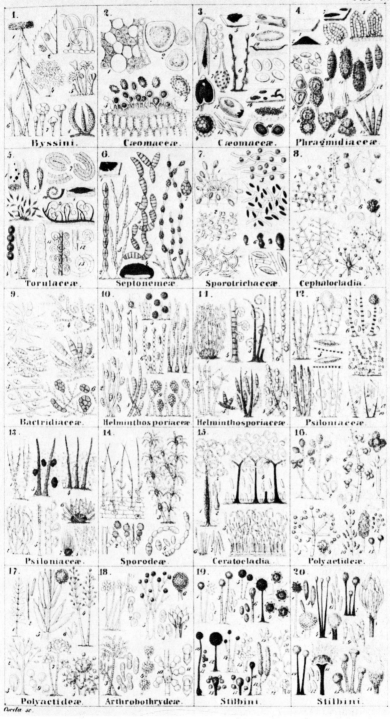

Fig. 30. Plate B (approximately natural size) of microfungi from
A. C. J. Corda's octavo *Anleitung*, 1842. Some plates of his folio
Icones have ninety or more such groups of figures.

fungi imperfecti where spore shape, septation, and colour were used by P. A. Saccardo in the arbitrary system he devised in the late eighteen-eighties for ordering these fungi in his *Sylloge* (see p. 242). Then followed a period during which much was done to widen the knowledge of imperfect fungi without any fundamental change in the approach to their spores. A new era opened in 1953 with the publication of the germinal paper by S. J. Hughes in which, by integrating views of such workers on hyphomycetes as Paul Vuillemin and E. W. Mason with the results of his own observations, he differentiated eight main sections of hyphomycetes on the basis of conidiophore and conidium development. This innovation stimulated a re-appraisal of the hyphomycetes (never more productively than in the splendid and comprehensive monograph on the dark-spored genera by M. B. Ellis of the Commonwealth Mycological Institute, Kew), and more recently still of the coelomycetes also, which is currently doing much to elucidate the fundamental structure of these microfungi and influence their taxonomy.

MICROSTRUCTURE OF LARGER FUNGI

Theophrastus (*Historia plantarum*, I, 5, § 3) described the stem of the fungus as being of uniform structure or evenness, without knots, prickles or divisions and there the matter rested until the advent of the microscope some two thousand years later but even then it took the best part of two centuries for the essential microscopic features and special organs of macrofungi to be established.

Robert Hooke gave the first account of the internal structure of mushrooms. In his *Micrographia*, in the section on sponges, he wrote:

Having examin'd also several kinds of Mushroms, I finde their texture to be somewhat of this kind, that is, to consist of an infinite company of small filaments, every way contex'd and woven together, so as to make a kind of cloth, and more particularly, examining a piece of Touch-wood (which is a kind of *Jews-ear*, or Mushrom...commonly call'd by the name of *Spunk*) I found it to be made of an exceedingly delicate texture...of an infinite company of filaments, somewhat like bushes interwoven with one another, that is, of bigger parts or stems, as it were, and smaller branches that grow out from them; or like a heap of Ropes ends, where each of the larger Ropes by degrees seems to split or untwist, into many smaller Cords, and each of these Cords into smaller Lines, and those Lines into Threads, &c and these strangely intangled, or interwoven with one another...The filaments I could plainly enough perceive to be even, round, cylindrical, transparent bodies and to cross each other every way, that is, there were not more seem'd to lie *horizontally* then *perpendicularly* and thwartway.

Fungal filaments were subsequently seen by Marchant *père* in 1678 and Tournefort (1707) in horse droppings (see p. 17) and both Marsigli (1714) and N. J. Necker (in his *Traité sur la mycitologie*, 1783) noted mycelial cords which they considered to be the transitional stage in the development of fungi from decaying plant tissue. Necker coined the term 'carcithium' for these structures and he illustrated a number of forms. It was not until 1830 that the Latin term *mycelium* was introduced by L. Trattinick (1804–6) for the *Schwammegewächs* in distinction to *Fruchtkorper* (fruit-body or 'encarpium'), another new introduction. K. L. Willdenow (1810) was the first to use *hypha* for an element of the mycelium.

A vegetative structure the elucidation of which gave much trouble was the sclerotium. The term *sclerotium* was first proposed as a generic name by Tode in 1790 for eight species but subsequent authors added many more. As Léveillé pointed out in his little monograph of 1843 on the sclerotium, in spite of the elimination of more than forty names, based on pathological products of plants such as insect galls (and here he included the ergots of Gramineae, including paspalum and maize), immature fungi, and species of other fungi and synonymous names, more than fifty species were compiled by Fries in the *Systema mycologicum*. As a result of his investigations of the range and structure of sclerotia, Léveillé concluded that mycelium occurs in four principal forms – '*filamenteux, membraneux, tuberculeux* et *pulpeux*' – and that the sclerotium 'n'est qu'une de ces formes; il ne doit pas être considéré comme une genre particulier'. The most notorious sclerotium was the 'fungus stone' (*Polyporus tuberaster*), a subject of speculation from early times (see pp. 32, 50), which was discussed by de Bary in his textbook when reviewing sclerotia (de Bary, 1887: 30-43) and set in perspective by C. Bommer (1894) in a study of diverse sclerotia and mycelial cords.

For the past hundred years medical mycologists have shown interest in the fungal 'grains' or 'granules' which characterise the group of human and animal mycoses known as mycetomas. These grains, which are analogous to sclerotia, develop in infected tissues as a result of host–pathogen interactions. The classical mycetoma is that described from India by Henry Vandyke Carter (professor of anatomy and physiology at the Grant Medical College, Bombay, in his monograph of 1874) as Madura foot which is typically caused by the black-grained *Madurella mycetomi*. Mycetomas are also caused by a number of other fungi and actinomycetes, characterised by grains of other colours, as reported in the doctoral thesis of Émile Brumpt (1906) and the recent monograph on mycetoma by Mahgoub & Murray (1973). Species of *Aspergillus* (especially *A. fumigatus*) give rise in the lungs and other organs to aspergillomas which, although not suppurating, have in the past often been classified as mycetomas.

Micheli (1729) figured asci and ascospores for lichens (tabs. 36, 52, 56), truffles (tab. 102, *Tuber*), and various pyrenomycetes. The details of asci in the published plates are not always as clear and accurate as in Micheli's original drawings which accompany the plates in Micheli's own copy of *Nova plantarum genera* now in the British Museum. The original of the asci of *Tuber* adds little to the published version (Fig. 31a). In Micheli's tab. 56 (*Lichen–Agaricus*), fig. 1 (which illustrates *Xylaria hypoxylon*), the rows of minute dots (labelled 's') representing the spores are shown in the original drawing as three vertical rows of six or seven spores, although the ascus wall is not apparent. The most striking series is that associated with tab. 56, fig. 1, illustrating a species of Micheli's genus *Lichenoides* – the lichen genus *Pertusaria* which is characterised by exceptionally large ascospores up to 300×100 μm. Part of the published illustration is reproduced here as Fig. 31b together with the relevant part of Micheli's original (Fig. 31c; a mirror image of the published figure) in which an ascus containing four ascospores is clearly drawn. But the most striking illustration is in the Micheli MSS. at Florence where in an early draft of his observations on *Lichenoides* he gives splendid and unequivocal sketches of asci and ascospores (Fig. 31d). It remained, however, for Hedwig, in 1788, to publish conclusive evidence of the existence in discomycetes of asci containing ascospores when he described and illustrated these organs (and also *paraphyses*, a term he introduced) (Fig. 32) by which he characterised his genus *Octospora*. He called the ascus a 'theca', a term he had previously coined for the sporangium of bryophytes. Subsequently, *ascus* was substituted for theca in mycology by Nees von Esenbeck

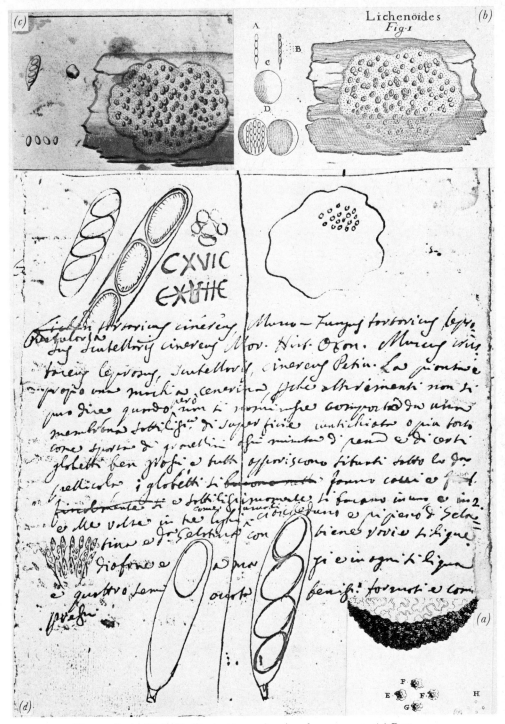

Fig. 31. P. A. Micheli's illustrations of asci and ascospores. (a) Part of *Tuber* fruit-body in section and detail of asci and ascospores (E, F, G) (Micheli (1729): pl. 102). (b)–(d), *Lichenoides* [*Pertusaria*]; (b), Micheli (1729), part of pl. 56, fig. 1; (c) Micheli's original drawing for the same illustration (British Museum); (d), page from the Micheli MS. Florence (vol. 50, facing p. 105) showing asci and ascospores of *Lichenoides*.

OCTOSPORA *scutellata.*

Fig. 32. Asci and ascospores of *Scutellinia scutellata* (J. Hedwig (1788); part of pl. 3).

(1816) – 'Thecae, den ich, da der Ausdruck doch noch öfter für die Moosfrucht gebraucht wird, mit der entsprechenderen Benennung: Asci, Schläuche, vertauschen will, bezeichnet zu werden pflegt.' – who restricted the use of theca to bryophytes. The origin of ascospores was thus established. That of basidiospores proved more troublesome.

Micheli, by viewing the gill surface of a dark-spored agaric such as *Coprinus*, noted the quaternary arrangement of the spores (Figs. 33, 34) but although he described cystidia both on the gill surface and on the gill edge he made no suggestion as to how the basidiospores were borne. Schaeffer (1759) was the first to describe sterigmata when he noted that examination with a

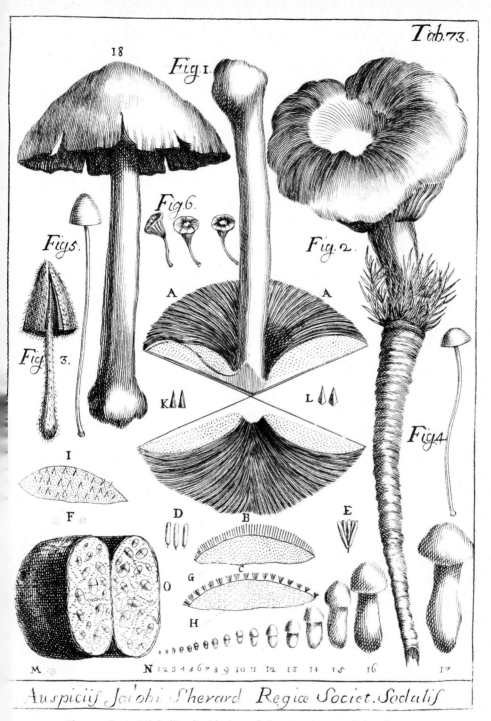

Fig. 33. P. A. Micheli's elucidation of the structure and development of the agaric fruit-body. The relevant illustrations are all included in fig. 1. B, C, Hairs on the gill edge (interpreted by Micheli as apetalous flowers); H, basidiospores arranged in fours; I, cystidia projecting from the gill surface; K, L, individual cystidia; N, 1–17, successive stages of fruit-body development. (P. A. Micheli 1729: pl. 73.)

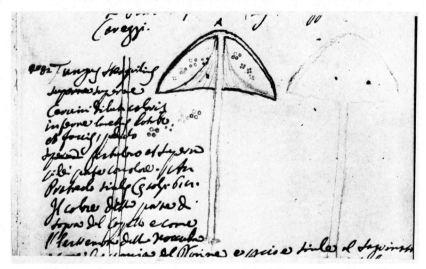

Fig. 34. Quaternate arrangement of basidiospores on the gill surface of *Coprinus*. (P. A. Micheli, MS., vol. 15:38, Florence.)

strong lens of an agaric, bolete or clavaria reveals 'round spores, like so many flowers, each of which is seated upon its own thread-like stem (fadennähnlichen Fusse und Stiele)' but it was not until 1780 that O. F. Müller when describing *Coprinus comatus* in *Icones florae Danicae* (Fasc. xiv, tab. 834) included a surface view of the hymenium which illustrated not only the quaternary arrangement of the spores but also the underlying basidia. He also illustrated paraphyses. In 1794 Persoon introduced the term *hymenium* and united Discomycetes and Hymenomycetes as his sixth order, Hymenothecium, characterised as having; 'Receptaculo ut plurimum carnoso membranae (*Hymenio*) ex thecis in seriem digestis constanti, toto, adnexo' and implied in his new concept of a hymenium-covered, spore-bearing organ that the spores were borne in asci. Two years later he described the quaternary arrangement of the spores of *Corticium caesium* which he apparently considered to be superficial (Persoon (1796–9) **1**: 15). Link (1809: 35–7) was more definite. In adopting Persoon's classification he stated unequivocally that the spores of agarics were borne in thecae, *Coprinus* having spores in fours because there are four series of spores in each theca. This view prevailed unchallenged for the next twenty-five years. Every expert (including Ditmar, 1813; Nees, 1816; Persoon, 1818; Fries, 1821, 1830; Desmazières, 1828; Krombholz, 1833; and Klotzsch,

1833) examining the hymenium of agarics described, and frequently clearly illustrated, the spores in asci as did R. K. Greville in his *Scottish cryptogamic flora*, 1823–8. In 1831, C. Vittadini, in *Monographia tuberacearum*, illustrated basidia, both unispored and tetraspored. For the former he illustrated internal spores and, for the latter, showed details of how in *Bovista* the internally developed spores pass to the exterior. Corda in J. Sturm's *Deutschlands Flora*, 1837 (**3**, Taf. 49), explained the quaternary arrangement as being due to the occurrence of the spores at the corners of a nework. Then there was a sudden change and between 1836 and 1838 no less than six investigators independently elucidated the problem of basidial structure.

F. M. Ascherson (of Berlin) in a note published in 1836 was the first to observe that the spores of several species of higher fungi are not borne in asci but at the ends of small stalks, usually four in number, which project from a cylindrical structure. He gave no illustration, and credit for first establishing the details of hymenium structure must be given to J. H. Léveillé, a medical man, who announced to the Société Philomatique de Paris on 17 March 1837 his findings which were published later in the year as his classical paper in which he coined the terms *basidium* and *cystidium* (Fig. 35). He recounted how he had repeated his experiments before Persoon who agreed with his conclusions and could not conceive how organs so constant and so easy to observe should have escaped notice by the many mycologists who had made microscopical investigations. At the same time in England the Rev. M. J. Berkeley, following the acquisition of a more powerful doublet than he hitherto possessed (his first compound microscope was a gift from J. D. Hooker, Director of the Royal Botanic Gardens, Kew, in 1868), came to similar conclusions regarding spore production in agarics (Fig. 35) (published in 1838 as a paper entitled 'On the fructifications of the pileate and clavate tribes of Hymenomycetous Fungi'), and the following year showed the same method to hold for gasteromycetes. It is interesting to note that Berkeley was on the verge of his discovery two years earlier when he wrote of *Agaricus prunulus*; 'Gills decurrent, narrow, more or less forked, covered with very minute conical papillae, ending in four spiculae. *Sporules* rose-coloured, elliptic, often seated upon the spiculae' (Berkeley, 1836: 76). Corda in 1837 correctly described

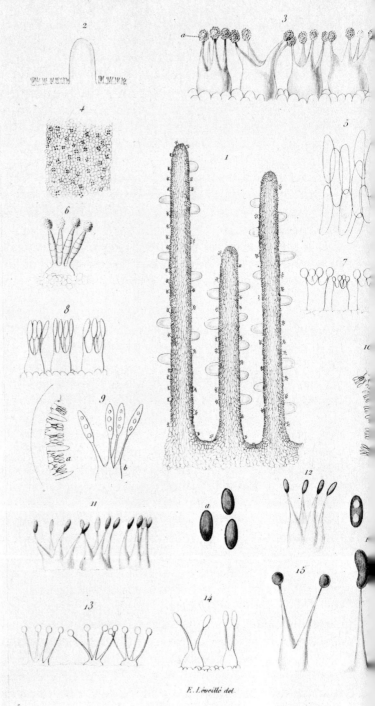

E. Léveillé del.

Structure de l'Hymenium des Champignons

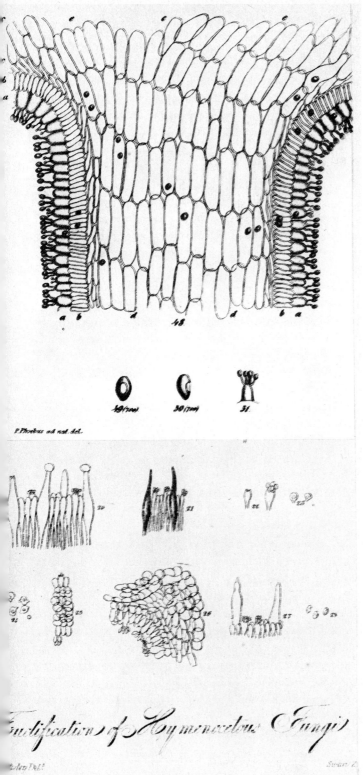

Fig. 35. Elucidation of basidial structure by Léveillé (1837: pl. 8, *left*), Berkeley (1838a: part of pl. 4, *bottom right*), and Phoebus (1842: part of pl. 56, *top right*).

and illustrated basidia and basidiospores of *Coprinus petasiformis* (Corda (1837–54) **1**: tab. vii, fig. 300) as did Klotzsch (1838) for no less than twenty-five basidiomycetes. The final and longest paper of this series, with 142 figures of basidia and basidial structure on two plates (Fig. 35), was that by P. Phoebus of Giessen, submitted for publication on 6 March 1838 but which did not appear in print until 1842. Phoebus had, however, in his *Deutschlands kryptogamische Giftgewächse* of 1838 correctly illustrated basidia (his 'Träger') and spores in tetrads.

This elucidation of the structure of the basidium is a good example of one way in which scientific discoveries are often made. Advance of knowledge, not only in science, is frequently retarded by the prevailing orthodoxy, by deference to authority, and – as every teacher knows – by the finding of what is expected. Correct observations not in accord with current ideas are explained away or their significance not appreciated. Finally, a new or improved technique enables several investigators by simultaneous unprejudiced observation to resolve a problem independently. Even then, older views linger on and the truth is not at once universally accepted. Fries in 1849[18] believed all genuine spores to be produced in evanescent asci. H. Schacht (1852) still considered the distinction between ascospore and basidiospore to be of little significance as did H. Hoffmann in 1856 and in the second edition of his textbook de Bary (1887: 341) was uncertain of the status of the asci of the parasite *Endomyces decipiens* which he had observed in the gills of *Armillaria mellea* in 1859. This led him to suggest that basidiomycetes were 'the gonidial [conidial] forms of perfect existing Ascomycetes'.

Following the establishment of the basic structure of larger fungi there was, as noted for microfungi, a period of widening knowledge during which little that was fundamental was added to the knowledge of the structure of these forms. This was particularly true of hymenomycetes. Thousands of additional species were described, frequently on external morphology and on the basic types of hymenium configuration adopted by Fries and other early nineteenth-century systematists. In marked contrast to that of his contemporaries was the approach of the French pharmacist Narcisse-Théophile Patouillard who in 1887 illustrated *Les Hyménomycètes d'Europe* by four plates devoted to microscopical details of fruit-bodies and not to illustrations of their external morphology, although at that time his classifica-

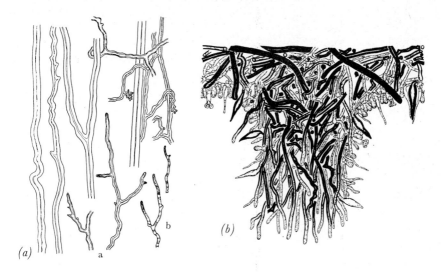

Fig. 36. Hyphal analysis. (*a*), *Polystictus xanthopus* 'Fragments of four skeletal hyphae with a few binding hyphae from the mature stem: a, the apices of binding hyphae from the primordial shaft; b, the apices of narrow generative hyphae from the pore-field.' (E. J. H. Corner (1932*a*): text-fig. 2.) (*b*), *Fomes levigatus*. Section of the pore-field of the developing dimitic fruit-body showing generative and skeletal hyphae. (Corner (1932*b*): part of text-fig. 9.)

tion was still basically Friesian (see p. 263). It was from such beginnings, and the researches of Fayod (1889), that the next major breakthrough had its origin. This was the technique of 'hyphal analysis' propounded by the Cambridge mycologist E. J. H. Corner in 1932 who showed that the fruit-bodies of polypores were made up of three main types of hyphae – *generative*, *skeletal*, and *binding* (Corner, 1932*a*) (Fig. 36*a*) – and he distinguished fruit-bodies composed of one, two, or three of these types as *monomitic*, *dimitic*, *trimitic*, respectively (Corner, 1932*b*) (Fig. 36*b*). Corner's approach and his terminology have been widely adopted and extended to other groups, as for example, by Corner himself to the Clavariaceae (Corner, 1950). Hyphal analysis was one of the criteria used by G. H. Cunningham in his studies from 1946 onwards of the aphyllophorales of New Zealand and Australia and the procedure is now standard practice. It has done much to draw attention to the microscopic details of fruit-body construction and to emphasise the importance of convergent evolution in the emergence of superficially

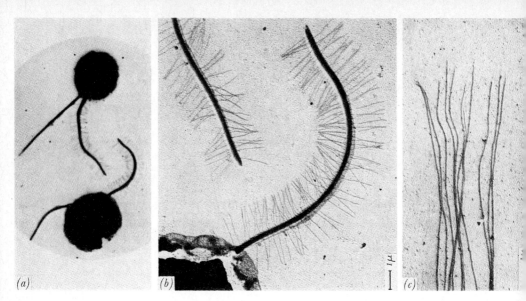

(a)   (b)   (c)

Fig. 37. Electron micrographs illustrating flagellum structure.
(a)–(b), *Saprolegnia ferax* flagella photographed (× 1200) with visual
light (a) and the corresponding electron micrograph (b). Note the
anterior whiplash flagellum and the posterior tinsel flagellum. (c),
dismembered *Olpidium brassicae* whiplash flagellum showing the
eleven component strands. (I. Manton *et al.* (1951): pl. 1, figs. 1, 2;
pl. 3, fig. 27.)

very similar fruit-bodies from diverse origins. As a result, the
taxonomy of the hymenomycetes, like that of the imperfect
fungi, is currently in a state of flux.

The introduction in 1942 (by the Dutch physicist F. Zernike, a
1932 Nobel prize-winner)[19] of phase-contrast microscopy for the
examination of unstained living preparations was a useful
development but it was the deployment of transmission electron
microscopy as a routine mycological technique from about 1950
(see Gregory & Nixon (1950), for example) and the scanning
electron microscope a few years later that allowed the investigation of fungus form and structure at a deeper level. Currently
this is a very popular field. In general, transmission electron
microscope investigations have shown the cells of fungi to have
an ultrastructure similar to that of other eukaryotic organisms.
One interesting outcome was the elucidation of the structure of
the pore in septa of basidiomycete hyphae which, unlike the
pore of ascomycetes, is a complex structure or *dolipore* (Moore &
McAlear, 1962), which while preventing nuclear migration
between adjacent cells allows continuity of the cytoplasm.
Another was the confirmation by Professor Irene Manton and

76

her collaborators (Manton *et al.*, 1952) at the University of Leeds that the anterior flagellum of the biflagellate zoospore of *Saprolegnia* is of the flimmer (tinsel or *Flimmergeissel*) type and the posterior of the whiplash (*Peitschgeissel*) type (Fig. 37*a,b*). They also demonstrated that the flagellum in fungi, whether flimmer or whiplash, like that of plants and animals, is composed of eleven longitudinally arranged fibres, a central pair surrounded by a ring of nine (Fig. 37*c*). Results obtained by the scanning electron microscope,[20] as exemplified by the monograph of D. N. Pegler & T. W. Young (1971) on basidiospore morphology in the Agaricales, should make increasingly significant contributions to taxonomic studies.

MORPHOGENESIS

For long the approach to both fungal form and structure was basically descriptive and, although this approach is still much used, studies on these aspects of fungi are becoming dominated by the more dynamic approach implicit in investigations on morphogenesis. Knowledge of morphogenesis is currently being gained at four levels. Instead of giving static descriptions morphologists and taxonomists are describing the developmental stages of ascocarps, basidiocarps, spores, and other fungal organs. Physiologists are investigating experimentally the diverse external factors which determine the expression of fungal form, geneticists are elucidating heriditary determinants, while, concurrently, chemists attempt to discover the underlying biochemical mechanisms. These investigations are being made by many workers, both as individuals and in teams, and while much detailed information is being accumulated few generalisations are as yet possible. Here only some of the more obvious main themes can be mentioned. For further detail the reader is referred to the many readily accessible reviews by which practitioners of this branch of mycological research are constantly updating their findings.[21]

Essentially a modern development, the study of fungal morphogenesis nevertheless has deep roots, for Micheli, in 1729, illustrated the developmental stages of several agarics (see Fig. 33) and 'fairy rings' have for long attracted attention and been a subject of much folklore.[22] Erasmus Darwin in 1791 believed fairy rings to originate from lightning strikes[23] but in the following year the English medical botanist William Withering seems to have been the first to recognise their fungal origin when he attributed rings associated with *Marasmius orcades* (*M. oreades*, the fairy-ring champignon) to the growth of the fungus.[24] During the eighteen-seventies and eighties the occurrence and chemistry of fairy rings was investigated by J. H. Gilbert and J. B. Lawes at Rothamsted and there have been a number of studies since those of Jessie M. Bayliss in 1911 of measurements of the growth rate of fairy rings. The best known is that of Schantz & Piemeisel (1917) in Colorado where they reported on centuries-old rings of great size and offered a detailed explanation for the well known effect of the ever expanding ring of mycelium on the grass or other vegetation (Fig. 38). More

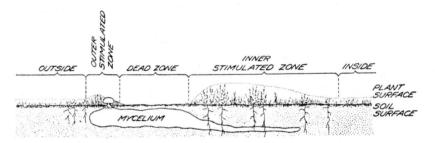

Fig. 38. Structure of a 'fairy ring' of *Agaricus tabularis* after H. L. Schantz & F. J. Piemeisel (1917): fig. 4.

recently, Burnett & Evans (1966) examined the genetical constitution of rings of *M. oreades* believed to be 100–150 years old and in no ring did they find more than two mating-type factors which they took to be good presumptive evidence that each mycelium was genetically homogeneous.

Current advances in fungal morphogenesis have largely stemmed from the work and views of Georg Klebs, professor of botany at Tübingen, as set out in his classic work *Die Bedingungen der Fortpflanzung bei einigen Algen und Pilzen* of 1896 and developed in later publications (e.g. Klebs, 1898–1900). Klebs recognised that living cells are influenced in three ways – by their specific structure (by which he understood their basic genetic constitution which ensures the constancy of expression shown by a specific organism under the same conditions); by internal conditions (the complex interaction between cells of the whole organism); and by external conditions (environmental factors). He did much of his experimental mycological work with *Saprolegnia* (see p. 138).

As already implied, knowledge of fungal morphogenesis has been acquired via many approaches using a great variety of techniques in attempts to elucidate an equally great variety of specific problems. The spore is an appropriate starting point because many fungi originate from single spores and the conditions which determine spore germination have been a favourite topic for reviews, at levels of increasing depth, since the accounts given by de Bary in the two editions of his textbook.[25] At first most attention was paid to environmental factors, that is to the external factors conditioning germination. These were succeeded by studies on internal factors and those associated with dormancy, longevity, and survivability are, as evidenced by the monograph of A. S. Sussman & H. O. Halvorson (1966), a prominent feature of recent investigations. Among other clarifications, these latter investigations proved self-inhibitors to occur in many spores and thus explained the observation made some fifty years earlier, and frequently confirmed, that for many different species spores in high concentrations germinate less well than at low concentrations. Such inhibitors may be volatile or non-volatile. A number have now been chemically defined but their bearing on ecological and evolutionary problems has yet to be assessed. A basic problem, and one still unresolved, is to elucidate the mechanism of the apical growth of hyphae and to correlate this with the chemistry of the cell wall and its ultrastructure; a topic illuminated by the work of E. S. Castle during 1927–42 on the growth of the sporangiophore of *Phycomyces* and the subsequent detailed study by M. Delbruck of the effect of light on the process. A related problem is that of dimorphism (see p. 30) – the mycelium (M) $\rightleftharpoons$ yeast (Y) transformation shown by many fungi under appropriate conditions – a transformation which, according to the convincing evidence obtained by the American microbiologist W. J. Nickerson and his collaborators during the nineteen-fifties from studies on *Candida albicans* (see the review by Nickerson, 1963), is associated with the oxidation–reduction state of certain sulphydryl (—SH) groups in a protein constituent of the cell wall which is in turn dependent on the supply of metabolic hydrogen for reduction in the presence of the enzyme protein-reductase. It is perhaps appropriate that attention should next be drawn to the dwarf form of *Coprinus fimetarius* observed by Quintanilha & Balle in 1940, the slow growing '*poky*' mutants of *Neurospora* described by Mary & H. K Mitchell in 1952, and the comparable '*petite*' mutants of *Saccharomyces cerevisiae* (Ephrussi, 1953) which are all examples of morphological expression controlled by extra-nuclear (cytoplasmic, non- chromosomal) inheritance.

Morphogenesis in the cellular slime moulds (Acrasiales) has received much attention because of its basic simplicity and the convenience of these slime moulds as experimental organisms. Knowledge of the aggregation of the myxoamoebae, due to the chemotactic effect of acrasin, into plasmodia and the subsequent development of the fruiting bodies (sorocarps) is well summarised in the two editions of the monograph by J. T. Bonner of Princeton University with whose name and that of Kenneth Raper so much of this work has been associated (Bonner, 1959). Water moulds, too, have proved favourite experimental subjects in this field. Much has been learned regarding the development of *Blastocladiella* from the productive researches of E. C. Cantino of Michigan State University at East Lansing (who like Bonner, the Raper brothers, and Ralph Emerson, among others, was a student of that inspiring American mycological teacher William H. Weston Jr of Harvard University), the results of which he has frequently reviewed. Equally fascinating knowledge of related fungi have been derived from the elucidation of the hormonal control of sexuality in *Allomyces*, a topic touched on in Chapter 5 (p. 138). Naturally, less progress has been made with the more complex ascomycetes and basidiomycetes on which there have, however, been notable studies including the monograph by A. F. M. Reijnders (1963) summarising his detailed anatomical investigations on fruit-body development in agarics and other larger basidiomycetes. Examples of a more experimental approach are provided by the work of S. D. Garrett at Cambridge and Gillian M. Butler at Birmingham on rhizomorphs and mycelial strands in basidiomycetes and that by B. E. Plunkett of Birkbeck College, London, on the basidiocarps of *Collybia velutipes* and *Polyporus brumalis*, while much attention has been paid to the genetic and environmental factors which control sporulation in both these and other groups of fungi (see p. 138). Finally, to return again to spores which are both a beginning and an end of fungal development, the studies on spore ontogeny in imperfect fungi, which are having a profound effect on the taxonomy of the groups (see p. 64), may be recalled.

4. Culture and nutrition

When and where the culture of fungi was first attempted is unknown. It was possibly some two thousand years ago in the Far East where the cultivation of *shii-take* (the edible *Lentinus edodes*) has long been practised in Japan. In Europe even as late as the mid-seventeenth century the guidance of classical writers on the cultivation of larger fungi was still being offered as is instanced by 'How Toad-stools may be generated' from the 1658 English translation of Giambattista della Porta's *Natural magic where in are set forth all the riches and delights of the natural sciences*[1] (a work first published in Italy in Latin in 1589):

Dioscorides, and others have written, That the bark of a white Poplar-tree, and of a black, being cut into small pieces, and sowed in dunged land or furrows, will at all times of the year bring forth mushromes or toad-stools that are good to be eaten. And another place he saith, that they are more particularly generated in those places, where there lies some old rusty iron, or some rotten cloth; but such as grow neer to a Serpents hole, or any noisome Plants, are very hurtful. But *Tarentium* speaks of this matter more precisely. If, saith he, you cut the stock of a black Poplar peece-meal into the earth, and pour upon it some leaven that hath been steeped in water, there will soon grow up some Poplar toad-stools. He added further; if an up-land or hilly field that hath in it much stubble and many stalks of corn, be set on fire at such a time as there is rain brewing in the clouds, then the rain falling, will cause many toad-stools there to spring up of their own accord: but if, after the field is thus set on fire, happily the rain which the clouds before threatened does not fall; then, if you take a thin linnen cloth, and let the water drop through by little and little like rain, upon some part of the field where the fire hath been, there will grow up toad-stools, but not so good as otherwise they would be, if they had been nourished with a showre of rain.

It was at about this time that the cultivation of mushrooms was being developed in France[2] and also in 1658 John Evelyn – diarist, man of letters, and later Fellow of the Royal Society – under the pseudonym 'Philocepos' published an English translation of *Le jardinier françois* by N. de Bonnefons (Paris, 1650) according to which to produce 'Bed-mushrooms' in your garden:

you must prepare a *bed* of *Mules*, or *Asses soyl*, covering it over four fingers thick with short, and rich *dung*, and when the great heat of the *bed* is qualified, you must cast upon it all the *parings*, and *offalls* of such *Mushrooms* as have been dressed in your *Kitchen*, together with the *water* wherein they were washed, and also such as are *old* and *Worm-eaten*, and a *bed* thus prepared, will produce you very good, and in a short space. The same bed may serve you two, or three years, and will much assist you in making another.

In 1707 a detailed and more technical account of the Parisian method of mushroom growing – a method which 'favorise la pensée de ceux qui croient que les Champignons naissent de grain de même que les autres plantes' – was given by de Tournefort to the French Academy in a paper in which it is stated that it was Marchant *père* who first demonstrated the development of mushrooms in horse dung to the Academy in 1678. Tournefort described the method (essentially the same as that practised in modern times) for preparing ridge beds in the open from turned stable manure, cased with soil, and inoculated with pieces of mouldy horse manure. He believed that stable manure always contained invisible mushroom 'seeds' which only revealed themselves after development into very delicate, white branched hairs or filaments which gave 'une odeur admirable de champignon', and from which the mushrooms developed by the stages indicated in an accompanying illustration (see Fig. 83).

Twenty-four years later, Philip Miller in the first edition of his *Gardener's Dictionary*, 1731, under 'Mushrooms', 'set down the Method practis'd by the Gardeners who cultivate them for Sale' near London. The details he gives of the preparation of the beds call for no comment but should you or your neighbours have no beds from which spawn can be obtained:

you should look abroad in rich Pastures, during the Months of August and September, until you find 'em...then you should open the Ground about the Roots of the *Mushrooms*, where you will find the Earth, very often, full of small white Knobs, which are the Off-sets or young Mushrooms; these should be carefully gather'd, preserving them in lumps with the Earth about them.

'Knobs' of this 'Mushroom Earth' are then inserted 'about six Inches asunder' in the beds of prepared and cased horse dung.

Mushroom growing, largely because of the difficulty of standardising spawn, and notwithstanding the preference for 'virgin spawn' (that is spawn collected from the wild and not from old mushroom beds) was an uncertain business and it was not until the closing years of the nineteenth century that, following the development of pure culture techniques, 'pure-culture spawn' became available and reduced this particular hazard. This uncertainty did not, however, hamper the development of a large commercial mushroom industry (see p. 207).

MICHELI'S OBSERVATIONS

A very different approach to the culture of fungi was made by Antonio Micheli who during the year 1718 made a fundamental series of experiments on the culture of agarics and various moulds from spores which he sowed on a variety of media under diverse environmental conditions. The results of these experiments summarised eleven years later in the *Nova plantarum genera* (pp. 136–9) as a series of 'Observations' are here reprinted in the English translation of Buller (1915*b*) together with relevant figures.

OBSERVATION I

On the tenth of June, 1718, I collected in the country around the town, many species of fungi which I had never seen coming up in the woods of the royal pleasure gardens commonly called Boboli. This was done to obtain the seeds of different species. Then I spread on the table in my room many leaves of Ilex, Quercus, Laurus, Fraxinus, and the like, which had already fallen some time but were not spoiled or rotten, keeping each kind separate. On the different heaps of leaves, I placed either erect or lying down, several of the fungi; on some, one species, but on others, several species, simply for the purpose of allowing each fungus to deposit its seeds on the leaves. When three or four hours had passed, the leaves were turned, not merely in order that each leaf should receive seeds, but also that the seeds should not be collected in a heap in one place. When that was done and as much of the seeds as there was need of had been collected, I threw away the fungi and divided the different kinds of leaves into two parts, of which I carried the one into a thicket of the Boboli gardens and the other outside the town to a forest of Mount Olivet. I laid all these leaves in a place suitable for producing fungi, i.e., shady, among semi-rotten leaves of kinds different from those which had been covered with the seeds of the fungi, and I so placed the different heaps that they could not be mixed either with one another or with the leaves lying around. From the twentieth of August to the fourth of September, there were several wet days, so, on the twentieth of September, I went to Mount Olivet and on the following day to the Boboli gardens to see whether I could find anything in particular on the heaps of leaves. In the Boboli gardens on some of the heaps which had been noted, and especially on those consisting of the leaves of the Ilex and Laurus, I saw that many of the seeds had increased to the size of a grain of Millet, and their margin went off into very white and thin down, and all were producing capillary and somewhat hairy roots, whence I conjectured that the fungus had already begun to grow. That the fact was really so, I found out after a few days when, in some of the heaps, I saw pilei beginning to break out from the down, and in others the whole form of the fungus plainly showing itself [Micheli (1729), tab. 77, G–K]. On the eighth of October, I visited the same places, not only for the sake of noting the progress of the fungi which had come into existence, but also of understanding, if it were possible, for what reason none had yet been observed growing on the other heaps. Not being able to see the reason easily, I therefore returned frequently to watch the progress of those which had arisen, observing them only with the eyes and not touching them at all by hand. And so toward the end of October, and after several days alternately fine and wet, they had at length become bigger, and projecting above the leaves were several of the fungi, which had been produced from the seeds scattered through the heaps. I repeated these observations several times at the beginning of summer and also in autumn; but since the growth of the seeds generally depends on the chance weather of nature, i.e., on the alternation of fine and wet days which does not always happen opportunely, I have observed almost innumerable cases where it befell that the fungi did not grow, or their formation was hastened or delayed, or they were brought forth in greater or less abundance. Hence it follows that on sowing the seeds, it is very difficult to determine anything with certainty, especially since I have not yet examined everything in detail; but in the meanwhile, it is sufficient for me that I have sown the seeds and have seen fungi arise from them. If anyone, however, desires a sure method of cultivating fungi in order to get profit from them, I am of the opinion that he should employ only that method which gardeners use, namely, oblong heaps of dung which by them are called by the French name of *Couches*. From these they raise meadow fungi, and indeed from the seed and not from the heating dung as they think. This seed is naturally mixed with the dung itself or with the earth or things of that kind which serve for the composition of the oblong heaps. The manner of preparing these heaps is shown by Quintinaeus in his book, lib. VI, pp. 292, 327, and 333, but far more exactly by Tournefort in Comm. Ac. R. Sc., An. 1707, pag. 72.

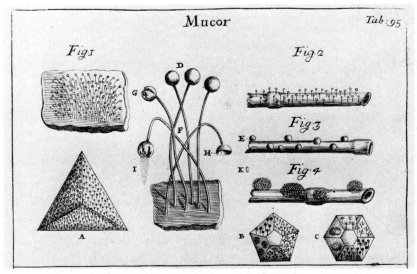

Fig. 39. *Mucor* (Fig. D, F, G, H, I) and cultures of *Mucor*, *Botrytis*, and *Aspergillus* on pieces of fruit. (P. A. Micheli (1729): part of pl. 95.)

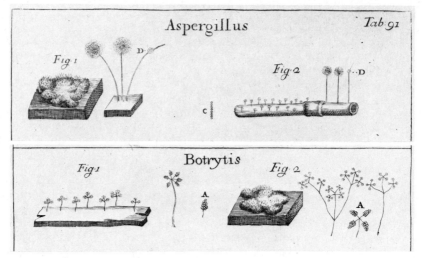

Fig. 40. *Aspergillus* and *Botrytis*. (P. A. Micheli (1729): parts of pl. 91.)

OBSERVATION II

On the fifth of November of the same year (1718) I took a piece of Melon (Pepo oblongus, C.B.Pin., 311) about four inches long and two inches wide and thick. Next, with a very soft brush, I collected from some other place the seeds from the dark sparkling heads of Mucor [Fig. 39, F]. I then smeared them on to the surface of the piece of melon on one side only, and put it in a place in no wise exposed to wind or sun. On the tenth of the same month, the infected part appeared everywhere white and strewn with a very thin down, like white cotton, which on the twelfth attained almost an inch in height and assumed a greyish colour; and some of the filaments of the down began to appear with white heads. On the fourteenth, the other filaments bore heads of the same kind. Finally, on the fifteenth, all the heads had become black, and after that the seeds came to maturity [Fig. 39, fig. 1].

OBSERVATION III

With the seeds which had been produced in the heads of the Mucor in the previous experiment, on the sixteenth day, I smeared another portion of the same melon on one side, and on the other side I placed the seeds of the capitate Aspergillus [Fig. 40] with glaucous heads and rounded seeds. On both sides, within the same interval of time as I have mentioned above, they sprang up, grew in the same manner as before, and produced seed after their kind. When I had done this several times, always using those seeds which each new crop of plants produced in its turn, I still observed no difference in the plants which sprang up.

OBSERVATION IV

On the sixth day of December, I took another piece of the melon of the same size and shape as before. In it I made five hollows, distinct from one another, and in them I placed a small piece of a fig infected with Mucor whose black and shining heads were already ripe. On the eighth day, many of the Mucor plants had bowed their heads; others were lowered around the holes where they had deposited seeds. Then, on the twelfth day, the whole surface of the piece of melon appeared covered over by the Mucor which, on the eighteenth day of the same month, brought its seeds to maturity. On the nineteenth day of the month, on certain parts of the surface of the above mentioned piece of melon, there appeared a white down, but in other places down of an ashen hue, arising from seeds which had fallen by chance from elsewhere on the melon. On the twentieth day, these grew up and one turned into Botrytis, branched, grey-coloured, with round seeds; the other into the capitate Aspergillus with glaucous heads and round seeds; but the black ones produced from seeds placed there by us forthwith perished.

OBSERVATION V

On the fourth day of November, I infected the sides of two pieces of the same melon with the seeds of the Botrytis [Fig. 40] which was branched and grey with round seeds. One of these pieces, I placed in a forcing house; the other in a small room with a window open. On these two pieces, on the seventh day, and on the same two sides which had been covered by seeds, granules appeared everywhere, like those which one may observe on the skin of what we commonly call shagreen (*sagri*), or rather a piece of pear which has been cut open for some days, but more sparsely than these. On the eighth day, these grains developed into very minute and almost imperceptible down, and especially those which were observed on the piece of melon which had been placed in the forcing house. On the evening of the ninth day, on both sides of the pieces, this down had much increased, so much so that they appeared as if covered with frost, or as it were by nitre which comes out on walls. The upper part of each piece of melon which had not been inoculated with the Botrytis seed was still intact. On the thirteenth day, the down on both pieces, at the infected spots, had produced heads which changed to a glaucous colour not differing from that of the heads from which I took the seeds, while not even a tiny plant of a second species of Botrytis or of Mucor had appeared. Howbeit, on the upper part of the piece of melon not placed in the forcing house, certain masses of white down had arisen, which on the eighteenth day developed into true plants, on the twentieth day came to maturity, and then Mucor was revealed with black and shining heads, no one having planted it, but the seeds having fallen there by chance.

OBSERVATION VI

On the first day of November, I infected another piece of the melon with the seed of the capitate Aspergillus [Fig. 40] which has spherical and glaucous heads. On the fourth day I observed no change. On the evening of the eighth day, the surface of the melon had a granular appearance similar to that already mentioned for the pear and the skin of shagreen (*sagri*), so that, on account of the abundance of the flakes of the investing down, it seemed like frost. On the thirteenth day, the down had grown and come to its final perfection, for it put forth heads and turned into plants of its own kind: and on the eighteenth day it disappeared. Howbeit, on a portion of the infected melon of very small extent which had not been inoculated with the seed of Aspergillus, another kind of plant was produced which showed itself clearly on the twenty-fourth day and came to maturity. It was indeed the branched, grey Botrytis with round seeds. On the twenty-seventh day, in certain places where the (capitate) Aspergillus already on the eighteenth day had come to perfect maturity, there appeared the very slender, white Aspergillus which is branched like Gramen Dactyloides, and has round seeds. Both these plants had sprung from seeds which had fallen on the piece of melon by chance.

OBSERVATION VII

On the thirtieth day of December, I took a piece of the melon and shaped it into a triangular pyramid. Then, choosing a piece of a quince and also of an almost ripe pear, commonly called *Spina*, I formed them into truncated pyramids, with their apices removed, giving the piece of quince a pentagonal, and the piece of pear a hexagonal base. On the individual faces of the pyramids, I sowed the seeds of Mucor, Aspergillus, and Botrytis, keeping each kind separate, so that on the piece of melon I had placed three kinds, on the quince on five sides five kinds, and lastly on the pear on six sides six kinds, just as is shown in the Plate [Fig. 39] of Mucor at the letters A, B, and C. All these species of seeds began to germinate from the fourth to the fifth or sixth day of the month, as I observed. They developed into plants according to their seed, of which some attained to their maturity on the tenth day, others on the twelfth, others on the thirteenth, and finally others on the fifteenth: and they produced the seeds of their kind. I kept these seeds separate, and again and again planted the seeds produced in like fashion from them; and then I always observed the same mode of growth in them, not in one trial only but however often and whenever I attempted it, without any difference whatsoever other than in the rate of growth or in the earlier or later ripening. These discrepancies could arise from various causes, perhaps from the difference in the time of the year, the place, or the structure of the substrata, or because the seeds were too ripe or immature.

It is clear from these observations that Micheli was both ingenious and perceptive. His experiments with microfungi were uniformly successful. Any contaminants of the freshly cut fruit surfaces were suppressed by the heavy inoculum; like produced like in both his primary cultures and subcultures and he recognised contamination by 'seeds' falling on his cultures by chance (Obs. IV, V, VI). His experiments with agarics (Obs. I) were naturally less successful – 'I have observed almost in-numerable cases where it befell that the fungi did not grow' – but he modestly concluded 'it is sufficient for me that I have sown the seeds and have seen fungi arise from them'.

SAPROBES, SYMBIONTS, AND PARASITES

The attempts of subsequent workers to repeat Micheli's cultural experiments met with varying success, and the controversies which these disagreements aroused have been touched on in Chapter 2 (p. 18). There was, however, no essential modification of the methods used by Micheli for some 125 years when the development of pure culture and the introduction of synthetic media of known composition enabled the cultural and the nutritional requirements of fungi to be investigated by greatly refined methods. In the meantime observational and descriptive work continued and it slowly emerged that while some fungi were agents of putrefaction of dead organic matter others developed on living organisms – in modern terminology, some were saprobes, others parasites.

The term parasite, according to *The Oxford English Dictionary*, was first introduced into biology in the early eighteenth century for ivy and other epiphytes and in the current sense of implying a nutritional dependence not until the nineteenth century. As late as 1857, Berkeley in his *Cryptogamic botany* (p. 235) described fungi as being 'hysterphytal or epiphytal mycetals (or more rarely epizoic or inhabitants of inorganic substances)' but de Bary (1866) in the first edition of his textbook distinguished saprophytic from parasitic fungi and in the second edition (1887: 356) differentiated 'pure saprophytes', 'facultative parasites' (a term introduced by van Tieghem), and 'obligate parasites' (including 'strictly obligate parasites' and 'facultative saprophytes'). The term *saprophyte* was a de Bary introduction and it was G. W. Martin in 1932[4] who suggested *saprobe* as more suitable for application to non-parasitic fungi, bacteria, protozoa, and other organisms for which an implied association with plants was undesirable.

Parasitism usually carries the implication of at least some degree of host impairment – that is, of pathogenicity. A. B. Frank (1877) from his studies of crustose lichens concluded that the algal and fungal components of the thallus exhibited a balanced relationship from which both partners benefited, an association he designated *symbiosis (Symbiotismus)*.[3] Later, Frank (1885) also coined the term *mycorrhiza* for a special class of symbiotic relationships.

Fungal parasitism is dealt with in Chapter 6. Here some impacts of the saprobic growth of fungi on the development of mycology and the recognition of symbiotic relationships of fungi with other organisms will now be considered.

TIMBER DECAY

Spoilage of man's food and property by moulds must always have occurred. Until recently most of these losses have been considered unavoidable and seem to have done little to stimulate advance in mycological knowledge. The main exception to this generalisation has been the decay of structural timber of which there are records from earliest times. The biblical account of 'leprosy of the house' (Leviticus 14. 33–48) possibly refers to dry rot (the control measures detailed are not inappropriate) and the Greek and Roman classical writers were well aware of the liability of wood to decay, From the seventeenth century the decay of the wooden ships of the navies of the European maritime nations became of increasing importance and reached a climax in the early years of the nineteenth century when in the United Kingdom the problem of decay in ships of the line was regarded as 'a national calamity'. The association of fungi with the decay was recognised. Leading mycologists were called in for consultation. A large literature was generated in both the popular and the scientific press. Mycological knowledge advanced, if slowly.

Decay in ships timbers was a recurring problem of long standing in the English navy.[5] It was in order to build ships of better quality than those supplied by contractors, which frequently proved unsatisfactory, that Henry VIII established royal dockyards at Woolwich, Deptford, and Portsmouth, but this did not eliminate the trouble. There was a major crisis after the Restoration when as Samuel Pepys reported (after recall in 1684 as Secretary for the Affairs of the Admiralty), most of thirty recently launched ships were in a shocking state of decay and Pepys, with his own hands, 'gathered toad-stools growing in the most considerable of them as big as my fists'. However, in spite of vigorous action the problem intensified. The normal life of a ship which was considered to be thirty years in the mid-seventeenth century and fourteen years a hundred years later, fell to eight during the Napoleonic war, and not infrequently, to nil at the end.

The basic reasons for these losses were twofold, the use of unseasoned timber and the lack of adequate ventilation of the ships. Merchantmen and coasting vessels were much less severely affected than ships of the navy because the former were

frequently loaded and unloaded while the latter were commonly
built in haste in large numbers at times of emergency (thus
aggravating the shortage of well-seasoned timber) and, during
intervals of peace, placed in reserve under conditions very
favourable for the development of wood-destroying fungi.
Another factor was the necessity to keep ships of the line at sea
during the winter months. As John Knowles, Secretary to the
Committee of Surveyors of His Majesty's Navy, pointed out in
1821 all those naval ships whose periods of durability had far
exceeded the ordinary course of things had been stationary in
harbours and he instanced the eighty-four-gun *Royal William*
which lasted ninety-four years and lay in harbour for ninety of
those years.

The importance of the use of properly seasoned mature
timber and adequate ventilation was early recognised. The
shortage of mature oaks capable of yielding the largest timber
with a high ratio of heart- to sap-wood is said to have begun at the
Reformation when the new owners made quick returns by
over-felling monastic woodlands and there was subsequently
much difference of opinion on the best time of year, and the
methods used, to fell trees for ship building. There were also
developments in the ventilation of ships in service by, for
example, wind-sails (invented in Denmark and introduced into
the British navy about 1740) and forced draughts induced by
devices such as the pneumatic machine (invented by Stephen
Hales in 1741) or strategically placed stoves. Some of these
methods were applicable to the drying out of recently con-
structed ships.

The famous case of the *Queen Charlotte,* a first rate of 100 guns,
marks a turning point. The keel of the ship was laid down in
October 1805. In the New Year of 1810 'stoves were placed in the
main hold, magazine, bread, and store rooms, and the orlop
deck' to dry the ship out before the launch in May. Fitting out
was completed by October and in May 1811 she was navigated to
Plymouth where the ship was found to have become so rotten
that 'all the planking within and without board together with
many of the timbers and beams' had to be replaced. At the
suggestion of Knowles, James Sowerby was called in to examine
the *Queen Charlotte* (possibly the first occasion on which a
mycologist was asked to advise a government department) and
he made three reports to the Honourable Commissioners of His

Majesty's Navy between August 1812 and October 1813.[6] He found many different fungi – and with his report submitted relevant plates from his *English Fungi* – including species of *Poria* (*Boletus hybridus* (289), *B. medulla-panis* (326)) and the dry-rot fungus *Serpula lacrimans* (*Auricularia pulverulenta* (358)) and he made recommendations, the implementation of which he subsequently supervised, for the stacking of timber supplies at the dockyard to ensure proper ventilation and the protection from the weather, the humidity being recorded by hygrometers. He emphasised the necessity to keep the wood dry while being worked and this was helped by the introduction of permanent roofs over the slipways on which the ships were built; the *Wellington*, of seventy-four guns, in 1816 being the first ship of the line to be launched from under such a roof. Sowerby was awarded 200 guineas by the Navy Board – less than he hoped. The *Queen Charlotte* was finally broken up in 1892.

In 1860 the Rev. Berkeley inspected the *Prince Regent* and the *Caroline* at Portsmouth and the *Arethusa* at Chatham when he found *Xylostroma giganteum* and *Polyporus hybridus* but not *Serpula lacrimans* or *Coniophora puteana* which so frequently cause decay in domestic structural timber.[7]

Ever since Noah was instructed to 'Make...an ark of gopher wood' and 'pitch it within and without with pitch' (Genesis 6. 14) there have been attempts to control timber decay by diverse treatments. In classical times cedar, larch, juniper, and other oils were used to prolong the life of wood and during the eighteenth and early nineteenth centuries common salt, alum, lime, sulphates of copper, iron, and zinc, mundic (an arsenic-containing ore from Cornish tin mines), and vegetable and coal tars were all advocated as panaceas for the decay of wood in ships.[8] Another widely used technique was to immerse the structural timber, or even to sink complete ships, for varying periods in water, either salt or fresh. In 1811 a seasoning-house was built at Woolwich in which timber could be exposed to the vapours derived from the destructive distillation of pitch pine sawdust but in December 1812 the plant exploded killing six men and doing extensive damage to the dockyard. At about this time many patents were taken out for the prevention of timber decay and in order to evaluate such treatments a 'fungus pit' was set up at Woolwich where samples of treated and untreated wood were stored in the closed and humid pit in contact with rotten timber. The most

famous patent was Mr Kyan's of 1832 which involved the
immersion of the timber to be treated in a solution of mercuric
chloride. This so-called 'Kyanization process' was the subject of
lectures by both Dr George Birkbeck (1835), the founder of
Birkbeck College of London University, and Michael Faraday
(1836), the latter devoting his inaugural lecture as Fullerian
Professor of Chemistry at the Royal Institution to the topic.

With the introduction of ironclad warships in the early
eighteen-sixties naval interest in timber decay waned and
attention concentrated on the preservation of wooden railway
sleepers, and later, of telegraph poles. 'Burnettizing'
(impregnation with zinc chloride, as set out in the patent of Sir
William Burnett, 1838) was at first popular in these fields but the
process which won the day was treatment with creosote under a
patent held by James Bethell and dated 1838.[9] Two years before,
Franz Moll had taken out a patent for the injection of wood in
closed cylinders with oils of coal tar and Moll called the
lighter-than-water fraction of these oils 'Eupion' and that
heavier than water 'Kreosot', The Bethell patent did not
specifically mention 'creosote' – originally and strictly the name
for the product obtained by the destructive distillation of
wood – but 'creosote' and 'creosoting' became the established
designations of the coal tar product and its still widely used
application as a wood preservative.[10]

The true relationship of fungi to timber decay proved as
troublesome to elucidate as that of pathogenic fungi to their
hosts (see Chapter 6). In 1833 Theodor Hartig, professor at
Brunswick and the discoverer of sieve tubes and aleurone
grains, believed that with age, or as the result of unfavourable
external conditions, a tree lost vitality; the contents of the wood
cells rounded off as little balls, or 'monads', which then aligned
themselves in rows to become fungus hyphae able to infect
sound wood and cause decay. Hartig called the fungus *Nycto-
myces* (*Nachtfaser*). He could only speculate on the relation of an
externally produced fruit-body to the decay as could Willkomm
more than thirty years later in his *Die mikroskopischen Feinde des
Waldes*, 1866–7, in which red rot and white rot were distinguished. In the light of an appreciation of the role of plant
pathogenic fungi Robert Hartig (the son of Theodor), in 1878, in
his well-illustrated and now classical *Zersetzungserscheinungen des
Holzes* opened the modern era of the understanding of timber

decay. He described the effects of a range of wood-rotting basidiomycetes and from microscopical observations and chemical analyses showed that some fungi such as the dry-rot fungus (*Serpula lacrimans*) attack cellulose but not lignin (to give brown rots) while others (e.g. *Stereum hirsutum*), which cause white rots, decompose all the components of the wood. Subsequently a large literature developed including monographs on the dry rot fungus by Robert Hartig (1885), Mez (1908) and Richard Falk (1912) among others. Constantin & Matruchot (1894) obtained the wood-destroying *Collybia velutipes* in pure culture and an important twentieth-century development, particularly associated with the names of Mildred K. Nobles (1948) in Canada and K. St G. Cartwright in England, has been the identification of wood-destroying fungi by their cultural characters rather than from the fruit-bodies.

Fig. 41. Erik Acharius (1757–1819). (E. Acharius, 1814, *frontispiece.*)

Fig. 42. *Usnea*. (J. J. Dillenius, *Historia muscorum*, 1741: pl. 12.)

LICHENS

The long history of lichenology has been well summarised by Annie Lorrain Smith (1921).[11] The common macrolichens, if of uncertain affinity, were, up to a hundred years ago, always considered independent organisms, as seemed so obvious. Tournefort (1694) first distinguished these forms as a distinct group, under the generic designation *Lichen*. Robert Morison, the Oxford botanist, in 1699 classified lichens as 'Musco-fungus' and thus emphasised their fungal nature and their association by early writers with bryophytes, while Linnaeus (1753) treated them as algae. The now universally recognised founder of lichenology was Erik Acharius (Fig. 41), a Swedish country doctor, who (Acharius, 1803, 1810) introduced such familiar terms as *thallus, apothecium,* and *perithecium* (which have proved so generally useful) together with *podetium, soredium, isidium,* and *cephalodium.* He also set lichenological taxonomy on a firm basis and treated Lichenes as one of the six groups into which he divided cryptogams.

Although Micheli devoted a series of plates to his thirty-eight 'Orders' of *Lichen* (Micheli, 1729: tabs. 36–54, 59) and Dillenius in *Historia muscorum,* 1741, had well illustrated the range of lichen form by beautiful copper-plates (Fig. 42), it was Wallroth (1825–7, **1**) who gave the first adequate description of the lichen thallus. He noticed the similarity of the colourless filaments to fungal hyphae and interpreted the green globose cells as reproductive organs which he accordingly named 'gonidia'. It was also Wallroth who characterised the arrangement of gonidia evenly throughout the thallus as *homoiomerous* and their limitation to a distinct zone as *heteromerous.* Wallroth noted the similarity of gonidia to unicellular algae and considered that free-living algae on the trunks of trees were 'unfortunate brood-cells' which originated from lichens but were unable to reform a lichen thallus. Tulasne (1852) concluded that the gonidia were budded off from hyphae in the gonidial layer, a conclusion which others confirmed and which even persisted well into the twentieth century. In 1866 de Bary expressed the opinion (de Bary, 1866: 291) that either lichens were the perfect fruiting forms of the Nostocaceae and Chroococcaceae or that the Nostocaceae and Chroococcaceae are typical algae which take the form of *Collema, Ephebe* etc. when attacked by parasitic

Fig. 43. S. Schwendener (1829–1919).

ascomycetes. The following year S. Schwendener (Fig. 43), then professor of botany at Basel, as a result of extensive studies on the anatomy of lichens, published his short epoch making paper (Schwendener, 1867) (Fig. 44) expressing the view that the green and blue-green gonidia really were algae and that every complete lichen represented a fungus living parasitically on an alga.

Schwendener's 'dual hypothesis' excited much resistance, in part from reasons of 'common sense', in part because lichenology has always been a somewhat esoteric pursuit. Leading contemporary lichenologists, including W. Nylander in Finland, Th. M. Fries in Sweden, J. Müller in Switzerland, and the Rev. J. M. Crombie in England, all bitterly opposed the hypothesis. Crombie characterised it as 'this sensational "Romance of Lichenology", or the unnatural union between a captive Algal damsel and a tyrant Fungal master'[12] and M. C. Cooke in 1879 asserted of the dual hypothesis that 'even if endorsed by the "Nineteenth Century" it will certainly be forgotten in the twentieth'.[13] The dual hypothesis was not forgotten. It became universally accepted but the problem of the taxonomic status of

97

the lichens remained. In 1911 the American lichenologist Bruce Fink on the basis of replies to a questionnaire he had sent, in 1909, to seventy-five American and an equal number of foreign botanists (including lichenologists, mycologists, and eminent general botanists) found that eighty-three per cent of all the botanists in his sample thought that lichens should be retained as a special group and not be distributed among the fungi, most of them (if lichenologists be excluded) on the grounds of convenience. This view is still prevalent but the current majority view is certainly that both the *phycobionts* and the *mycobionts*[14] should be treated taxonomically as Algae and Fungi, respectively.

As already noted, Schwendener considered that the association of fungus to alga was one of parasitism. Reinke (1873) pointed out that parasitism was inconsistent with the long-continued healthy life of the associated organisms and he suggested the use of 'consortium' as a term to express the relationship. Frank suggested the term 'homobium' (which implies complete interdependence of the two partners) and in the following year he (Frank, 1877) introduced 'symbiosis' which after being taken up by de Bary (1879) came into general use. Elenkin (1902) believed that the fungus obtained its nourishment from dead or dying algal cells, in other words, the relationship of fungus to alga was a saprobic one for which he coined 'endosaprophytism'. The claim by F. Moreau (1928: chap. 9) that the algal component parasitises the mycobiont is apparently unique.

The phycobiont frequently shows no apparent differences from its free-living counterpart and when isolated grows normally, although *Trebouxia*-like gonidia, which commonly occur, have lost the ability to live in the absence of the appropriate mycobiont. Mycobionts are more dependent than the phycobionts and if isolated do not produce spores.

From shortly after Schwendener's discovery there have been many attempts to synthesise lichens but Ahmadjian (1967) in his recent review concluded that among the early attempts only the claims of Stahl (1877) for *Endocarpon*, and Bonnier (1886) for *Xanthoria* and other genera, appear to be substantiated and that the first convincing experimental lichen synthesis was that by E. A. Thomas (1939).

Since 1950 there has been a very marked revival in physiological investigations on both whole lichens and their component

Protokoll

der

botanischen Sektion.

Sitzung: Dienstag den 10. September 1867,
Morgens 8 Uhr.

Die Sektion constituirt sich, indem sie zum Präsidenten Herrn Prof. Heer aus Zürich, zum Sekretär Herrn Rothenbach, Lehrer in Basel, wählt.

1. Herr Prof. Schwendener aus Basel begründet in einem längeren Vortrag, Bezug nehmend auf seine früheren Publikationen über den Bau des Flechtenthallus, die ihm in Folge neuer Untersuchungen zur Ueberzeugung gewordene Ansicht, dass bei einer ganzen Gruppe von Flechten Gonidien und Fasern nicht in einem genetischen Zusammenhange stehen, sondern dass letztere als Wucherungen von Pilzfaden auf Algenformen zu betrachten seien. Es wird an der Hand tabellarischer Abbildungen und kleinerer Farbenskizzen nachgewiesen:

1) dass der Flechtengattung Ephebe: eine Stigonema,
2) der Ephebella Hegetschweilerii: eine Scytonema,
3) den Collomaceen: Nostoc-Colonien,
4) den Omphalariaceen: Gloeocapsen,
5) den Racobloemaceen: Rivularien
zu Grunde liegen, und dass ferner ein Zusammenhang zwischen Cystococcus humicula und den gewöhnlichen Strauch- und Laubflechten, sowie zwischen Graphideen und den Algen aus der Chroolepus-Gruppe zu vermuthen sei.

Die jetzige Ansicht des Dozenten geht also dahin, dass die Flechten nicht selbstständige Pflanzen, sondern Pilze in Verbindung mit Algen seien, welch' letztere der erstern Nährpflanze bilden. Betont wird namentlich die Beobachtung des Eindringens von Flechtenfasern in junge Nostoc-Colonien, wie die durch zahlreiche Uebergänge an Exemplaren vom Rheinfalle zur Gewissheit gewordene Verwandtschaft von Rivularien mit Racobloemaceen. Ferner wird die von Famietzin in der botanischen Zeitung mitgetheilte Thatsache hervorgehoben, dass verschiedene Strauch- und Laubflechten, welche in fliessendem Wasser sich von ihren Faden befreiten, in ganz anderer Weise fortvegetirten und Schwärmzellen bildeten, endlich, dass die von de Bary, Tulasne und Anderen vorgenommenen Keimungsversuche mit Flechtensporen nicht gelungen seien. Von der Cultur von Flechtensporen auf Algen verspricht sich der Dozent ein besseres Resultat; es bleiben übrigens noch verschiedene entwicklungsgeschichtliche Thatsachen zu studiren, besonders für diejenigen Flechten, deren Gonidien nicht in Reihen liegen.

Das Präsidium verdankt den Vortrag des Herrn Prof. Schwendener, verhehlt aber dessen Ansicht gegenüber seine auf die grosse Verbreitung der höheren Krusten- und Laubflechten und auf deren chemische Zusammensetzung (Stärkemehlgehalt) gegründeten Bedenken nicht; die Apothecien würden also als Pilzfrüchte zu betrachten sein.

Diese Frage wird von Herrn Prof. Schwendener bejaht. In der grossen Verbreitung der Krustenflechten sieht

Fig. 44. Schwendener's announcement of the dual nature of lichens in *Verh. Schweiz. Naturf. Ges. Aarau* 51: 88-9 (1867).

parts which have shed much light on the nutrition of lichens and the physiological relationships of phycobionts to mycobionts. These developments are outside the scope of this review. It is however clear that whatever status they are accorded, in the words of Hale (1967:71):

Lichens are undeniably more than the sum of their parts, for lichenization is accompanied by structural modifications (e.g. thalloid exciple, soredia) new to the plant kingdom and physiological activities (production of lichen acids) different from those of either component.

MYCORRHIZA

The symbiotic association of fungi with phanerogams and vascular cryptogams is the reverse of the relationship found in lichens. In the latter the fungal component visually dominates the partnership while in the former it plays an inconspicuous part; in many instances having been unsuspected until revealed by careful microscopical investigation. As so often with new developments, the initiation of the serious study of the phenomenon can be attributed to the worker who introduced the terminology, in this case A. B. Frank who in 1885 coined 'Mycorhiza' for the fungus roots (*Pilzwurzeln*) he found on cupuliferous trees (Fig. 45). Two years later he distinguished the mycorrhiza of Cupuliferae, conifers, and other trees as *ecto-trophic* from the *endotrophic* mycorrhiza of Orchidaceae and Ericaceae (Frank, 1887).

Frank, as is frequently recalled, began these studies at the instigation of the German State Forestry Department in connection with a scheme for the development of truffle culture in Prussia. Truffles were known to be associated with certain broad-leaved trees and this association was thought to be parasitic in nature. Frank did not advance truffle culture but he did study the relationship of the fungi which he found regularly infecting and frequently distorting the roots of forest trees. He came to the conclusion that the association was one of mutual benefit; a conclusion that was subsequently abundantly substantiated.

There are a number of earlier records of tree mycorrhiza. In 1840 Theodor Hartig noted the hyphal sheath and the weft of hyphae between the cortical cells of coralloid root tips of conifers but he mistook the hyphae for intercellular canals. The 'Hartig net' was later described by O. Nicolai in 1865 and van Tieghem,

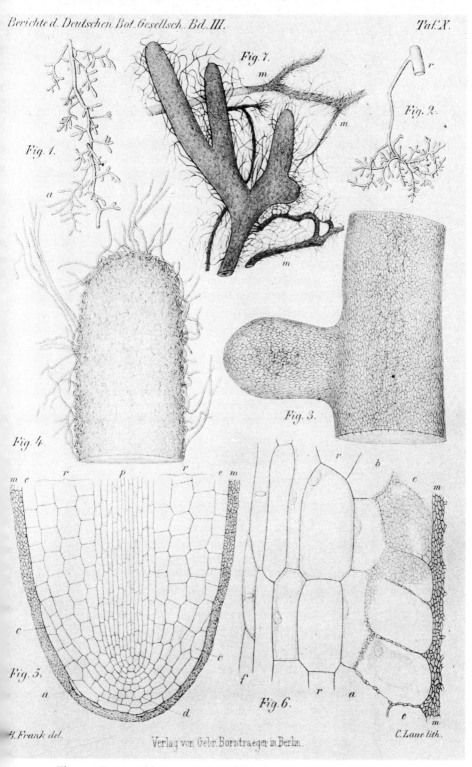

B. Frank del.   Verlag von Gebr. Borntraeger in Berlin.   C. Laue lith.

Fig. 45. Ectotrophic mycorrhiza of hornbeam (*Carpinus betulinus*) (1–6) and beech (*Fagus sylvatica*) (7). (A. B. Frank (1885): pl. 10.)

1871, and Reinke (1873) established its fungal nature. Also, Boudier (1876) described the association of *Elaphomyces* with the roots of birch, oak and chestnut.

Even earlier, a controversy during the eighteen-forties on the nutritional status of *Monotropa hypopitys* (the Yellow Bird's Nest) led E. Lees (1841) to describe 'a hirsuture that appears to be a byssoid fungus' on *Monotropa*. The next year T. G. Rylands (1842) illustrated both intra- and intercellular mycelium[15] in the roots of the same plant and concluded that the 'byssoid substance is really fungal, and performs no essential function in the economy of *Monotropa*'. It was F. Kamienski (1882) who elucidated this association and also studied mycorrhizas of other plants. He came to the conclusion that there was no evidence to support the view that the fungi were parasitic and claims have therefore been made to regard Kamienski as the first to recognise the true nature of the mycorrhizal association, but as Rayner (1927: 15–16) has pointed out it is in fact Pfeffer to whom such credit is due for in 1877 he ascribed to the orchid fungi a physiological role analogous to that of root hairs.

The mycorrhiza that attracted most attention during the nineteenth century was that of orchids.[16] Link, in 1840, was the first to illustrate, apparently unknowingly, fungi in the cells of an orchid, in the protocorm of *Goodyera procera*.[17] S. Reissek (1847)

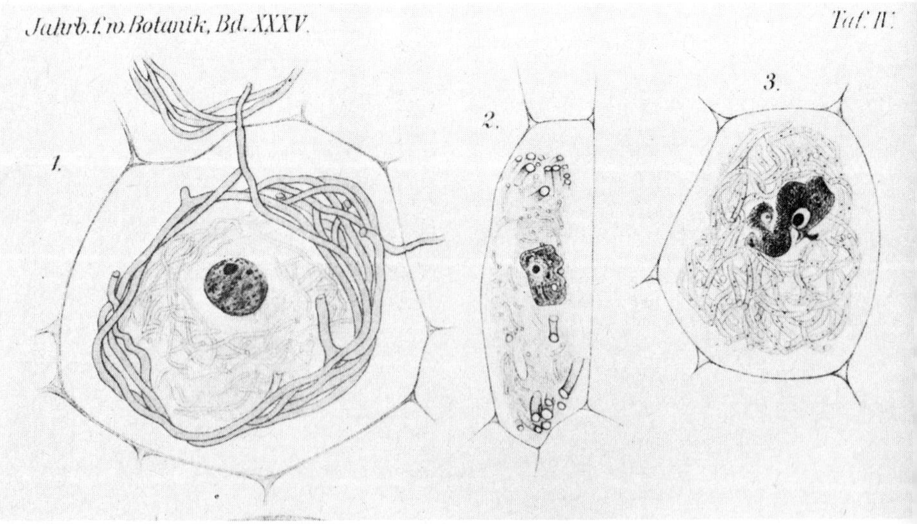

Fig. 46. Mycorrhiza of *Neottia nidus-avis*. Two 'host cells' (1–2) and one 'digesting cell'. (W. Magnus (1900): pl. 4.)

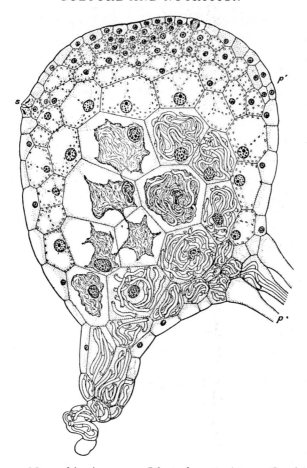

Fig. 47. Mycorrhiza in young *Odontoglossum crispum*× *O. adriane.*
(N. Bernard (1909): part of pl. 2.)

made extensive studies of mycorrhiza of orchids and other
plants and even attempted to isolate the associated fungi but
with the techniques available his isolates were only common soil
fungi and it was W. Magnus (1900) who distinguished between
the 'host cells' (*Pilzwirthzellen*) and the 'digesting cells' (*Ver-
dauungszellen*) in *Neottia* (Fig. 46). A climax to the elucidation
of orchid mycorrhiza resulted from the prolonged investigations by Noel Bernard and H. Burgeff who both made their
major contributions in this field by monographs published in
1909. Both these workers isolated mycorrhizal fungi in pure
culture and Bernard in a brilliant series of researches demonstrated that the *Rhizoctonia* species (for which Burgeff proposed

a new genus *Orcheomyces*) he had isolated were able to induce the germination of orchid seed and subsequently infect the seedlings (Fig. 47). Bernard also found that orchid seeds would germinate on sterile sugar-containing media and this discovery, supplemented by the investigations of the American L. Knudsen (1922–5), provided the basis of the commercial methods for raising orchids from seed.

Mycorrhizal associations of the Ericaceae have attracted much attention since the paper by Charlotte Ternetz in 1907 and the topic, including her own researches in this field, is reviewed by M. C. Rayner (1927: esp. chap. 6).

G. Gallaud in his doctoral thesis of 1904 characterised four series of endotrophic mycorrhiza by that of *Arum maculatum*, *Paris quadrifolia*, liverworts, and orchids, respectively, and in so doing described in detail 'vesicular–arbuscular' mycorrhizas. He introduced the term *arbuscule* for the terminal, tree-like branching systems (Fig. 48) of the intracellular mycelium which on degeneration become what Janse in 1897 had called 'sporangioles' (Fig. 48) and he also gave excellent illustrations of both *vesicles* and the *pelotons* (Fig. 48) or tangled masses of intracellular hyphae.

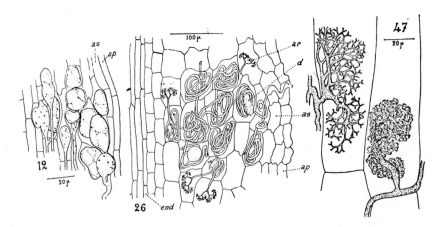

Fig. 48. Vesicular-arbuscular mycorrhiza. Arbuscule (47, *left*) and sporangiole (47, *right*) in *Allium sphaerocephalum*; vesicles in *Parnassia palustris* (12); peloton in *Tamus communis* (26). (G. Gallaud (1904): pls. 4, 1, 2, in part.)

PURE CULTURE

The essentials of pure culture technique are to inoculate a sterile substratum, usually a nutrient medium of defined composition, with the required organism and to prevent subsequent contamination from the environment. Sterilisation of liquids liable to putrefaction by boiling was, from the middle of the eighteenth century, standard practice among investigators of 'spontaneous generation' and sterilisation of nutrient media by boiling or to obtain higher temperatures by immersing the media-containing vessels in water baths of boiling solutions of common salt or calcium chloride, or an oil bath, became common procedures by the mid-nineteenth century. It was in 1877 that the English chemist John Tyndall and the German bacteriologist Ferdinand Cohn discovered independently 'discontinuous sterilisation' (sterilisation by discontinuous heating or 'Tyndallisation'). By this technique, as Tyndall concluded, 'Five minutes of discontinuous heating can accomplish more than five hours' continuous heating'[18] at boiling point in spite of the fact that the discontinuous heating is below the boiling point of the medium. In the early eighteen-eighties commercial availability of the 'autoclave',[19] a development of a seventeenth-century invention by Denis Papin of a 'digester or engine for softening bones', enabled advantage to be taken of the sterilising effects of superheated steam. And it was at this time too that Robert Koch and Gustav Wolffhügel[20] drew attention to the value of hot air as a sterilising agent.

At first the vessels containing sterilised material were hermetically sealed. In the early nineteenth century air, after being strongly heated, was sometimes allowed entrance and Pasteur developed the flask, which still bears his name, with an open-ended, elongated, swan-like neck of fine bore which acted as a spore trap. Starting from a clue provided by the chemist Loewel in 1853 that after filtration through cotton wool air would no longer cause crystallisation of supersaturated solutions, H. G. F. Schröder and T. von Dusch in 1854 demonstrated that such air no longer induced putrefaction in freshly boiled solutions[21] and it appears to be the Manchester physician William Roberts (1874) who, in his studies on spontaneous generation, first used plugs of cotton wool to seal the glass vessels containing sterilised infusions. A few years later Pasteur reported that he routinely plugged the open ends of his flasks with sterilised asbestos but he emphasised that this was merely an extra precaution mainly directed against the entry of mites, etc. (Pasteur, 1876: 27).

No major progress in methods for the isolation and purification of fungi and bacteria was possible without solid media. Vittadini in 1852[22] seems to have been the first to solidify media with gelatine when attempting the culture of the muscardine fungus (*Beauveria bassiana*). Gelatine was also later employed by Oscar Brefeld (1872), who was so influential in introducing pure culture methods into mycology,[23] and in the following year by the bacteriologist Klebs. Brefeld (1872) also devised the dilution method for single-spore culture by which drops of a progressively more dilute suspension of spores were examined under the microscope until a drop was found containing a single spore to which a drop of culture medium was then added. (Six years later Joseph Lister, by means of a special syringe, adapted Brefeld's method for making single-spore cultures from bacterial cells.)[24] In 1883 the poured-plate method for the isolation and purification of bacteria, and one that has proved so useful in mycology, was published by Koch[25] and shortly afterwards agar–agar was introduced into bacteriology at the suggestion of Frau Hess, the wife of one of Koch's co-workers.

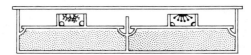

Fig. 49. The 'van Tieghem cell'. (P. van Tieghem & G. Le Monnier (1873): pl. 20, fig. 1.)

Finally, two minor but most useful technical innovations must be recalled. In 1873 the 'van Tieghem cell' for the microscopical examination of hanging drop cultures was introduced (van Tieghem & Le Monnier, 1873) (Fig. 49) and in 1887 R. J. Petri, one of Koch's assistants, published details of 'Eine kleine Modification des Koch'schen Plattenverfahrens' – the plates of glass on which gelatine and agar cultures had hitherto been made. He had invented the 'Petri dish'.

Jules Raulin, Phillippe van Tieghem, and Émile Duclaux were the first three students to work with Pasteur in his laboratory at the École Normale in Paris in the early eighteen-sixties and each in his turn produced a doctoral thesis on a topic related to Pasteur's studies. Van Tieghem's thesis 'Sur la fermentation de l'urée et de l'acide hippurique' appeared in 1864 and Duclaux's 'Sur l'absorption de l'ammoniaque et la production des acides volatile pendant la fermentation alcoolique' the next year. Raulin's, 'Études chimiques sur la végétation', on which he continued to work after leaving Paris for teaching posts at Brest and Caen, did not appear until the end of the decade (as a paper in the *Annales des Sciences naturelles* in 1869 and as his thesis on 22 March 1870) but it was a brilliant achievement able to take its place with such mycological landmarks of the eighteen-sixties as the Tulasnes' *Carpologia* (1861–5) and the first edition of de Bary's textbook (1866). A new standard for the physiological investigation of fungi was set.

Raulin's study was designed to elucidate the mineral nutrition of *Aspergillus niger* by quantitative measurement of the growth obtained on chemically defined media under standardised conditions. By preliminary studies the optimum conditions for the growth of the mould were determined and the final experiments were made, on series of paired cultures in un-covered shallow porcelain dishes ($16 \times 28 \times 4$ cm), each containing 1.5 l. of culture fluid, maintained in a specially designed incubator (Fig. 50) at 35 °C and at a humidity of 70° as registered on the de Saussure hygrometer. The general method adopted was to inoculate each culture by brushing the surface of the liquid with spores from a young pure culture and after three days to remove the mycelial mat, re-inoculate and incubate for a further three days, by which time the medium was too exhausted to allow more growth. The mycelial mats were squeezed free of culture fluid and dried in a Gay-Lussac oven, first at 50 °C, then at 100 °C. In each experiment the yields from a pair of dishes containing the complete medium served as a control by which the effects of experimental treatments could be compared and finally Raulin was able to obtain from a control culture after six days a yield of 25 g of dry mycelium, fifty times that of preliminary experiments, and to reduce the variation between paired control cultures in one experiment from 1 : 1.8 to 1 : 1.05.

Raulin's starting point was a medium, containing water 1000, sugar 20, ammonium bitartrate 2, ashes of yeast 0.5, which Pasteur had devised for the culture of *Penicillium* and other moulds. By trial and error Raulin evolved the synthetic medium which still bears his name that he found best suited for the growth of *Aspergillus niger*. Raulin's original formula for this medium was:

'Eau	1.500
Sucri candi	70
Acide tartrique	4
Nitrate d'ammoniaque	4
Phosphate d'ammoniaque	0,60
Carbonate de potasse	0,60
Carbonate de magnesie	0,60
Sulfate d'ammoniaque	0,25
Sulfate de zinc	0,07
Sulfate de fer	0,07
Silicate de potasse	0,07'

By omitting each ingredient in turn or by substituting equivalent amounts of related compounds Raulin was able to explore the effects of different compounds or elements on the growth of the mould, and he expressed the effects observed in two ways: by the ratio of the mycelial dry weights yielded by experimental and control cultures, and by calculating the dry weight of mycelium resulting from the utilisation of one unit of the nutrient or element, values which, he emphasised, should not be taken in an absolute sense as they could only be used to show relative effects. The omission of sugar reduced the yield 65-fold, and that of ammonium nitrate 24- to 153-fold. One gram of sugar was utilised in the production of one third of a gram of mycelium; 1 g of ammonium-nitrogen in the production of 17 g. The comparable values were for phosphoric acid 182, potassium carbonate 21, magnesium carbonate 91, and sulphuric acid 11.4 while 1 g of phosphorus, potassium, magnesium, and sulphur permitted the development of 157, 64, 200, and 346 g of mycelium, respectively.

Raulin was able to show that nitrates (but not nitrites or cyanate) were equally effective nitrogen sources and that any sulphate could be used provided the base did not have an intrinsic toxic effect. He also understood that in the absence of tartaric acid the development of infusoria which prevented the growth of the mould was a pH effect and found that small quantities of other organic acids and even mineral acids such as sulphuric acid could be used as a replacement.

Subsequently it became apparent from the many investigations by later authors that fungi could use a remarkably wide

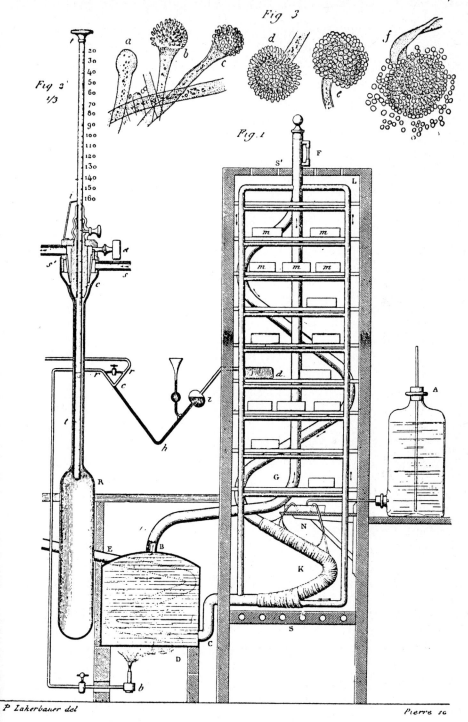

Fig 1 Étuve   2 Thermorégulateur   3. Aspergillus niger

Fig. 50. Raulin's incubator. (J. Raulin, 1869.)

range of carbon sources and as regards nitrogen nutrition Robbins (1937) was able to classify fungi into a series of four increasingly specialised groups according to whether they could utilise elemental nitrogen, nitrate-nitrogen, ammonium-nitrogen, and organic nitrogen, or only the last three (the majority of fungi), two, or one of the categories of nitrogen compounds.

One of the most interesting aspects of Raulin's investigation was that the techniques employed were sufficiently refined to enable the effects of trace elements to be discovered. Raulin noticed that as he took greater and greater care to prevent his medium being contaminated by foreign substances the growth yields became strangely smaller and that a three- to five-fold increase of these low yields could be induced by the addition to the medium of powdered baked earth ('terre cuite pulverisée') powdered porcelain, or wood ashes. He demonstrated that this improvement was not caused by the addition of any macro-nutrient but by traces of zinc and iron. He found that these two elements could not replace one another and he calculated the ratio of organic matter to the weight of zinc or iron contributing to its formation to have reached values as high as 953 for zinc and 857 for iron. His attempts to replace iron or zinc by manganese were inconclusive and he concluded that growth was enhanced by the addition of silicon as sodium silicate. Finally, the addition of powdered soil, etc., to his complete medium caused no improvement of growth but this still left unsolved the problem of whether 'le miellieu formé de la réunion de ces douze éléments, *supposés purs*, est la réalisation la plus parfaite d'un sol fertile'. Raulin appreciated that the solution depended on more refined methods of chemical purification, but after 1870 his mycological interests faded.

It was, however, the development of such methods that enabled the American R. A. Steinberg, some fifty years later in a notable series of papers (Steinberg, 1919 *et seq.*), to confirm Raulin's findings and show that traces of both molybdenum (Steinberg, 1936) and gallium (Steinberg, 1938) were also essential micro-nutrients. Earlier, Bertrand & Javillier (1911), in France, had established a requirement for manganese and Mulder (1938) demonstrated that in the absence of minute traces of copper cultures of *Aspergillus niger* were sterile. The need for trace elements in the nutrition of many fungi became generally accepted.[26]

The requirement of some fungi for 'growth factors' (variously known as growth-substances, accessory growth-factors, and vitamins) took longer to elucidate although the question had been unwittingly raised by Pasteur. In his famous controversy with Leibig, Pasteur claimed that beer yeast would grow and multiply in a synthetic medium composed of water, sugar, ammonium tartrate, and ashes of yeast, Leibig maintained that it would not. Both were careful experimenters and the disagreement might have been resolved if Pasteur's offer in 1871 to 'prepare in a mineral medium, in the presence of a commission chosen for the purpose, as great a weight of ferment as Leibig could reasonably demand' had been taken up; but Leibig made no move and died in 1873. It remained for the Belgian E. Wildiers in 1901 to explain the discrepancy. He showed that an inoculum of a few yeast cells sufficient to initiate fermentation of beer wort would not grow in a sugar-containing synthetic medium while a heavier inoculum would. Pasteur must have used a larger pinhead-sized inoculum than did Leibig. Wildiers was able to demonstrate that, among other characters, the growth factor, which he called 'bios', was water soluble, absent from the ashes of yeast, and that yeast by multiplication and fermentation did not produce new bios. Bios was found to be present in Leibig's meat extract, commercial peptones, and in the boiled wort of germinated barley before seeding with yeast. He also showed that bios was not one of a range of known organic compounds. Later biochemical studies showed bios to be a complex mixture.

Eastcott (1928) was the first to identify a fraction of bios – bios I as i-inositol – but more components have since been characterised. They include thiamin (aneurin, vitamin B_1), biotin (vitamin H), pyridoxine (vitamin B_6), and pantothenic acid, all compounds for which many yeasts and mycelial fungi have been found to have requirements.[27] Thiamin, first identified by Schopfer (1934) as an essential metabolite for *Phycomyces blakesleeanus*, is the growth factor most generally required by fungi, the needs of some fungi being satisfied by one or other or both moieties – thiazole and pyramidine – of the thiamine molecule while others need the complete molecule. An example of an interesting and rare requirement is that for oleic acid by the yeast, *Pityrosporum ovale*, a normal inhabitant of human skin, particularly of the scalp (Benham, 1939). It is probable that given appropriate nutrients all fungi, even obligate parasites, could be grown *in vitro*. The powdery mildews (Erysiphaceae) and rusts (Uredinales) are frequently cited as two typical groups of obligate plant parasitic fungi. In-vitro culture of the former is still limited to spore germination but, since the description in 1966–7 by the Australians Williams, Scott & Kuhl of the successful urediniospore to urediniospore and teliospore development of *Puccinia graminis* f. sp. *tritici* on yeast-extract Czapek-Dox agar, several other rusts have been successfully cultured and the earlier claim made by the American V. M. Cutter (1959) for the axenic culture of *Gymnosporangium juniperi-virginianae* substantiated.

METABOLISM AND METABOLITES

During the past twenty-five years the elucidation of the biochemical pathways underlying basic fungal metabolism constitutes a major mycological development of this century. These discoveries were both preceded and overlapped by studies on the many and chemically diverse fungal metabolites; an aspect of mycology that has attracted the interest of chemists and biochemists for more than a hundred years.

The first chemical investigations on fungi were analyses of the ash of mushrooms and the fruit-bodies of other larger fungi. After about 1870 many organic chemical constituents of fungi were defined and by 1896 Zopf, in his textbook, gave a more comprehensive account of the chemical constituents of fungi than has appeared in any subsequent traditional textbook of mycology. In addition to inorganic constituents, his account (Zopf, 1890: 116–86) included reviews of the knowledge (mostly of German origin) of, among others, sugars, fats, organic acids, pigments, proteins (*Eiweisstoffe*), and enzymes (*Fermente*) found in fungi, as well as a section on lichen acids. Investigations continued and by 1907 Zellner was able to monograph at book length the chemistry of the larger fungi. Subsequently, a major contribution to the knowledge of fungal metabolites was made by Harold Raistrick, J. H. Birkinshaw, and their colleagues (including the self-taught mycologist George Smith, author of the popular *Introduction to industrial mycology*, 1938) who between 1931 and 1964 published the 116 parts of their famous 'Studies in the biochemistry of micro-organisms'. This series was almost exclusively devoted to fungi and had an emphasis on secondary metabolites, that is, on the multitude of synthetic by-products which have no obvious role in the economy of the organism as have such primary metabolites as proteins, DNA, and various storage products. Current knowledge of the secondary metabolites of fungi has been well summarised, according to their biosynthetic origins, in the recent compilation by W. B. Turner (1971).

Those wide ranging studies of fungal metabolites and the many technological advances which accompanied the commercial exploitation of the metabolic activities of fungi (see Chapter 8) did little to elucidate the basic dynamic metabolic mechanisms of fungi. Determinations of the 'carbon balance sheet', that is of the ratio of the carbon removed from a synthetic medium to the amount of the resulting fungal growth (which allows the calculation of 'economic coefficients') were in 1931 published by Raistrick's team for more than 200 fungi. These results showed that, under optimum conditions, fungi utilise carbon compounds very efficiently but it was not until shortly after Foster, in 1949, made his masterly survey of the chemical activities of filamentous fungi that the biochemical pathways by which carbohydrates are metabolised in fungi were shown, as a result of investigations by teams of biochemists in a number of laboratories, to be similar in pattern to those of other organisms. It was established, first for yeast and later for filamentous fungi that the carbohydrate source after having been transformed, if necessary, into glucose is then converted into pyruvate by either the Embden–Meyerhof–Parnas (EMP) or the hexose monophosphate (HMP; pentose phosphate pathway) pathways, or, more rarely, by the Entner–Douderoff (ED) pathway. The next step is the incorporation of the pyruvate into the tricarboxylic acid cycle (TCA cycle; citric acid cycle; Krebs cycle) which occurs in both animals and plants and which plays an important role in providing material for the biosynthesis of essential cell constituents. Finally, by the coupling to the respiratory chain considerable energy is made available to the cell during the process, mediated by cytochromes, of hydrogen (electron) transfer to molecular oxygen.

5. Sexuality, cytology, and genetics

The first observation of fungal spores by della Porta in 1588 and the later experimental demonstration by Micheli that these spores functioned as 'seeds' were described in an earlier chapter (see pp. 14, 84). Micheli and his successors suspected that fungal spores had a sexual aspect but this suspicion was difficult to confirm. The elucidation of sexuality in flowering plants had given much trouble and it was not until 1694 that R. J. Camerarius published the results of experiments demonstrating the function of pollen and a century and a half later, in 1846, that gametic union was first unequivocally described by Amici, in orchids. Sexuality in fungi proved even more troublesome to clarify.

Among early writers there was no lack of speculation on sexuality in both plants and fungi and as *Dryopteris filix-mas* and *Athyrium filix-femina* were distinguished on imaginary sexual characters as the male and female ferns so herbalists differentiated the 'male agarick' (*Fomes officinalis*) from the 'female agarick' (*Phellinus igniarius*). The first observational approach to the problem of sex in fungi was to seek for structures comparable to the sexual organs of higher plants in the larger fungi, particularly hymenomycetes in which sexuality is more difficult to interpret than in any other group. The result was more than a century of erroneous explanations based on preconceptions and on 'minute microscopic analysis where', as the Rev. Berkeley wrote in 1838(*a*), 'there is so much room for the exercise of imagination'. Two of these false trails, those based on marginal gill hairs and cystidia, respectively, will first be traced.

It was Micheli (1729) who interpreted the hair-like projections from the gill edge, which are a not uncommon feature of agarics, as 'apetalous flowers'. The clearest description and illustrations of these structures are those which he gave for the genus *Fungus* (agarics with a central stipe) in which 'at the margin of the laminae are developed apetalous flowers, naked, composed of

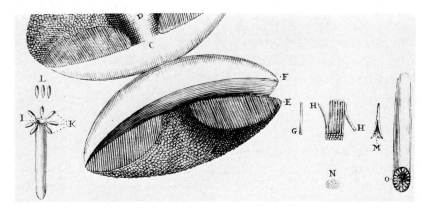

Fig. 51. Flower of *Suillus*. (P. A. Micheli (1729): from pl. 68.)

nothing but a cylindrical filament; in some species solitary, in others arranged as a mass or tuft' (see Fig. 33). The flowers of *Agaricum* (various dimidiate hymenomycetes) are described as 'apetalous, monostemous (consisting of a single filament), sterile, naked, and lacking calyx, pistil, and stamens' (Fig. 33) as are those of *Suillus* (*Boletus*) which are found 'in the mouths of the tubes and upper part of the stipe while the pileus is expanding' but the accompanying illustration (Fig. 51), showing seven 'flowers' arranged round the mouth of a tube, is obscure and has never been satisfactorily explained. Schmidel (1762) also figured gill hairs (in *Coprinus comatus*) and, inspired by Micheli, Gleditsch (1753) expecting to find what he sought, described and illustrated for boletes the five or more 'stamens', each consisting of a 'filament' and terminal 'anther', which surrounded the mouth of each tube, and similar stamens for agarics in which they occur singly or in fascicles of five or more as fructifications on the gill margin. Schaeffer (1759), who claimed the spores of fungi to be asexually produced gemmae, could not find these structures (and said that he had written to Gleditsch to learn how to observe them); neither could von Haller ('etsi nonnunquam minutias rerum viderit, quas post eum nemo reperit, ut fungorum antheras').[1]

Although these reverses encouraged some to follow Schaeffer and claim that fungi lacked sexuality – as did J. Gaertner who wrote in his *De fructibus et seminibus plantarum*, 1788 (p. xv), 'Among purely gemmiparous or asexual plants Fungi deserve the first place... Every method of fertilization whatsoever that is attributed to fungi should certainly be considered a mere deception, since they increase not by true seeds but by buds'[2] – they did little to stem interest in the sexuality which as Linnaeus had implied by his nomenclature would be discovered in the diverse members of his twenty-fourth class (Cryptogamia) of the plant kingdom.

Attention next turned to cystidia, Micheli's 'corpori diaphani' which Bulliard (1791–1812, **1**) equated with pollen and designated 'vesicules spermatiques'. Then, the details of pollination were unknown and pollen grains were widely believed (as discussed at some length by Bulliard) to contain a spermatic fluid. Bulliard believed cystidia to contain a similar fluid and he illustrated cystidia squirting their seminal contents onto the spores. This view of cystidial function persisted for the next seventy-five years. Corda (1834) called Micheli's 'corpori diaphani' 'fungus anthers' and in Sturm's *Deutschlands Flora* (1837, **3**: tab. 49) he showed an anther of *Coprinus micaceus* discharging its contents. Klotzsch (1838) believed that a juice excreted through the walls of the 'anthers' fertilised spores which settle on them while in their classical and objective accounts of hymenial structure Léveillé (1837), who introduced the term cystidium (see p. 52), and Berkeley (1838a), who referred to the cystidia as 'utricles', were both noncommittal as to the function of these organs. Because Corda came to the conclusion that his 'anthers' were analogous to pollen, in his *Icones fungorum* (1837–54, **1**), he substituted the designation 'pollinaria' (*Pollinarien*) but he was uncertain about the details of fertilisation. Montagne in his *Equisse* (1841) accepted Corda's conclusions and explained away the lack of pollinaria in some hymenomycetes by a comparison with mosses where though the presence of organs considered as endowed with the property of fecundating sporidia is averred there are a great number of species in which they cannot be found. The theory of the sexual function of cystidia was exploded by H. Hoffmann (1856) in a detailed account of his thorough investigation of cystidia which also includes the first description and illustrations (Fig. 56) of *clamp-connections* ('Schnallenzallen', buckle-cells). Hoffmann showed cystidia to be sterile structures more reminiscent of the glandular hairs on the leaves and stems of higher plants than organs of sex. Not everybody, however, was convinced, for example, G. Sicard as late as 1875;[3] and in the same year the climax of misinterpretation was attained by Worthington G. Smith[4] who in a study of the reproduction of *Coprinus radiatus* claimed that when both spores and cystidia fall on to moist ground fertilisation occurs and he illustrated the motile spermatozoids (presumably contaminating protozoa), which he thought were liberated from the cystidia and fertilised the spores. The first true observation of sexuality in fungi must now be intercalated.

In 1818 Christian Gottfried Ehrenberg, Geheimer Medicinal-rath and professor at Berlin University, named, in a new genus, a mould which he had found growing on the pileus of *Agaricus aurantius* Persoon as *Syzygites megalocarpus*. Two years later (Ehrenberg, 1820 *b*) he described, with beautiful illustrations (see Fig. 52), the stages by which, as a result of the conjugation of two filaments, a black fruit-body (*Fruchtwarze*) is formed in this species. After discussing whether the organism was animal or plant, alga or fungus, he concluded that it was a fungus comparable with a conjugate alga and he classified it in the family Ascophorae which he had proposed for mucors (fungi designated 'Fadenpilze' or 'Luftalgen' by Nees von Esenbeck in 1816). Ehrenberg regarded the conjugation as a sexual process but, as for conjugation in *Spirogyra* first observed by O. F. Müller in 1782 and only recognised as a sexual phenomenon twenty years later (by Vaucher in 1803), the significance of Ehrenberg's observations was not generally recognised until mid-century when de Bary (see de Bary & Woronin (1864–81) **2**) published the results of a re-examination of *Syzygites megalocarpus* made in 1860, emphasised the sexual nature of the conjugation, and named the fusion spore a *zygospore*, a term he had already coined for the analogous spore in *Spirogyra*. Two years later, de Bary & Woronin (1864–81, **2**) described zygospores in *Rhizopus stolonifer*, and the Tulasnes (1867) did the same for *Mucor fusiger*.

Earlier, Pringsheim (1857–60) had published a series of papers in which the oogonia and antheridia of Saprolegniales (previously observed by Schleiden, 1842; C. Naegeli, 1847; and A. Braun, 1851) were described in detail and de Bary (1852) had discovered oogonia, antheridia, and rediscovered zoospores (first observed by Prévost (1807), see p. 62) in *Albugo* and *Peronospora*. It was again de Bary who must be credited with the discovery of sex organs in ascomycetes by describing the male

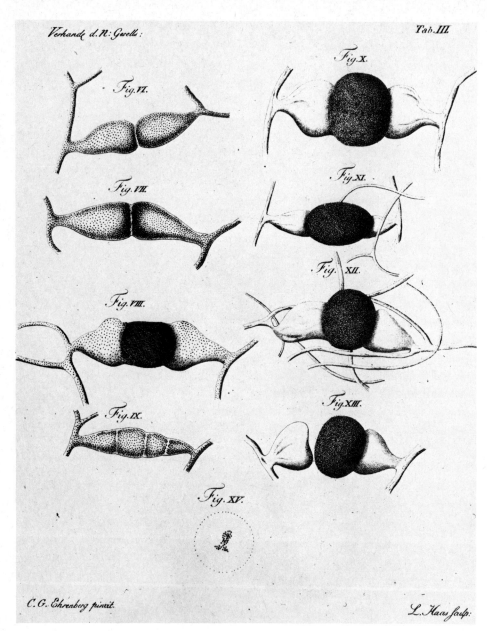

Fig. 52. Zygospore formation in *Syzygites megalocarpus*. (C. G. Ehrenberg (1820 b): pl. 3.)

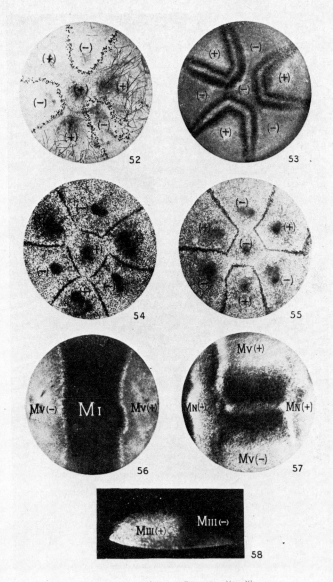

Fig. 53. Heterothallism in *Mucor*. (A. F. Blakeslee (1904): pl. 4.)

and female structures in *Pyronema confluens* and *Erysiphe cichoracearum* (de Bary, 1862 a). De Bary's findings for *Pyronema* were confirmed by the Tulasnes (1867) who distinguished the ascogonium and antheridium as the 'macrocyst' and 'paracyst'.

The discovery of sexual organs in ascomycetes and lower fungi initiated the third, and final, unsuccessful approach to the problem of sex in the hymenomycetes. Attention hitherto paid to the fruit-body now turned to the mycelium. In 1856 H. Hoffmann had noted small cells abstricted from the mycelium at the base of the stipe of *Agaricus metatus*. As these cells would not germinate he considered them to correspond to the spermatia of lichens and other fungi and concluded that they possibly fertilise the young mycelium. Ørsted (1865) described egg cells and long slender antheridia growing up from their base on the mycelium of *Agaricus variabilis* and Karsten (1866) described sexual organs reminiscent of those found in ascomycetes in *Agaricus campestris* and *A. vaginatus*. In 1873 Richon & Roze could not confirm these results but they noted hyphal fusions in the mycelium of *Coprinus ephemerus* (see Richon & Roze, 1885–9: 78–80). In 1875 Max Rees described carpogonia and spermatia in *Coprinus stercorarius* (see Rees, 1876) and in the same year van Tieghem published two papers in the first of which (1875 a) he claimed that the mycelia from spores are of two sexes and bear either male (spermatia) of female (carpogonia) organs and that fruit-body development followed fusion of spermatia with carpogonia; a claim quickly 'confirmed' by de Seynes[5] and Eidam.[6] In the second paper (van Tieghem, 1875 b) he retracted these findings and stated that as the so-called 'spermatia' can germinate to produce 'spermatia'-bearing mycelium they cannot be spermatia. He observed mycelial anastomoses and that fruit-bodies can be produced on mycelium lacking 'spermatia'. He therefore concluded fruit-body production to be sexual. It was left to Brefeld (1876) by pure culture studies to describe how he had followed the development of various agarics, in uncontaminated culture fluids, from spore to spore and had seen no evidence suggesting a sexual origin for fruit-bodies. These observations he amplified and published more fully in his *Botanische Untersuchungen über Schimmelpilze* (1877, **3**:13) which includes a detailed account of the life history of *Coprinus stercorarius* in which the fruit-body develops from a single hypha. Spermatia were considered by Brefeld to be conidia which had lost their powers of development.

The studies which finally elucidated sexuality in hymenomycetes were made possible by developments in cytology already in progress and by the discovery of the phenomenon of heterothallism in the opening years of the twentieth century. A digression, to summarise these advances, must therefore be made.

CYTOLOGY

The 'cell theory', the concept of the cell as the unit of organic life, which was developed during the fourth decade of the nineteenth century was a major biological generalisation. Its basis was laid by the English botanist Robert Brown who in 1831 announced to the Linnean Society of London his discovery of what he called the *nucleus* in the cells of orchids and Asclepidiaceae.[7] The theory was elaborated for plants by M. J. Schleiden, professor of botany at Jena, in his paper 'On Phytogenesis', 1838, and for animals by Theodor Schwann, professor at Louvain, in *Mikroscopische Untersuchungen über die Wachstum der Thiere und Pflanzen*, 1839. Among others who made important contributions were von Mohl and Karl Naegeli while Robert Remak, between 1842 and 1854, established that animal cells multiply by division and not, as supposed by Schwann and others extracellularly.[8] The subsequent extension of the cell theory and its developments can conveniently be recalled by merely noting the dates at which a number of then new, and now universally accepted, descriptive terms were introduced. J. E. Purkinje, professor of physiology at Breslau, coined the term *protoplasm* in 1839 and in 1882 E. Strasburger, professor of botany at Bonn, *cytoplasm*, as opposed to *nucleoplasm*. Two years later Strasburger introduced *prophase, metaphase,* and *anaphase* for the stages of *mitosis* (W. Fleming, 1882; who also coined *chromatin*) and it was J. B. Farmer and J. E. Moore, in 1905, who applied the term *meiosis*[9] to the reduction division which Strasburger in 1882 had shown to occur in both pollen mother cells and the embryo sac.

W. Waldeyer, in 1888, was responsible for *chromosome*. Subsequently there was a gradual recognition of the individuality and constancy of chromosomes and the significance of *chromosome number*. In 1905 Strasburger, who had previously distinguished the gametophyte generation from the sporophyte generation according to nuclear status, introduced *haploid* and *diploid*[10] as terms applicable in both plant and animal kingdoms to genera-

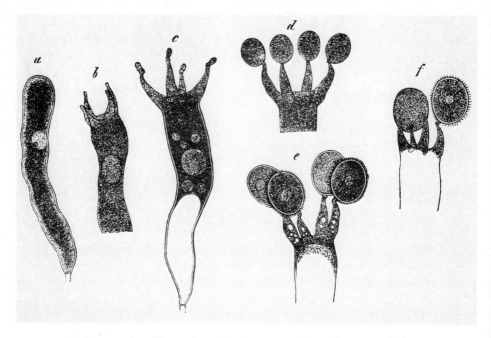

Fig. 54. First illustration of a fungus nucleus – in young living basidium of *Aleurodiscus amorphus*. (A. de Bary (1866): fig. 45.)

tions characterised by either a single or a double set of chromosomes. These most useful terms were seized on and applied not only to generations but also to nuclei, cells, tissues, and even individuals. (For further details on these and related topics general histories of botany or biology should be consulted.)

Nuclei were first recorded for fungi by de Bary (1866) in the first edition of his textbook where he described the single large nucleus he had seen in the young living basidium of *Corticium amorphum* [*Aleurodiscus amorphus*] (Fig. 54) and Schmitz (1880), by haematoxylin staining, concluded that nuclei were present in all fungi. Strasburger (1884), by alcohol fixation followed by alum haematoxylin staining, demonstrated the presence of nuclei in the hyphal cells of *Agaricus campestris*. It was Rosen (1892) who first observed nuclear fusion in the basidium and in the next year the same observation was made by Wager (1893), who in the previous year had made the first observation of mitosis in the basidium of *Stropharia stercoraria* (*S. semiglobata*). Like Rosen, Wager believed that more than two nuclei may fuse to form the secondary nucleus in the basidium, an error corrected by Dangeard (1895)[11] whose findings were confirmed by René Maire (1902) in his doctoral thesis (the historical introduction to which reviews the early studies on the cytology of hymenomycetes and other basidiomycetes). Two years before, Maire (1900) had shown that the cells of both hymenomycetes and gasteromycetes from the development of the fruit-body to the basidium are binucleate, the two nuclei in each cell dividing simultaneously by conjugate divisions ('mitoses conjugées'). Perhaps the only further point to which attention need be called is the subdivision by Juel (1898:385) of the hymenomycetes into two series 'Les Stichobasidiés' and 'Les Chiastobasidiés' according to whether at nuclear division in the basidium the spindles are longitudinally or transversely arranged in relation to the main axis.

Schmitz (1880) also detected nuclei in ascomycetes and Dangeard (1894) described and illustrated nuclear fusion in the ascus[12] of *Exoascus* and *Peziza*. Shortly afterwards the American R. A. Harper, then working in Strasburger's laboratory at Bonn, described nuclear fusion in *Sphaerotheca castagnei* (Harper, 1895) and *Pyronema confluens* (Harper, 1900) and it was Harper who showed that in these fungi there is an alternation of a haploid phase (the vegetative mycelium) and a diploid phase (the fertilised ascogonium and the ascogenous hyphae) (see Harper, 1905).

During the second half of the nineteenth century the problem of sexuality in the rusts had been attracting much attention. Meyen (1841) had suggested that the spermogonia and the aecia were the male and female organs. Early workers found the spermatia (produced in the spermogonia) to lack the power of germination until Tulasne (1875) found that some germination of spermatia occurred in nutrient solutions. Cytological studies by Poirault & Račiborski (1895) showed that the cells of the Uredinales frequently contained two nuclei ('noyaux conjugués', conjugate nuclei) which divided in pairs ('division conjugée', conjugate division) and Dangeard's student, Sapin-Trouffy (1896), described how in rusts exhibiting all spore forms the teliospore is always uninucleate and gives rise to four uninucleate sporidia each of which on germination yields a mycelium of uninucleate cells. Further, he found the mycelium bearing the aecia to be binucleate, as were the aeciospores, and that this condition was also that of the uredinospores (if any) and the young teliospore. Nuclear fusion, which he interpreted as sexual, occurred in the mature teliospore. Sapin-Trouffy considered the uninucleate sporidia to be conidial. V. H. Blackman (1904), then assistant in the department of botany at the British Museum (Natural History) (he later became professor of botany at the Imperial College of Science) thought it obvious 'that the critical point for the investigation in relation to sexuality is the early development of the aecidium' and as a result of a careful and thorough cytological study on *Phragmidium violaceum* and *Gymnosporangium clavariiforme* he confirmed Sappin-Trouffy's findings and illustrated the nuclear migration between cells of the developing aecia to give the binucleate condition (see Fig. 55). He concluded that the spermatia were functionless male cells.

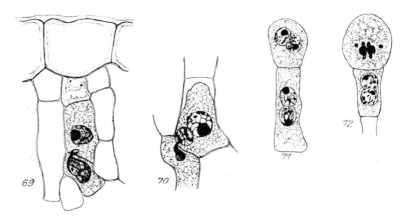

Fig. 55. Nuclear migration between cells of developing aecia of *Phragmidium violaceum* (69, 70) to give the dikaryotic condition (71) in the developing aeciospore (72). (V. H. Blackman (1904): part of pl. 23.)

To summarise, by the opening years of the twentieth century sexual organs had been described for both ascomycetes and lower fungi. Nuclear fusion in the oogonia and developing asci and basidia had been established, and the idea of alternating asexual and sexual phases introduced. The distinction between the dikaryotic and diploid condition had been observed (although the modern terminology applied to these states had yet to be introduced) while as Blackman had realised the three morphological phases observed in connection with the nuclear cycle – 'nuclear association within the same mass of cytoplasm . . . nuclear reduction (fusion), and chromosome reduction' may either take place together or be separated by a considerable number of divisions. Further progress in the elucidation of sexuality was to stem from the discovery of heterothallism.

HETEROTHALLISM

Heterothalism was discovered by the young American mycologist Albert Francis Blakeslee (Fig. 104) at Harvard University and the essentials of Blakeslee's findings may appropriately be given in the author's own words taken from the note in *Science* of 3 June 1904 summarising his more important results.

Zygospore production in the Mucorineae is conditioned by the inherent nature of the individual species and only secondarily or not at all by external factors.

According to their method of zygospore formation, the various species among the Mucorineae may be divided into two main categories, which may be designated as *homothallic* and *heterothallic*, which correspond respectively to monoecious and dioecious forms among the higher plants.

In the homothallic group, zygospores are developed from branches of the same thallus or mycelium and can be obtained from the sowing of a single spore. Although it has been currently assumed that all mucors belong to this class, it comprises a very small percentage of species and contains the only forms from which heretofore it has proved possible to obtain a constant production of zygospores...

In the heterothallic group, comprising a large majority of the species, zygospores are developed from branches which necessarily belong to thalli or mycelia diverse in character and can never be obtained from the sowing of a single spore. Every heterothallic species is, therefore, an aggregate of two distinct strains through the interaction of which zygosporic reproduction is brought about. If inoculations of these two opposite strains of a given species are so disposed that their mycelia can grow together, there will be developed, at the region of contact, a distinct dark line produced by the accumulation of zygospores formed between filaments of opposite strains [Fig. 53].

Rhizopus nigricans [*R. stolonifer*], the common bread mold which is used in nearly every elementary class in cryptogamic botany, may be taken as the type of this group. An accidental mixture of its two strains has been kept under cultivation for nearly ten years and as the 'Harvard strain' has furnished zygospores for class work to many botanical laboratories in this country.

In an individual species these sexual strains show in general a more or less marked differentiation in vegetative luxuriance, and the more and less luxuriant may be appropriately designated by the use of (+) and (−) signs respectively. In a few forms, no differentiation has yet been detected...

In heterothallic species, strains have been found which from their failure to react with (+) and (−) strains of the same form have been called 'neutral', and a similar neutrality may be induced by cultivation under adverse conditions...

In all species of both homo- and heterothallic groups in which the process of conjugation has been carefully followed, the swollen portions (*progametes*) from which the gametes are cut off do not grow towards each other, as curently believed, but arise as the result of contact between more or less differentiated hyphae (*zygophores*) and are from the outset always normally adherent.

In some species the zygophores have been demonstrated to be mutually attractive (*zygotactic*).

In the *heterogamic* subdivision of the homothallic group, a distinct and constant differentiation exists between the zygophoric hyphae and the gametes derived from them, but in the remaining homothallic forms and in all heterothallic forms no such differentiation is apparent...

A process of imperfect hybridization will occur between *unlike* strains of different heterothallic species in the same or even different genera, or between a homothallic form and *both* strains of a heterothallic species, and distinct white lines are produced in many cases at the regions of hybridization...

Two months later the results were published more fully in the *Proceedings of the American Academy of Arts and Sciences* and a number of additional papers followed during the next twenty-six years. In 1906 Blakeslee showed that zygospores of the homothallic *Sporodinia grandis* (*Syzygites megalocarpus*) yielded mycelia carrying only homothallic sporangiospores. Those of the heterothallic *Mucor mucedo* gave sporangia in which the spores were either (+) or (−), thus indicating segregation for 'sex' before the formation of sporangiospores, while in the heterothallic *Phycomyces nitens* (+) and (−) spores occurred in the same sporangium and in addition there were some homothallic spores which gave rise to abnormal homothallic mycelia which were unstable and developed (+), (−), and homothallic sporangiospores. In the course of other studies, when attempting to show serological differences between (+) and (−) strains, it was incidentally found that an aqueous extract of dried *Rhizopus stolonifer* was highly toxic to the rabbit on intravenous injection (Blakeslee & Gortner, 1913).

During the succeeding decade Blakeslee's findings for the Mucorales were amply confirmed.

Hymenomycetes

The first extension of heterothallism to other groups was to the Hymenomycetes. In 1908–9 Elsie M. Wakefield, as a post-graduate student, when making cultural studies on hymenomycetes in the laboratory of Professor Karl von Tubeuf at Munich found that only three of thirteen single-spore cultures of *Schizophyllum commune* produced fruit-bodies (Wakefield, 1909).

Ten years later, in order to investigate this inconsistency Hans Kniep of Würzburg repeated and confirmed Miss Wakefield's findings and showed that, when appropriate sterile strains were mated by being grown as adjacent colonies on one culture plate, fruit-bodies developed. He also observed that the mycelium derived from fertile matings exhibited clamp-connections (Fig. 56) which a few years before he had shown by cytological

Fig. 56. First illustration of a clamp-connection. (H. Hoffman (1856): pl. 5.)

investigations on *Corticium varians* and *C. serum* to be associated with conjugate nuclear division (Kniep, 1915). Further, he established that the mycelium of a fruit-body produced from a single-spore culture lacked clamp-connections and that meiosis was replaced by mitosis in the basidia which bore spores of one mating type only. On 27 November 1919 Kniep announced his discovery of heterothallism in Hymenomycetes at a meeting of the Physikalisch-medizinischen Gesellschaft of Würzburg (Kniep, 1920). Because of the recent war, what Kniep did not know was that this discovery had already been reported by Mathilde Bensaude to the French Académie des Sciences in 1917 and published in the *Comptes Rendus* for that year and in detail as a doctoral thesis in 1918. Neither did Mlle Bensaude know that her discovery of the association of clamp-connections with conjugate nuclear division had been forestalled by Kniep. Mlle Bensaude experimented with two monosporous cultures of *Coprinus fimetarius* (*C. cinereus*) – the survivors of four obtained by the use of a van Tieghem cell. Both these cultures, which she distinguished as α and β, lacked clamp connections and remained sterile indefinitely but when mated always gave a clamp-connection bearing mycelium on which fruit-bodies developed. Mlle Bensaude concluded that *C. cinereus* exhibited heterothallism analogous to that in the Mucorales and she designated her α strain (+) and the β strain (−).

These discoveries stimulated much research, work particularly associated with the names of Buller and his students (including Irene Mounce, W. F. Hanna, Dorothy E. Newton (Fig. 57) and H. J. Brodie) in Canada, H. Brunswik in Austria, R. Vandendries in Belgium, A. J. P. Oort in Holland, A. Quintanilha in Portugal, and Nils Fries in Sweden (who established heterothallism in the Gasteromycetes); and also by Kniep himself – Mlle Bensaude had turned her energies to studies on the plant diseases of Portugal, the Azores, and elsewhere. A complex situation was revealed.

Fig. 57. Staff of the Dominion Rust Research Laboratory, Winnipeg on the occasion of the retirement of Professor Buller in 1936.
Back row (left to right): F. Greaney, J. E. Machacek (1902–70), W. A. F. Hagburg.
Middle row: J. M. Welsh, R. Peturson, D. B. Waddell, T. Johnson, A. M. Brown, W. Popp.
Front row: B. Peturson, W. F. Hanna (1892–1972), Margaret Newton (1887–1971), A. H. R. Buller (1874–1944), J. H. Craigie, C. H. Goulden.

It became firmly established that among hymenomycetes fruit-bodies may bear one, two, or four types of basidiospores. Those of the first category are typically homothallic. That is, fruit-body development occurs on a monosporous, clamp-connection bearing mycelium as demonstrated by Irene Mounce (1921) for *Coprinus sterquilinus*, *C. stercorarius*, and *C. narcoticus*. Less frequently the fruit-body bearing one type of spore is produced on a mycelium derived from a single spore of one mating type of a heterothallic species. Then the mycelium bearing the fruit-bodies lacks clamp-connections, as shown by Kniep for *Schizophyllum commune*. Further, it was found, first by B. O. Dodge (1928) for the ascomycete *Neurospora tetrasperma* and two years later by Sass (1929) for *Coprinus ephemerus* f. *bisporus*, that a second type of homothallism ('secondary homothallism') may be distinguished in which the fruit-bodies are derived from a single spore which contains not one diploid nucleus (as in 'primary homothallism') but two haploid nuclei of different but compatible mating types.

Fruit-bodies of the second category, that is with basidiospores of two mating types (*A* and *a*) determined by one pair of Mendelian factors (*Aa*) are characteristic of *bipolar heterothallism* which was discovered independently during 1923–4 by Vandendries in *Annellaria separata* (*Panaeolus separatus*) (Vandendries, 1923), Brunswik (1924) in *Coprinus comatus*, and Dorothy Newton (1926) in *Coprinus rostrupianus* (*C. flocculosus*).

In the third category, *tetrapolar heterothallism*, mating is determined by two pairs of factors (*Aa*, *Bb*) at two loci. The four types of spore then possible have constitutions *AB*, *ab*, *Ab*, *aB* and mating can only take place between mycelia of the first and second and the third and fourth classes. Kniep (1920) was the first to demonstrate tetrapolar heterothallism in *Schizophyllum commune* and *Aleurodiscus polygonius* and he was also the first to discover (by crossing single-spore cultures of fruit-bodies from different sources) an additional complication caused by the presence of multiple alleles at both loci (A_1, A_2, A_3 etc.; B_1, B_2, B_3 etc.). Within a few years the occurrence of multiple alleles (up to a hundred or more in some cases) was established for many examples of both tetrapolar and bipolar heterothallism in diverse hymenomycetes and the presence of multiple alleles explained why single-spore cultures of heterothallic species from different sources (the so-called 'geographical races' of Brunswik (1924) and others) were completely compatible when paired.

Kniep was also the first to investigate the individual spores in a tetrad. The four spores of the basidium of the tetrapolar *Aleurodiscus polygonius* are sometimes shot away as one clump which can be separated by shaking into its components which are then grown on an agar culture medium. Two types of spores were found on each basidium, either two A_1B_1 and two A_2B_2 or two A_2B_1 and two A_1B_2, from which Kniep deduced that segregation for sex occurred at the first reduction division. Funke (1924), in Kniep's laboratory, confirmed this by micro-manipulation for tetrapolar species of *Hypholoma* and *Collybia* and also found some basidia bearing all four possible types of spore (indicating segregation for sex at the second reduction division) as did Hanna (1925) for *Coprinus lagopus* (*sensu* Buller = *C. radiatus*) by a technique which he and Buller devised involving the pressing of a cover slip against the surface of a young gill when a number of the spore tetrads adhered to the cover slip from which they were removed individually by means of a dry needle, transferred to a suitable agar and the resulting mycelia tested for mating type (Hanna, 1924).

In 1930 Buller coined the useful term *diploidisation* for the process by which a haploid (or monokaryotic) mycelium is converted into a diploid (or dikaryotic) mycelium; see Buller (1941) for a review. Diploidisation by the fusion of one hypha of a haploid mycelium with another of a compatible mating type was first observed at Kniep's laboratory, in cytological preparations of *Corticium serum* [*Hypodontia sambuci*] and *Typhula ery-thropus*, by Lehfeldt (1923) who also studied nuclear migration and the development of the first clamp connection. The diploidisation of one haploid mycelium by another is the normal process but in 1930 Buller reported diploidisation of the haploid mycelium of *Coprinus radiatus* by fusion with a diploid (dikaryotic) mycelium (Fig. 58). This process was subsequently designated the 'Buller phenomenon' by Quintanilha (1937).

Another method of diploidisation is by oidia. Oidia were described for numerous hymenomycetes by Brefeld who con-cluded that they were vestigial structures and although oidia were germinated by, among others, van Tieghem (who consi-dered them to be male cells), their function remained unknown. It was Bensaude (1918) who first noted that oidia were produced only on haploid mycelia and that after diploidisation oidial production ceased. She also observed that occasionally in paired cultures one of the haploid colonies became diploid before

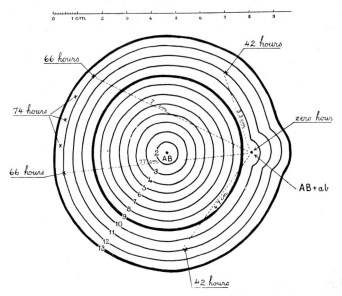

FIG. 5.—The diploidisation of a large haploid mycelium (*AB*) by a *diploid* mycelium (*AB*)+(*ab*). The (*AB*) mycelium was inoculated with a tiny hyphal mass of the (*AB*)+(*ab*) mycelium after 9 days of growth (periphery shown by heavier inner circle, No. 9) at the zero hour. The diploid mycelium diploidised the haploid mycelium in a little more than three days. The crosses show where clamp-connexions were observed at particular times. The (*ab*) nuclei must have travelled more than 7·7 cm. or 77 mm. through the (*AB*) hyphæ in about 64 hours, or more than 1·2 mm. per hour. Two-thirds the actual size.

Fig. 58. The 'Buller phenomenon', the diploidisation of a monokaryotic mycelium by a dikaryotic mycelium (A. H. R. Buller (1930): fig. 5.)

fusion had occurred with the opposite strain which she suggested was because oidia from the opposite strain had crossed the gap between the two mycelia in a water film on the surface of the agar, germinated, and fused with a hypha of the other strain. Brodie (1931) established that oidia can effect diploidisation in *Coprinus radiatus* and, in the light of Craigie's experiments on rusts which are noted below, showed that such diploidisation could be brought about by the agency of flies.

Ustilaginales and Uredinales

In 1919, Kniep reported that the anther smut of Caryophyllaceous plants (*Ustilago violacea*) exhibited bipolar heterothallism and this condition, lacking multiple allelomorphs, was shown by subsequent workers to be typical for smuts in general. Sexuality in rusts was more difficult to elucidate.

In July 1927 J. H. Craigie (Fig. 57) of the Dominion Rust Research Laboratory, Winnipeg, announced in *Nature* his demonstration by monosporidial infections of sunflower that *Puccinia helianthi* was heterothallic (Craigie, 1927a). He found that isolated monosporidial infections gave rise to pycnia which produced pycniospores and nectar but no aecia developed as they frequently did when two or more pycnial pustules were adjacent to one another or coalesced. Craigie concluded that the sporidia were of two types (+) and (−) and that when adjacent monosporidial infections were of opposite type, hyphal fusions occurred, and aecial development followed. In November of the same year, in a second letter to *Nature*, Craigie (1927 b) elucidated the function of the pycnia by reporting that aecial development followed the transference of spores from one pycnium to another of opposite mating type.

In 1933 Craigie illustrated the fusion of a pycniospore with a flexuous hypha projecting from a pycnium in *Puccinia helianthi* and in 1938 Buller recorded and illustrated the same process in *P. graminis* while A. M. Brown (1932) noted the Buller phenomenon (the diploidisation of a haploid by a diploid mycelium) in *P. helianthi*.

Ascomycetes and other fungi

The first case of heterothallism in the Ascomycetes was that of the discomycete *Ascobolus magnificus* described by B. O. Dodge (1920). C. L. Shear & B. O. Dodge (1927) showed several species of *Neurospora* to be heterothallic while the four-spored *N. tetrasperma* was secondarily homothallic with occasional heterothallic haplonts. F. L. Drayton (1932, 1934) found that in *Sclerotinia gladioli* the microconidia borne on monosporous mycelial cultures, together with receptive bodies, functioned as spermatia which when transferred to the receptive bodies of other compatible monosporous cultures induced the formation of apothecia.

Heterothallism in sporogenous yeasts took many years to elucidate. As long ago as 1891, E. C. Hansen reported that there was fusion between ascospores of *Saccharomycodes ludwigii* on germination. A few years later his assistant H. Schiønning (1895) observed the copulation of cells before ascus formation in *Schizosaccharomyces octosporus* (Fig. 59) and at the turn of the century it was established by Hoffmeister (1900), and subse-

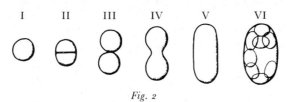

Fig. 2

Formation d'ascus. I, cellule ronde peu avant la formation de la cloison transversale; II, au bout d'une heure; III, au bout de 3; IV, au bout de 6; V, au bout de 10; VI, au bout de 17. Les indications du temps sont comptées à partir du commencement de l'observation. Grossissement linéaire 1000 fois.

Fig. 59. Copulation of cells before ascus formation in *Schizosaccharomyces octosporus*. (H. Schiønning (1895): fig. 3 (redrawn).)

quently confirmed by others, that such plasmogamy is followed by karyogamy. In 1935 Ø. Winge reported the alternation of haploid and diploid generations in yeasts but it was not until 1943 that Lindegren & Lindegren demonstrated heterothallism in *Saccharomyces cerevisiae* in the laboratory and in 1952 that L. J. Wickerham and K. A. Burton established the occurrence of the same phenomenon in nature.

Many more examples have been recorded and the term heterothallism has also been applied to laboulbeniomycetes and many phycomycetes where the two types of thalli can be distinguished by the presence of gametes or gametangia of opposite sex. H. L. K. Whitehouse (1949a) in the first of his excellent reviews of heterothallism, distinguished this type of heterothallism as *morphological heterothallism* (or haplo-dioecism) as opposed to *physiological heterothallism* (or haploid incompatibility) when there are no significant morphological differences between the two thalli associated with sexual reproduction.

GENETICS

One of the major developments following the elucidation of fungal sexuality was the use made of fungi by geneticists in experimental studies on problems of formal genetics. Species of *Neurospora, Saccharomyces,* and *Aspergillus* were taken over as most convenient subjects for such studies, just as the fruit-fly *Drosophila* had been, and a large new literature was generated. Most of the results obtained, many of which are post-1950, belong to the history of genetics rather than mycology and will

133

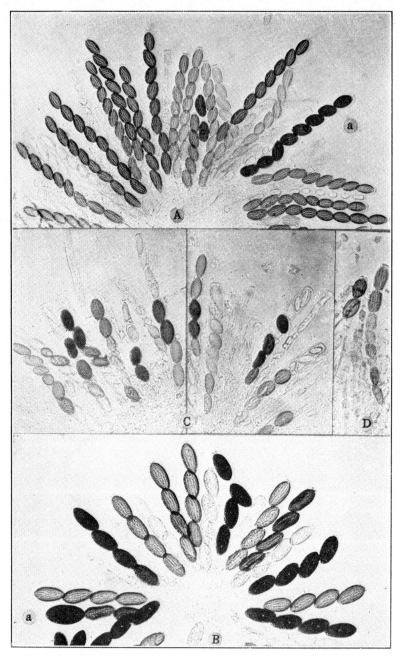

Asci of Neurospora sitophila, A; N. tetrasperma, B; and sitophila × tetrasperma, C, D.

Fig. 60. First experimental hybridisation in *Neurospora* by B. O. Dodge (1928): pl. 1.

not be reviewed here (see Fincham & Day (1963) and Esser & Kuenen (1965) for details). Only a few points need to be recalled.

B. O. Dodge (1928) was the first to report fertile hybrids in *Neurospora* (Fig. 60). Following this lead Carl C. Lindegren was in 1936 able to establish the first linkage map for the fungal chromosome; a six-point map for the sex-chromosome of *Neurospora crassa* published in the April issue of the *Journal of Genetics* and as a slightly revised version in July in the *Journal of Heredity* (Fig. 61).

Lindegren: Sex-Chromosomes of Neurospora   255

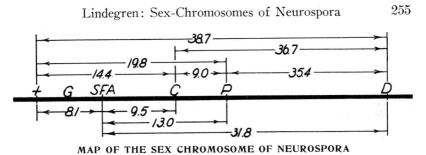

MAP OF THE SEX CHROMOSOME OF NEUROSPORA
Figure 4

Genetical distances in the sex-chromosome determined by data submitted in this paper. All are direct measurements and none is the result of addition or subtraction of two other values. The gene *gap* (G) has been previously (Lindegren[9]) shown to be about 4 units to the left of the spindle-fiber attachment (SFA). Although there are discrepancies due to sampling errors concerning actual distances, all data agree on the seriation of the six loci: the sex-differentiating allels (+/−), the gene *gap* (G), the spindle-fiber attachment (SFA), and the genes *crisp* (C), *pale* (P), and *dirty* (D).

Fig. 61. First map for a fungus chromosome. (C. C. Lindegren (1936): fig. 4.)

For any pair of alleles the eight ascospores in the ascus occur in six different patterns according to whether segregation occurs at the first or the second reduction division and Lindegren, working with the sex determining factor and five other single factors which determined morphological characteristics, was able from considerations derived from segregation of gene pairs at the second reduction division, which is a function of the amount of crossing-over which has occurred between the loci in question and the spindle fibre attachment (centromere), to calculate the distance of each of the six loci from the centromere.

The important modern development of 'biochemical genetics' has two main aspects, both of which have deep roots. The first, in which fungi have played a minor role, concerns the chemical nature of the genetic material. This began with the description by Johan Friedrich Mieschner of 'nuclein' in pus cells (1868) and in the spermatozoa of fish (1874). At the turn of the century, Albrecht Kossel, professor of physiology at Marburg and, in 1910, a Nobel prize winner for medicine, differentiated two types of nucleic acid as 'thymus nucleic acid' (deoxyribose nucleic acid, DNA) and 'yeast nucleic acid' (ribose nucleic acid, RNA) and a climax was reached by the elucidation of the chemical structure of DNA by J. D. Watson and F. H. C. Crick in 1953.

The second concerns the biochemical effects of genes. In 1941 Beadle and Tatum looking for a convenient organism capable of yielding many mutants to extend work begun with the fruit-fly *Drosophila*, selected *Neurospora* which has simple nutritional requirements, is haploid so that the analysis is not complicated by dominance and recessiveness, produces large numbers of asexual spores of identical genetic constitution, and yet has a sexual phase (B. O. Dodge, 1928) which can be utilised for the genetic analysis of mutant strains. Beadle & Tatum (1945) reported that by testing *Neurospora* cultures after treatment with X-rays or other radiation they had obtained 380 strains of altered nutritional requirements (for example, mutants which unlike the wild type would only grow on the 'minimal medium' after supplementation with pyridoxine, thiamine, *para*-aminobenzoic acid, or other nutrient, respectively) and in each case analysed the mutant showed a single-gene difference from the parent type. By undertaking breeding experiments with such strains it is frequently possible to recognise progeny carrying genes of particular types simply by plating large numbers of spores on test media of appropriate composition. Variations in spore colour too can provide most useful genetic markers. This technique was widely adopted and extended to other fungi, including *Saccharomyces* and *Aspergillus*, and the results obtained have greatly extended and deepened the knowledge of metabolic pathways in fungi, and of gene action.

G. Pontecorvo and J. A. Roper, geneticists of Glasgow University, discovered that genetic recombination occurred in *Aspergillus nidulans* in the absence of sexual reproduction by a process which they called the 'parasexual cycle' (Pontecorvo & Roper, 1952). To use Pontecorvo's own words,

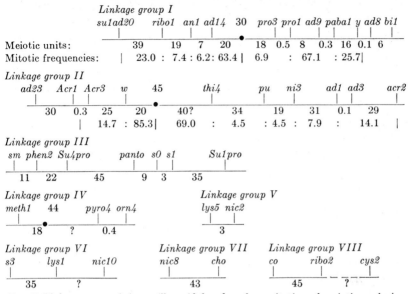

FIG. 2. Linkage maps of *Aspergillus nidulans* based on mitotic and meiotic analysis.

Fig. 62. First maps for the chromosome complement of a fungus (*Emericella nidulans*). (Etta Käfer (1958): fig. 2.)

The steps in the cycle are the following: (a) Fusion of two unlike haploid nuclei in a heterokaryon. (b) Multiplication of the resulting diploid heterozygous nucleus side by side with the parent haploid nuclei in a heterokaryotic condition. (c) Eventual sorting out of a homokaryotic diploid mycelium which becomes established as a strain. (d) Mitotic crossing over occurring during multiplication of the diploid nuclei. (e) Vegetative haploidisation of the diploid nuclei (Pontecorvo, 1956).*

A. nidulans has a sexual phase but Pontecorvo and his co-workers soon established the presence of the parasexual cycle in both *A. niger* and *Penicillium chrysogenum* for which sexual reproduction is unknown and since then the phenomenon has been extended to additional fungi by other workers. The parasexual cycle has clearly many potentially useful applications in breeding new strains of imperfect fungi but such potentialities are still under investigation.

A climax in this direction was the set of chromosome maps published in 1958 by Etta Käfer, of Pontecorvo's laboratory, for the eight chromosomes of *A. nidulans*. This was the first occasion on which maps for a complete chromosome complement of any fungus had been offered and as can be seen from Fig. 62 the data on which the maps were based were obtained by two independent methods – by meiotic analysis of the perfect state (*Emericella nidulans*) and by mitotic analysis of the imperfect state (*A. nidulans*).[13]

PHYSIOLOGY AND BIOCHEMISTRY OF SPORULATION AND SEXUALITY

During the past fifty years, research on sporulation and sexuality, like that already noted on morphology and nutrition, has been increasingly physiological and biochemical. These investigations, on which few generalisations are possible, have been dominated by studies on the effects of diverse environmental factors on sporulation. The first aspect to attract attention was the effect of nutrition, a topic which came into prominence after the turn of the century following the demonstration by Klebs in 1896 that the transfer of Saprolegnia from a nutrient medium to distilled water induced the formation of zoosporangia and zoospores. Later P. Claussen (1912) made the experiment, now so familiar to students, in which a culture of *Pyronema confluens* growing on a plate of solid, inulin-containing medium formed a ring of apothecia as it overgrew onto a medium lacking carbohydrate. Subsequently it was found, first by W. H. Schopfer in 1931 for *Phycomyces blakesleeanus* (see Schopfer, 1934), that some fungi have, in addition to requirements for trace elements (see p. 110), specific growth factor requirements for sporulation; for example, *Sordaria fimicola* (studied in detail, as *Melanospora destruens*, by Lilian E. Hawker and her students in the decade following 1936) which requires biotin for assimilative growth and the addition of thiamin (or its pyramidine moiety) for sporulation. These and many other examples are covered in the reviews by Hawker (1950: chaps. 5, 6; 1957; and 1966 in Ainsworth & Sussman (1965–73) **2**: chap. 14), as are studies on the effects on sporulation of such external factors as temperature, water, hydrogen-ion concentration, and radiation. One aspect of the diurnal effect of daylight on cultures early noted was the phenomenon of zonation – apparently first studied experimentally by G. G. Hedgcock in 1906 – in which the growth shows alternating bands of differing degrees of sporulation, while the observations of F. L. Stevens in 1928 on the stimulatory effect on sporulation of ultra-violet radiation was an inspiration for the current routine use of 'black light' (near ultra-violet radiation of 320–420 nm) for inducing spore production in sterile culture for identification purposes.[14]

One of the major mycological advances of the twentieth century was the experimental proof of hormonal control of sexuality in fungi, a discovery associated with the studies of John R. Raper at Harvard University. The possibility that sexual hormones occurred in *Saprolegnia* was suggested by de Bary in 1881 but it was not until 1924 that Burgeff announced his experimental proof of chemotropism (or 'zygotropism' as he termed it) in *Mucor mucedo* in which he observed the beginning of the development of zygophores in both (+) and (−) strains before hyphae of the two strains had come into contact and he demonstrated the development of zygophores (and their mutual attraction) in both strains when grown adjacent to one another but separated by a membrane. Later M. Plempel (1960) showed that two hormones were involved, one responsible for zygophore induction, the other for the chemotropism exhibited by the zygophores. In the meantime, in a brilliant series of papers published between 1939 and 1950 J. R. Raper had offered experimental evidence that sexuality in *Achlya* – particularly in the heterothallic *A. ambisexualis* and *A. bisexualis* – is subject to complex hormonal control (see Raper (1952) for a review; also Machlis, 1966). He showed that four specific hormones (the A complex), two derived from the female thallus and two from the male, were responsible for the development of antheridial hyphae and that other hormones controlled the production of oogonial initials, the growth of the antheridial hyphae to the oogonial initials, and the formation of the cross-walls which delimit the antheridia and oogonia. Subsequently, proof of sex hormones in ascomycetes was given by Levi (1956) who published experimental evidence in support of suggestions made by earlier workers that conjugation in *Saccharomyces cerevisiae* is under hormonal control, and in 1956–7 G. N. Bistis demonstrated multihormonal control of sexuality in *Ascobolus stercorarius*.

6. Pathogenicity

Although advances in scientific knowledge may in part be attributed to man's curiosity and desire for explanation, more practical ends and economic drives have also given impetus to advances in most branches of science. What navigation was to horology and cartography fungal disease in man, animals, and, particularly, plants has been to mycology.

PATHOGENICITY TO PLANTS

There are references to plant diseases in the Indian Vedas, Sumerian clay tablets, and the Greek and Roman classics extending back some 2000 years before the Christian era and if some of these often imprecise records were of depredations by pests, others certainly referred to fungal infections.

A frequent explanation of outbreaks of disease in plants was Divine intervention; particularly as a punishment for wrong-doing as when the Hebrew prophet Amos, on behalf of the Almighty, declaimed:

I have smitten you with blasting and mildew: when your gardens and your vineyards and your fig trees and your olive trees increased the palmerworm devoured them: yet have ye not returned to me saith the LORD [*Amos*, 4. 9].

The Romans (traditionally during the seventh century B.C. under Numa Pompilius, the second legendary king of Rome) instituted an annual festival, the Robigalia, which was celebrated on 25 April when, to propitiate Robigus, the rust god, a procession left Rome by the Flamian gate, crossed the Milvian bridge, and proceeded to the fifth milestone on the Claudian Way where in a sacred grove prayers were offered and the priest (*flamen quirinalis*) sacrificed a reddish dog (the colour being possibly symbolic of the disease to be averted) and a sheep.[1]

Alternative explanations were the evil influence of the stars (but not, according to Theophrastus, the sun), witchcraft, and, more rationally, environmental factors: for example, Pharaoh, recounting his dream to Moses, told how he beheld 'Seven ears, withered, thin, *and* blasted with the east wind' (Genesis, 41. 23).

It was not until the seventeenth century that the association of fungi with disease in plants was established but even then it was usual to account for the fungi as being a result rather than the cause of the disease. Only as recently as the last quarter of the eighteenth century was the elucidation of pathogenicity of fungi for plants begun on a sound experimental basis but its final establishment at the turn of the century did not prevent the causal role of environmental factors and heterogenesis receiving intermittent support for another fifty years.

Today, the study of the aetiology, cure, and prevention of infectious diseases, non-parasitic disorders, and pest infestations of plants shows less coherence than veterinary or human medicine of which it is the counterpart. There is a widespread dichotomy into 'pests' and 'diseases' and 'plant pathology' (or phytopathology) is restricted to infectious diseases and non-parasitic disorders, although it does at times cover nematode infestations. Fungal diseases constitute the major branch of plant pathology, so much so that, for long, plant pathologists were known as 'mycologists' and in the United Kingdom it was not until the National Agricultural Advisory Service (N.A.A.S.) was established in 1946 that the designation of the Ministry of Agriculture's 'advisory mycologists' was changed to 'plant pathologists'. Many practical aspects of the treatment and prevention of disease in plants belong to the history of phytopathology rather than mycology and will be given little emphasis here where the significant mycological development of the gradual acceptance between 1750 and 1850 of the ability of fungi to parasitise plants and induce disease, and the effect this has had on the growth of mycology, will be stressed. But first a glance at the changing views on the classification of disorders of plants may help to set the topic in perspective.

Classification of disease in plants

The first classifications proposed were comprehensive. J. S. Elsholz (1684) considered plant diseases to be caused by meteors, animals, and pests.[2] In 1705 J. P. de Tournefort in his 'Observations sur les maladies des plantes' distinguished two groups of diseases; those due to internal and external causes, respectively. C. S. Eysfarth in his doctoral thesis *De morbis plantarum,* 1723, after an introduction on the nature of plants grouped plant disorders according to the time of occurrence: at germination, during growth, or at maturity. Linnaeus, in *Philosophia botanica,* 1751, distinguished mildew, rusts, smuts, and ergot and various types of insect damage (Fig. 63) while Adanson (1763, **1**:4), like de Tournefort, differentiated internal and external categories of disease. Plant disease classifications during the last quarter of the eighteenth century and the first

312. Morbofas plantas, vel etiam ætates in Nomini-
bus varietatum affumere, fæpius fuperfluum eft.

Morbofæ plantæ a Botanicis receptæ variæ funt, prout earum morbi.

Eryfiphe Th. eft Mucor albus, capitulis fufcis feffilibus, quo folia afperguntur, frequens in *Humulo, Lamio* Fl. fuec. 494. *Galeopfide* Fl. fuec. 491. *Lithofpermo* Fl. fuec. 152. *Acere* Fl. fuec. 303.

Rubigo eft pulvis ferrugineus, foliis fubtus adfperfus, frequens in *Alchemilla* Fl. fuec. 135. *Rubo faxatili* Fl. fuec. 411. *Efula degenere* R. & præfertim in *Senecione* five *Jacobæa fenecionis folio incano perenni.* Hall. jen. 177. inprimis in folo fylvatico uftulato.

Clavus cum Semina enafcuntur in cornicula majora extus nigra, ut in *Secali* & *Caricibus.*

Uftilago cum fructus loco feminum farinam nigram proferunt.

Uftilago Hordei C. B. Uftilago Avenæ. C. B.
Scorzonera pulverifloraH.R.P.Tragopogon abortivumLœf.

Nidus infectorum cauffatus ab infectis, quæ ova depofuere in plantis unde excrefcentiæ variæ.

Gallæ *Querci e Glechoma, Cifti, Populi tremula, Salicum Hieracii myophori.*

Bedeguar *Rofæ.*

Folliculi *Piftaciæ, Populi nigræ.*

Contorfiones *Ceraftii, Veronicæ, Loti.*

Squamationes *Abietis, Salicis rofeæ.*

Infecta fæpe caufant plenitudines & prolificationes florum.

Matricaria Chamæmelum vulgare Fl. fuec. 702. ab infectis minimis fit prolifera.

Carduus caule crifpo Fl. fuec. 658. cura infectorum gerit flofculos majores, grifeos plenos vel potius prolifero frondofos, piftillis in folia enafcentibus.

Fig. 63. Linnaeus's classification of plant diseases. (C. Linnaeus (1751): 243.)

half of the nineteenth were much influenced by medical nosology. Johan Baptista Zallinger (*De morbis plantarum*, 1773) recognised five classes of plant disease – *Phlegmasiae* or inflammatory disease; *Paralyses seu debilitates*, paralysis or debility; *Cachexia*, bad constitution; etc. – as did the Danish naturalist Johan Christian Fabricius who, in the following year, attempted to 'deduce the classes and genera according to the apparent cases, but species according to the causes of the disease'. For example, his Class I, *Rendering unproductive* included five genera 'Super-abundance of sap', 'Smut', 'Barrenness', 'Unseasonable deciduity', and 'Doubling'; species of the third genus being Barrenness due to rain, cold, smoke, bastardised plants, etc. A climax in this direction was the system devised by the Italian Philippo Ré who in the second edition of his *Saggio teorico-pratico sulle malattie delle piante* of 1817 listed sixty-seven genera of diseases (in five classes) from 'Bulbomania' and 'Antheromania' to 'Contagio' and 'Morbo del gelso' (disease of the mulberry tree), together with many species (e.g. *Necrosi solare*, *Necrosi elettrica*, etc.). Ré's system attracted the Rev. M. J. Berkeley who adapted it as a framework for his famous exposition of 'Vegetable pathology' in a series of 173 articles in the *Gardeners' Chronicle* during 1854–7. Berkeley's 'Genus XLII Parasitae' covered both phanerogams, cryptogams (fungi, mosses, lichens, algae), and insects.

By this time the reality of fungal pathogenicity for plants was becoming appreciated and the textbook by Julius Gotthelf Kühn, *Die Krankheiten der Kulturgewächse, ihre Ursachen und Verhütung*, 1858, which, with the one exception of ear cockles of wheat (caused by the nematode *Tylenchus* [*Anguillulina*] *tritici*) was confined to fungal diseases, set the pattern for later texts. As the number of recognised fungal diseases increased it became the custom to group them taxonomically into diseases caused by phycomycetes, ascomycetes, basidiomycetes, and fungi imperfecti; a pattern that persisted. As late as 1928 *Plant diseases* by F. T. Brooks was a treatise on plant pathogenic fungi rather than on diseases. Although this approach certainly has didactic advantages, and monographs on particular groups of pathogens such as powdery mildews, rusts, and smuts are very valuable adjuncts for plant pathologists (as the many contributions by plant pathologists to this aspect of taxonomic mycology demonstrates), most practising plant pathologists, like most agriculturalists and horticulturalists, are crop orientated. This has resulted in the publication of many texts on the diseases of particular crops, a trend that began in forestry with Robert Hartig's *Wichtige Krankheiten der Waldbäume*, 1874.

As already implied, there was little progress in the understanding of infectious disease in plants (and also in animals and man) from classical times to the sixteenth century herbalists. In the mid-seventeenth century Robert Hooke, in his *Micrographia*, 1665 (Obs. xix), wrote of rose rust (*Phragmidium mucronatum*):

The generation of this Plant seems in part, ascribable to a kind of *Mildew* or *Blight*, whereby the parts of the leaves grow scabby, or putrify'd, as it were, so as that the moisture breaks out in little scabs or spots.

From this putrify'd scab breaks out this little Vegetable; which may be somewhat like a *Mould* or Moss; and may have its *equivocal* generation much after the same manner as I have supposed *Moss* or *Mould* to have, and to be a more simple and uncompounded kind of vegetation.

J. P. de Tournefort (1705), although without proof, believed that fungi were produced from 'seeds', that these seeds germinated under moist conditions and caused mouldiness of plants, particularly during the winter months under glass, to prevent which he recommended keeping greenhouses drier. These conclusions were advanced and more typical of the period were the views of the Rev. Stephen Hales, perpetual curate of Teddington on the Thames near London, on hop mildew in his *Vegetable Staticks*, 1727, where he wrote:

in a rainy moist state of air, without a due mixture of dry weather, too much moisture hovers about the hops, so as to hinder in good measure the kindly perspiration of the leaves, whereby the stagnating sap corrupts, and breeds moldy fen, which often spoils vast quantities of flourishing hop grounds. This was the case in the year 1723.

But Hales does not appear to have been a consistent heterogeneticist for he concluded:

The planters observe, that when mold or fen has once seized any part of the ground, it soon runs over the whole...

Probably because the small seeds of the quick growing mold, which soon come to maturity, are blown over the whole ground: Which spreading of the seed may be the reason why some grounds are infected with fen for several years successively; *viz.* from the seeds of last year's fen. Might it not then be adviseable to burn the fenny hop-vines as soon as the hops are picked, in hopes thereby to destroy some of the seed of the mold?

There were two main approaches to the elucidation of the pathogenicity of fungi to plants between 1750 and 1850. One, which was largely responsible for perpetuating explanations based on heterogenesis, was to make a-priori deductions from observations or reports. The other was to limit deductions to those justified by the results of experiments. As might be expected, the investigations which contributed most to the reality of fungal pathogenicity were those on diseases of such staple crops as cereals; studies on bunt or stinking smut of wheat (*Tilletia caries*) and on black rust (*Puccinia graminis*) being of particular importance.

Bunt of wheat

Woolman & Humphrey (1924) and G. W. Fischer & Holton (1957: chap. 2) have reviewed the early records of stinking smut of wheat which was, even if at times confused with rusts and other smuts, known to the Greeks and Romans and also at first hand by the herbalists, including Jerome Bock (1552) and the Bauhin brothers. The most popular explanation of bunt was unfavourable environmental factors, an explanation still current in the eighteenth century when Jethro Tull in the chapter 'Of Smuttiness' in his *The horse-hoing husbandry*, 1733, attributed the disease to cold wet summers and

was confirm'd in this by several Plants of Wheat, taken up when they were in Grass in the Spring, and Plac'd in Troughs in my Chamber-window, with some of the Roots in Water. These Wheat-plants sent up several Ears each; but at Harvest, every Grain was smutty; and I observed that none of the Ears ever sent out Blossom; This Smuttiness could not be from any Moisture that descended upon it, but from the Earth, which always kept very moist.

Tull's contemporary, Richard Bradley, professor of botany at the University of Cambridge, in 1725 thought otherwise. In his opinion 'all Blights proceed from insects'. Microscopic observation on 'the Black in corn... plainly discover it to be the Eggs of Insects'.[3] Both these authors held that bunt could be prevented by steeping the seed in solutions of brine before sowing.

A major advance, not only in the understanding of bunt but also in phytopathological field experimentation, resulted from the investigations of Mathieu Tillet, Director of the Mint at Troyes, who devoted his spare time to the improvement of agriculture, as detailed in his 150-page *Dissertation sur la cause qui corrupt et noircit les grains de bled dans les épis; et sur les moyens de prevenir ces accidens*, 1755, which gained for him the prize, offered in 1750 by the Académie Royale des Belles-Lettres, Sciences & Arts de Bordeaux, for the best dissertation on the cause and cure

EXPERIENCES DE 1751-52.

N.	QUATRIEME DIVISION. FUMIER DE CHEVAL & DE MULET.						CINQUIEME DIVISION. TERRE NATURELLE.					
2. Nov.	16. Oct.	22. Oct.	27. Oct.	3. Nov.	10. Nov.	22. Nov.	16. Oct.	22. Oct.	27. Oct.	3. Nov.	10. Nov.	22. Nov.
Moucheté. & noirci. 69	Noirci. 73	Mouchété. 77	Noirci. 81	Mouchete. 85	Noirci. 89	Mouchete. & noirci. 93	Noirci. 97	Mouchete. 101	Nourci. 105	Mouchete. 109	Noirci. 113	Mouchete. & noirci. 117
Mariné. 70	Mariné. 74	Mariné. 78	Mariné. 82	Mariné. 86	Mariné. 90	Mariné. 94	Mariné. 98	Mariné. 102	Mariné. 106	Mariné. 110	Mariné. 114	Mariné. 118
Pur. 71	Pur. 75	Pur. 79	Pur. 83	Pur. 87	Pur. 91	Pur. 95	Pur. 99	Pur. 103	Pur. 107	Pur. 111	Pur. 115	Pur. 119
Chaulé. 72	Nitré. 76	Chaulé. 80	Nitré. 84	Chaulé. 88	Nitré. 92	Chaulé. 96	Nitré. 100	Chaulé. 104	Nitré. 108	Chaulé. 112	Nitré. 116	Chaulé. 120

108. Pieds. 108. Pieds.

... ... entier .... 540. pieds ....

Fig. 64. Part of the plan of Tillet's first field experiment on bunt of wheat. (M. Tillet, 1755.)

of blackening of wheat. After preliminary experiments by which Tillet refuted Tull's explanation that bunt resulted from too much moisture, the first of his two main experiments on winter wheat was set up in the autumn of 1751 when a plot of land suitable for wheat, 24 by 540 feet, was divided into 120 subplots on which he compared the effects of unmanured soil with four manurial treatments (pigeon manure, sheep manure, fecal matter, horse and mule manure) on clean seed, seed naturally contaminated or experimentally inoculated with bunt spores, and seed left untreated or after treatment with several formulations of salt, lime, and saltpetre (see Fig. 64). In addition, a comparison was made of six dates of sowing on each of which seed was sown on all five manurial treatments. During the second season (1752–3) on a similar but slightly smaller plot, again divided into 120 subplots, the effects of various methods of inoculation, different types of seed treatment, and the effects on clean seed of the addition to the soil of contaminated wheat straw, manure obtained from a Spanish horse which had eaten (somewhat unwillingly) straw heavily contaminated with bunt spores, and various other types of manure before or after admixture with contaminated straw. Further seed treatments were also included as well as a series of plots of darnel, rye, and barley. Finally, additional trials were made during 1753–4.

Tillet's results made it abundantly clear 'that' – in his own words (in H. B. Humphrey's translation) – 'the common cause, the abounding source of bunted wheat plants resides in the dust of the bunt balls of diseased wheat'; and further 'that the treatments I employed have protected the most heavily infested seed against the effects of the contagion'.

Tillet was equally convinced that insects played no role in the aetiology of bunt (although he did believe, but without any detailed experimentation, that ergot was caused by the sting of an insect). He distinguished loose smut (*Ustilago tritici*) from bunt but was not able to identify the smut spores as being fungal in nature. Although he likened bunt spores to the spores of *Lycoperdon*, he held, to continue the first of the above quotations,

that the clean healthy seed, inoculated with this dust, receives through rapid contagion and a very intimate communication the poison peculiar to it; that it transmits the poison to the kernels of which it is the origin; that these kernels, once infested become converted into a black dust and become for others a cause of disease; that the culms, themselves, that bore the bunt heads contain something pestilential for the seed that lies near them and on which they germinate.

The critical step was taken by Bénédict Prévost fifty-two years later. In the intervening period Fabricius edged nearer when he concluded in 1774 that the 'symptoms of smut can never be better explained than by assuming something organised to be the cause' and in 1783 the Abbé Tessier's *Traité des maladies des grains* gave very well-illustrated accounts of both bunt and loose smut.

Isaac-Bénédict Prévost, who will always be remembered as the first to offer experimental proof of the pathogenicity of a micro-organism, was born in Geneva of Swiss parents and in 1777 at the age of 22 became a private tutor at Montauban, Départment du Lot, France, where he spent the rest of his life; from 1810 as professor of philosophy at the Faculté de Theologie Protestante. Prévost was a self-taught scientist and his one mycological investigation originated from an invitation in 1797 by the scientific section of the Society of Montauban to its members to occupy themselves with the problem of bunt in wheat. Ten years later saw the publication of *Mémoire sur la cause immédiate de la carie ou charbon des blés*. . .in which Prévost gave an account of his historic experiments which demonstrated that 'la cause immédiate de la carie est un plante du genre des uredo ou d'un genre très-voisin', a conclusion he had reached by 1804. Prévost had also made observations on various rusts and mildews (in which he was the first to observe germination by zoospores, see p. 62) and a few lines later he generalised his discovery by suggesting that plants of this class should be designated 'plantes parasites intestines' or simply 'plantes intestines'. Prévost complemented Tillet and set a new standard for phytopathological laboratory experimentation. He established the nature of the 'grains' filling the bunted kernels by a series of careful and rigorous experiments on the external factors determining their germination (and illustrated sporidial production, see Fig. 65), including the effects of toxic agents (see

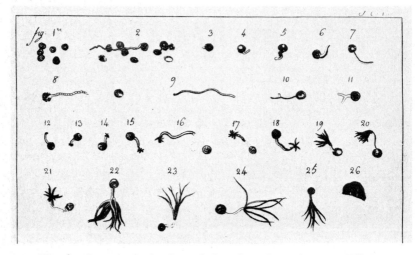

Fig. 65. Spore germination of the wheat bunt fungus (*Tilletia caries*). (B. Prévost (1807): part of pl. 1.)

p. 161), as well as age, previous treatment, and concentration of these spores. He undertook extensive inoculation experiments and defined factors determining infection. He observed germinating spores in soil and on the surface of young wheat plants and described the development of spores in the infected ovary, and correctly believed, though he could not prove it, that infection of the seedling spread through the developing plant to sporulate in the ovaries. Also, incidentally, he noted the phenomenon later known as 'Brownian movement'. Prévost's concepts were precise. He claimed the parasite as the 'immediate' or 'direct' cause of bunt because he realised that infection occurs only when the environmental conditions are favourable.

The publication of Prévost's memoir caused little stir. The Montauban Academy submitted the published memoir to the Institute in Paris where the Abbé Tessier was the reporter of a commission appointed to examine it. The study was praised but its significance not appreciated, and only after Prévost's death was the true worth of his investigations recognised, among others by the Tulasnes, in their monograph on smuts (Tulasne, 1847) in which they proposed the genus *Tilletia* (typified by the bunt fungus, *T. caries*). This delayed acceptance was in part due to the inaccessibility of the memoir which even de Bary (1853) could only cite from secondary sources.

The final link in the life history of the bunt fungus was completed in 1859 when Kühn observed the penetration of the coleoptile of wheat by the bunt fungus and in 1896 Frank Maddox in Tasmania recorded blossom infection of wheat by the loose smut (*Ustilago tritici*).

Black rust of wheat

A new era in the understanding of rusts, which like smuts were known to the ancients, began in 1767 with two independent publications by the Italians Giovanni Targioni-Tozzetti and Felice Fontana.

G. Targioni-Tozzetti, the son of a Florentine physician, after qualifying in medicine at Pisa returned to Florence where he studied botany under the guidance of Micheli. After Micheli's death in 1737 Targioni-Tozzetti acquired Micheli's manuscripts (as noted in Chapter 3) with a view to their publication but little was achieved because of other heavy literary commitments. One of Targioni-Tozzetti's long-term interests was the improvement of agriculture by rationalised practice and he planned a comprehensive survey of the subject as a contribution towards increasing agricultural productivity. The fifth and final chapter of the first (and only completed) volume of this work, *Alimurgia o sia modo di render meno gravi le carestie, proposto per sollievo de' poveri* ('The Alimurgia or means of rendering less serious the dearths. Proposed for the relief of the poor'), published in 1767, was devoted to a survey of diseases of plants and contained much original observation. A wide range of diseases was covered, and supplemented by a lithographed plate illustrating the micro-scopical appearance of fungal spores and other structures. Cereal diseases, particularly rusts and smuts, were given prom-inence and Targioni-Tozzetti will now always be remembered for his conclusion that 'la ruggine è una intiera planta parasitica picolissima, la quale non masie sermochè fra pelle è pelle del grano' (the rust is an entire, very tiny, parasitic plant that grows nowhere except inside the skin of the wheat). Targioni-Tozzetti described the subepidermal origin of the uredia (and the forcing of the cuticle of the wheat outward to make 'it swell and finally crack in a very delicate spangle') at the base of which he observed hyphae ('certain very short and delicate rootlike hairs' which 'steal, and suck up for themselves the nutriment prepared, and destined for the wheat'), illustrated urediniospores and telios-pores, concluded that the dispersion of the rust was by wind, and that, under appropriately humid atmospheric conditions, infec-tion occurred via the stomata. In one thing he was mistaken. He took each spore to be a little plant which 'ought to be born of its own proper seed' but 'The smallness...in its entirety of a plantlet of the Rust, even of the greatest rankness, forces the belief that these seeds are of a scarcely credible smallness, such as they could not be seen with even the most acute microscopes'. It was these 'seeds' which he considered must be 'scattered about by the air, without our eyes being able to discern them, even to a distance, on the wings of the Winds'.

In 1766, Felice Fontana, physicist, naturalist, and professor of philosophy at Pisa, was, like Targioni-Tozzetti, induced to make a microscopical study of wheat rust because of a severe rust epidemic in Tuscany. His observations, which he published as a little book, *Osservazioni spora la ruggine del grano*, in the following year, agreed with those of Targioni-Tozzetti. On 10 June 1766, he discovered that the rust was 'a grove of plant parasites that nourish themselves at the expense of the grain', not insects as was widely believed. He gave descriptions and excellent illustra-tions of the urediniospores (to demonstrate that they were pedicellate gave him much trouble) and teliospores of *Puccinia graminis* which he believed to be two distinct rusts, recognised their fungal nature, and compared them with *Pilobolus*, on the basis of a recent record from England.[4] Like Targioni-Tozzetti, Fontana interpreted individual spores as plants, which behaved as 'semi-parasites', that is to say 'they can attach themselves on the grain plant only when there is a rupturing of the vessels and a diffusing of the humours'.

These results were received with interest. Fontana's observations became generally known (there was an English translation of his book in 1792) but Targioni-Tozzetti's were soon forgotten and remained unnoticed by de Bary and subsequent writers until 1943 when Professor G. Goidànich was instrumental in arranging for a reprint of the *Almurgia*, and in 1952 an English translation of the chapter on plant diseases, together with a biographical commentary by Goidànich, appeared in the *Phytopathological Classics* series.

The relationships of the different types of spores exhibited by black rust and the bearing of heteroecism on the epidemiology of the disease proved to be very troublesome problems. Sir Joseph Banks in his famous pamphlet, *A short account of the cause of the disease in corn*, of 1805, with its two attractive colour plates by Francis Bauer, 'Botanical Painter to His Majesty' King George III ('Farmer George'), took it for granted that the rust was a parasitic plant, acknowledged that Felice Fontana was the first to give 'an elaborate account of this mischievious weed', and like Fontana and others believed the teliospores to be little plants which he figured bursting and shedding 'seeds'. Banks also believed the urediniospores and teliospores to be genetically related as did de Candolle (1807). Prévost (1807) saw the 'uredo' and 'puccinias' of rose rust (*Phragmidium mucronatum*) in the same sorus but the controversy on the relationship of the uredinial and telial states continued for a number of years, some authors holding that the urediniospores were parasitic on the teliospores. It was not until 1854 that Tulasne proved (for *Phragmidium*) that the two states were different stages of the same species.

Proof of heteroecism took even longer. It was farmers who first became convinced that barberry bushes growing in the neighbourhood of wheat induced disease in the latter and in 1755 the famous Barberry Law of Massachusetts, 'An Act to prevent Danger to English Grain arising from Barberry Bushes', was passed. This first legislative measure against a plant disease began:

Whereas it has been found by experience, that Blasting of Wheat and other English Grain is often occasioned by Barberry Bushes, to the great loss and damage of the inhabitants of this province:–

Be it therefore enacted by the Governour, Council, and House of Representatives, that whoever, whether community or private person, hath any Barberry Bushes standing or growing in his or their Land, within any of the Towns of this Province, he or they shall cause the same to be extirpated or destroyed on or before the thirteenth Day of June Anno Domini One Thousand Seven Hundred and Sixty.[5]

and continued with the penalties to be applied to those who did not comply with the Act.

Similar legislation was enacted in Connecticut and Rhode Island and in France and by 1805 Banks was writing: 'It has long been admitted by farmers, though scarcely credited by botanists, that wheat in the neighbourhood of a barberry bush seldom escapes the Blight. The village of Rollesby in Norfolk, where barberry bushes abound, is called by the opprobious appellation of Mildew Rollesby.' He continued, that it is 'notorious to all botanical observers, that the leaves of the barberry are very subject to the attack of a yellow parasitic fungus, larger, but otherwise much resembling rust in corn', and asks 'Is it not more than possible that the parasitic fungus of the barberry and that of the wheat are one and the same species, and that seed is transferred from barberry to the corn?' The next year L. G. Windt (1806) in Germany independently came to a similar conclusion, particularly regarding the effect of barberry on rye.

In August 1805 the English horticulturalist Thomas Andrew Knight dusted the leaves of a wheat plant with *Puccinia graminis* urediniospores and after five days he began sprinkling the plant with water three or five times daily. One week later infection was observed. He also applied aeciospores from barberry to three other plants and obtained urediniospores on two of them, but not only on the leaves inoculated and the experiment terminated when his 'villanous jack ass' broke in and 'swallowed Rust, Wheat, and all together'. Knight considered his experiments to be inconclusive and was inclined to think that 'much Moisture, applied to the stem and Leaves when the Roots are ill supplied with it, will in almost all cases, bring it [rust] on'. A Danish schoolmaster of Hammel in Jutland, Niels Pedersen Scholer, was more successful. From 1807 he made observations on the association of black rust of rye with barberry bushes and in 1816 he succeeded in experimentally infecting rye by the inoculation

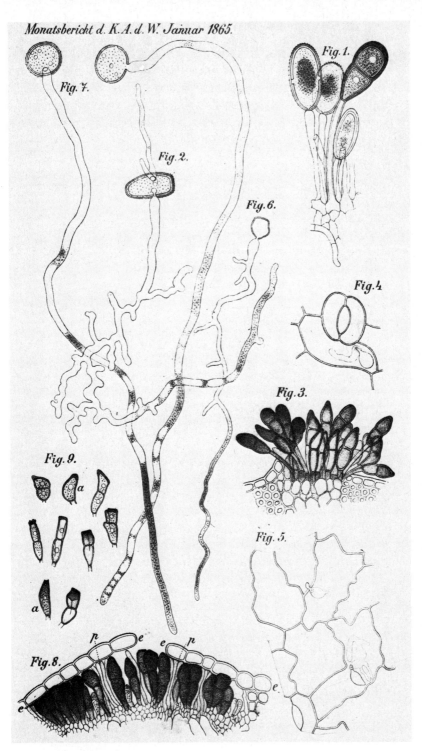

Fig. 66. *Puccinia graminis*. 1, 3, Urediniospores and teliospores on wheat; 2, germinating urediniospore; 4, 5, sporidia infecting barberry leaf; 6, germinating aeciospore. (A. de Bary, 1865–6.)

of healthy plants wet with dew with aeciospores from barberry. This result was received with controversy and it was de Bary who finally established beyond all doubt the reality of *heteroecism* (a term he introduced; de Bary, 1865: 32) by the experimental infection of barberry with sporidia from teliospores in 1865 and of rye with aeciospores from barberry the following year (see Fig. 66).

The autogenicists and predispositionists

While these developments were taking place, old and erroneous beliefs on the role and nature of plant pathogens persisted, and lingered on until the mid-nineteenth century. Elias Fries in the Introduction to the *Systema* (1821) declared *Uredo antherarum* [*Ustilago violacea*] to be nothing but pollen in a morbid state ('Pollen in statu morbosa'), that the Mucedineae 'sunt tantum *pili plantarum in statu morbosa*'; while in the third volume of the same work (1832) he expressed his belief in the heterogenic origin of rusts and smuts (Hypodermii):

Coniomycetes 'Ordo IV HYPODERMII (Entophyti) CHAR. Vegetatio propria nulla. Sporidia ex anamorphosi telae cellulosae *plantarum vivarum* orta; sub earum epidermide enata et per hanc erumpentia.
    Obs. Ex altera parte sunt exanthemata plantarum; ex altera simul vegetationem propriam offerunt.'

Franz Unger in his important monograph *Die Exantheme der Pflanzen*, 1833, which was written under the influence of Naturphilosophie, elaborated similar beliefs to those of Fries and concluded that parasitic fungi (or 'entophytes' as he called them) arose from morbid sap and so were the signs rather than the cause of disease. In spite of this opinion Unger, like Fries, treated fungi as though they were independent organisms and made notable mycological contributions including the proposal of *Protomyces* and *Ramularia* (= *Ovularia*) as new genera. Léveillé, Turpin, and de Candolle in France, Ré in Italy, and Unger's fellow Germans Theodor Hartig (1833) (see p. 93), Meyen (1837–8), and Wiegmann (1839) were prominent among those who supported heterogenesis.

Raspail attributed the origin of parasitic fungi to insect punctures. In his *Histoire naturelle de la santé et de la maladie*, 1846, he wrote:

L'insecte qui produit les *erineum, uredo, aecidium, xyloma, puccinia*, n'est donc

Fig. 67. Rev. Miles Joseph Berkeley (1803–89).

pour nous un insecte inconnu, mais un *acarus* (grise), un *aphis* (puceron) ou un *thrips*, qui produit au printemps une deviation.

The conflict of opinion on the role of plant pathogenic fungi at that time is well brought out by the reaction to the devastating epidemic of potato blight in 1845 which resulted in the Irish famine and caused untold economic and social distress.[6] Among the many experts called in for consultation was the Rev. M. J. Berkeley (Fig. 67) who stated in his classic paper, 'Observations, botanical and physiological, on the potato murrain', which apeared in the *Journal of the Horticultural Society of London* for January 1846: 'The decay is the consequence of the presence of the mould, and not the mould of the decay'. This conclusion regarding the origin of the potato blight was not original. As Berkeley records, it was held by the American, J. E. Teschmacher, the Belgian, Dr C. J. E. Morren, and in France by Monsieur Antoine de Payen and also by Dr J. F. C. Montagne who, on 30 August 1845, was the first to name the potato blight fungus which he described and designated *Botrytis infestans*. It was however, a minority opinion. Most authors including the French mycologist J. B. H. J. Desmazières (who also described and named the potato blight fungus), and Dr John Lindley, editor of the *Gardeners' Chronicle* and professor of botany at University College London, considered the mould to be the result of the decay and attributed the primary disease to diverse environmental and physiological factors. The controversy continued for the next twenty years and it was again de Bary, in his two memoirs of 1861 and 1863, who finally silenced the heterogenicists.

The expansion of plant pathology

After the general acceptance of the importance of plant pathogenic fungi, that is from about 1870, the development of phytopathology was rapid. Many more fungal diseases of all the major economic crops, both temperate and tropical, were recognised, many new plant pathogenic fungi were described, and much detail regarding the life histories and general biology of the parasites was elucidated. Even after the discovery of bacterial diseases of plants in the eighteen-eighties and of virus diseases at the turn of the century – and, later still, deficiency disorders – fungal infections maintained their position as the most important diseases of plants. Textbooks,[7] monographs, and journal articles multiplied. Special journals to report the results of phytopathological research were begun (the *Zeitschrift für Pflanzenkrankheiten*, started in 1891, was the first) and phytopathologists associated together in professional societies.[8] This activity did much to expand interest in mycology and extensive additions to mycological knowledge resulted. Most of this work was 'filling in', the extension of known principles to additional particular cases, and need not be referred to here; but major new advances were made particularly in the understanding of host–parasite relationships and in the control of fungi by chemical fungicides.

Host–parasite relations

The main development in the knowledge of host–parasite relationships was the recognition of the interaction between resistance of the host to infection on the one hand and specialisation of the parasite on the other. Varietal differences in susceptibility to disease had long been noted. R. Austen in his *Treatise of fruit-trees*, 1657, recorded that 'Crab trees...are usually free from the canker' and it appears that T. A. Knight in 1806 was among the first to suggest the use of resistant varieties to combat rust of cereals. The major British contribution in this field resulted from the work of Rowland H. Biffen in the early years of the present century at Cambridge, where he later became professor of agricultural botany and served as the first director of the Plant Breeding Institute there. Biffen demonstrated that the inheritance of susceptibility and resistance of wheat varieties to yellow rust (*Puccinia striiformis*) was subject to Mendel's laws of heredity (Biffen, 1904). Two of the best known wheat varieties developed by Biffen were Little Joss and Yeoman.

J. Schroeter (1879) was the first to suggest that rust fungi showed host specialisation and in 1894 Jakob Eriksson in Sweden proved experimentally the existence of five morphologically similar races of *Puccinia graminis* exhibiting specialised parasitism for wheat (*Triticum vulgare*); oats; rye barley, and *T. repens*; *Aira caespitosa*; or *Poa compressa*. He called these races 'formae speciales' and found that they retained their own characteristics after passage through barberry. These results were generally accepted and by the turn of the century had been extended to other rusts and the powdery mildews (Erysiphaceae) and subsequently formae speciales of many other fungi were demonstrated.

Biffen's results on breeding cereals for rust resistance were at first received with a certain amount of caution (as by E. J. Butler, 1905) because it was not then appreciated that cereal rusts showed important differences in host specialisation within one forma specialis, a discovery that will always be associated with the name of Elvin Charles Stakman, a graduate of the University of Minnesota with which his life work has been so closely associated. Stakman's interest in cereal rusts began with his doctoral thesis, submitted and published in 1913, and he continued his investigations in this field for the next half century. It was in 1917 that Stakman & Piemeisel showed that *Puccinia graminis* f.sp. *tritici* comprised several biological forms which could be differentiated by the host reactions of a series of varieties (cultivars) of wheat.

By 1922, thirty-seven of these biological forms (or physiologic races as they are now designated)[9] had been recognised and Stakman & Levine published details of the technique (which became standard practice among cereal pathologists) for their differentiation by the inoculation of twelve varieties of wheat and recording the type of host response obtained (see Fig. 68).

TABLE IV

VARIATIONS AND CONSTANTS IN REACTION OF DIFFERENTIAL HOSTS OF WHEAT TO BIOLOGIC FORMS OF STEM RUST

Biologic forms	Ranges and means of infection																							
	Ltl.Club		Marquis		Kanred		Kota		Arnautka		Mindum		Snt.Marz		Kubanka		Acme		Einkorn		Emmer		Khapli	
	Range	Mean	Range	Mean	Range	Mean	Range	Mean	Range	Mean	Range	Mean	Range	Mean	Range	Mean	Range	Mean	Range	Mean	Range	Mean	Range	Mean
I.......	3+ 4+	4	3+ 4+	4	0 0;	0	3- 4+	3+	0. 1	1=	0 2+	1	0. 1+	1=	3- 4	3+	3 4	3+	3- 3+	3	0 1	0;	0 1+	1=
II......	4- 4+	4	1= 2+	2=	0 2+	2=	0 2	2=	0 1+	1-	0 2+	1	0 1+	1=	0 2	1+	3- 4+	3+	3= 4	3+	0; 1	1-	0 1	0;
III.....	3+ 4+	4	3+ 4+	4	3= 4+	4=	3= 4	3+	0 1+	1=	0 2+	1=	0 2+	1-	0 2	1+	3= 4+	3+	3= 4	3+	0 2-	1=	0 1+	0;
IV......	4 4+	4+	1 2+	2-	0 2+	1-	1 2	2=	3+ 4	4=	3 4	3+	3 4	3+	0; 2+	2	3 4	3+	3- 4	3+	0. 1+	1=	0 1	1=
V.......	3+ 4+	4	3 4+	4	0 2	0;	3= 3+	3	3 4	4=	3 4	3+	3- 4	3+	0; 2+	1+	3= 4	3+	3 3+	3	0 1-	0;	0 1-	0.
VI......	4 4+	4	0; 2+	2	0 1-	1=	0 1	0;	3 4	3+	0; 2.	2=	1 2=	2=	1= 2+	1	3 4+	3+	3 4	3	0 0.	0.	0 1	0;
VII.....	3+ 4+	4	0 2+	2=	3- 4+	3+	0 1+	1=	0. 1	1=	0 2	1+	0 2	1-	0 2	1	3 4	3+	3= 3	3-	0 2	1	0 1+	1-
VIII....	3+ 4+	4	3+ 4+	4	0 0;	0.	3= 4	4	3+ 4	4=	3 4	3+	3 4	4=	0 0;	0.	3- 3+	3	3 3+	3	3+ 4	4	0 1-	0;
IX......	3+ 4+	4	3 4+	4	0 0;	0	3= 4	3+	3- 4+	4-	3 4+	4=	3- 4+	4=	3- 4	4=	3= 4+	3+	3- 4	3+	3- 4+	4=	0 1+	1-
X.......	3+ 4+	4+	1- 2+	2-	3 4	3+	1= 2	2	3+ 4	4	3 4+	4	4 4+	4	3 4	3+	3+ 4	4	3- 3+	3+	0; 1-	1=	0; 1-	1=

Fig. 68. Part of tabulation of the reactions to differential varieties of physiologic races of *Puccinia graminis* f. sp. *tritici*. (E. C. Stakman & M. N. Levine (1922): table 4.)

They also distinguished the different races by numbers and subsequently it was from Stakman's laboratory at St Paul, Minnesota, that numbers were allocated for the new races recognised by cereal pathologists in all parts of the world. By 1962, no less than 297 races had been differentiated (see Stakman, Stewart & Loegering, 1962) and a number of these have been subdivided into 'biotypes' which are usually designated by letters e.g. *P. graminis* f.sp. *tritici* 15B.

A notable deepening of the understanding of host–parasite relationships resulted from the investigations of the American H. H. Flor on flax rust (*Melampsora lini*). By 1940, Flor had recognised twenty-four physiologic races of flax rust and from the results of a subsequent series of brilliant breeding experiments on the flax plant and its rust it became evident that the genetic systems of both organisms were involved in determining host–pathogen interactions. He noted that the genes for resistance in flax cultivars and those for virulence in the rust race were numerically equal and from this concluded that for each locus conditioning rust reaction in the host there was a specific, related locus conditioning virulence and avirulence in the rust (see the summary by Flor, 1956). Person in 1959 used Flor's hypothesis predictively and extended the phenomenon to potato and potato blight while others have since established gene-for-gene relationships for a number of additional major diseases (see Person, 1968).

Physiology of parasitism

Research on fungal parasitism for plants – the physiological study of host–parasite relationships – began, like so many other mycological topics, with de Bary who in 1886 reported that the expressed sap of carrots attacked by *Sclerotinia sclerotiorum* contained a thermolabile substance able to disorganise the host tissue *in vitro*. Marshall Ward (1888) extracted a similar principle from ground-up mycelial mats of two- to six-week-old cultures of *Botrytis* and in 1915 William Brown, working at the Imperial College of Science, London, as reported in the first instalment of 'Studies on the physiology of parasitism' (a series that continued for the next forty years), obtained much more active preparations from the hyphal mass of recently germinated *Botrytis cinerea* spores and confirmed Ward's view that the active principle was an enzyme which, he demonstrated, was able to disintegrate plant tissue by using the middle lamella as its substrate. Brown called the enzyme (or enzyme mixture as it proved to be) 'cytase', a name subsequently replaced by 'pectinase', 'protopectinase', and, later still, by the more specific appellations 'polygalacturonase' and 'polymethylgalacturonase'.

The state of knowledge in this field in the mid-thirties, as regards both facultative and obligate parasites, was well reviewed by Brown (1936), and again fifty years after his first paper, and need not be re-iterated here, while R. K. S. Wood, also of the Imperial College, has summarised recent developments in his textbook, *Physiological plant pathology*, 1967. Current interest in the physiology of parasitism and related topics necessitated the founding of a new journal, *Physiological plant pathology*, in 1971.

The discovery that metabolites of certain plant pathogenic fungi are phytotoxic is considered briefly in Chapter 7 (p. 191).

Fungicides

Some of the earliest attempts to mitigate the effects of plant disease by chemical treatments such as the brining of seed wheat against bunt (p. 145) and the diverse seed treatments employed by Tillet (p. 146) against the same disease have already been noted. So too have chemical treatments against decay in structural timber (p. 92). The development of effective fungicidal treatment, both therapeutic and prophylactic (or 'eradicative' and 'protective'), for controlling disease in plants, which constitutes the main stream for the development of fungicides in general, has yet to be considered. This development dates from the beginning of the nineteenth century and the first fungicide to gain popularity was sulphur.

Sulphur was employed in dermatological preparations by Dioscorides in the first century A.D. and there are a number of early references to its use against infestations. It was William Forsyth, gardener to King George III, who, in 1802, introduced lime sulphur for the control of peach mildew and T. A. Knight in 1834 reported the control of both leaf curl (*Taphrina deformans*) and red spider mite on peach by sprinkling the trees in early summer with 'water holding in solution or suspension a mixture of lime and flowers of sulphur' (Knight, 1842). The same remedy was successfully used by Edward Tucker, the Margate gardener, who first recorded powdery mildew of the grape vine in England in 1845 and who was commemorated by Berkeley in 1847 by the name *Oidium tuckeri* given to the pathogen *Uncinula necator*. Subsequently, the use of sulphur continued and it still finds application today, particularly as lime sulphur preparations and as dusts containing wettable sulphur (introduced in the nineteen-twenties).

For the half century preceding 1935 copper fungicides predominated. Schulthess in 1761 was apparently the first to use copper sulphate as a treatment for wheat bunt and copper was also used (but in very high concentration) for the same purpose by Tessier (1783) but rejected, along with lemon juice, mint brandy, ether, etc., as being of no use. Undoubtedly the discovery of the high toxicity of copper to fungal spores must be attributed to Prévost (1807) who, during his investigations on wheat bunt (see above, p. 147), noted that germination of bunt spores was inhibited in water that had been distilled in a copper alembic; just as sixty years later Raulin (1869) failed to obtain growth of *Aspergillus niger* cultures made in a silver dish which also led him to investigate the effects of highly toxic chemicals.[10] Shortly afterwards Prévost had his attention drawn to the relative freedom from bunt of two local farms where the seed wheat was given pre-sowing treatment by lime in copper vessels. This led him to perform a series of elegant experiments on the effects of copper sulphate on the germination of bunt spores when he found that a 1:280000 (w/w) solution of copper sulphate prevented germination and that 1:600000 or even 1:1000000 retarded it perceptibly. He also noted the diminution in toxicity of dilute solutions of copper sulphate made in hard water after filtering off the precipitate formed. On the basis of these results he made practical recommendations for seed treatment. Some farmers followed Prévost's advice, but his recommendations were not widely applied and copper did not come into general use in plant protection until after 1885 when P. M. A. Millardet, professor of botany at Bordeaux, announced the formulation which became known as 'bouillie bordelaise'. Details of the discovery of Bordeaux mixture following the recognition of downy mildew of the vine (*Plasmopara viticola*) in France in 1878 are familiar but it is perhaps of interest to recall Millardet's own account of the circumstances (in Schneiderhan's translation).

Towards the end of October, 1882, I had occasion to traverse the vineyard of Saint-Julien en Médoc. I was not a little surprised to see, all along my way, that the vines still bore their leaves, while everywhere else they had long ago fallen. There had been some mildew that year, and my first reaction was to attribute the persistence of the leaves along the way to some treatment that had preserved them from the malady. Examination, indeed, permitted me to confirm the fact that these leaves were covered in great part on the upper surface by a thin coating of a pulverized bluish white substance.

After arriving at the Chateau Beaucaillon, I questioned the manager, M. Ernest David, who told me that the custom in Médoc was to cover the leaves with verdigris or with copper sulphate mixed with lime, when the grapes were ripening, in order to keep away thieves who, on seeing these leaves covered with copperish spots, would not dare to taste the fruit hidden underneath for fear of its having been blemished in the same way.

I called the attention of M. David to the fact of the preservation of the leaves under immediate consideration and shared with him the hope, engendered in my mind, that the salts of copper might from the basis of the treatment of mildew.

During the next two years Millardet, with the help of the chemist U. Gayon, perfected and tested his treatment. His final recommendation reads:

In 100 litres of water (either well, rain, or river) are dissolved 8 kilos of commercial copper sulphate. Then, from 30 litres of water and 15 kilos of rich rock lime, milk of lime is made and mixed with the solution of copper sulphate. It forms a bluish paste. The workman pours part of the mixture, while stirring it, into a watering pot, which he takes in his left hand, while, with the right, by the aid of a small brush he wets the leaves, taking care constantly not to touch the grapes. One need fear no harm, even to the most tender organs.

At M. Johnston's, 50 litres of the mixture sufficed, on an average, for treatment of 1,000 plants, which, for one *hectare* (10,000 plants), places the total expense (cost of materials and labour) at not more than 50 francs.

Bordeaux mixture was an immediate success. The threat of downy mildew to the French wine-growing industry was averted. The mixture proved equally successful against potato blight and on Millardet's death in 1902 a statue was erected to his memory at Bordeaux and in 1935 'Le cinquantenaire de la bouillie bordelaise' was celebrated by a symposium in Paris.[11]

This commemorative symposium coincided with the end of an era for although Bordeaux mixture and fixed copper compounds, such as the basic sulphates and chlorides and oxides (introduced in the nineteen-twenties and thirties), continued to be used (as they still are) they began to be replaced by a wide range of organic fungicides. The first of these was the dithiocarbamate series (derivatives of dithiocarbamic acid), which was covered by an American patent of 1931. The first and most famous of the dithiocarbamate fungicides was Thiram (tetramethylthiuram disulphide). This was followed by Ferbam and Ziram and, later still, by Nabam, Zineb, and Maneb. Other organic fungicides introduced commercially in the decade following the Second World War include the quinone derivatives Chloranil (tetrachlorol, 4-benzoquinone) and Dichlone (2,3-dichloro-1,4-naphthoquinone) and the trichloromethyl-dithiocarboximides, Captan and Folpet.[12] Antifungal antibiotics, which have figured prominently in medicine, have found few applications for the control of plant diseases (see p. 221).

The intensive search for new and better fungicides has been accompanied by many studies on techniques for the laboratory testing of fungicides and much investigation on the action of fungicides on fungi. For reviews of this work the two monographs of J. G. Horsfall (1945, 1956) may be consulted.

A major advance in both fungicide research and its application is the current development of systemic fungicides for the control of fungal diseases of crop plants, a development well documented by the comprehensive multi-author, book-length review edited by R. W. Marsh (1972) in which over half of the more than 1100 references cited are to papers published since 1967. Systemic therapeutants have been successfully employed against mycoses of man and animals since potassium iodide was first used for the treatment of sporotrichosis in 1903 (see p. 181) and there are a number of early references to sporadic attempts to employ systemic fungicides in plants (see Marsh (1972) for details), and even more to discover systemic pesticides.[13] The recent spectacular advance stems mainly from three later developments: work on the therapy of deficiency diseases by the injection of suitable solutions into the plant, such as that undertaken by W. A. Roach in England at the East Malling Research Station during the nineteen-thirties; the deployment of antibiotics in medicine and the concurrent advent of the antibiotic industry; and the introduction of a wide range of organic fungicides which provided a great diversity of chemical compounds to test for their potentialities as systemic fungicides of low phytotoxicity. One systemic fungicide now commercially available is Benomyl (methyl(1-butylcarbamoyl)benzimidazol-2-yl-carbamate) which has been widely used for the control of diseases of fruit, vegetable, and glasshouse crops.

PATHOGENICITY TO ANIMALS AND MAN

During the eighteenth century, as noted in Chapter 2 (p. 25), there was much speculation on the relationship of entomogenous fungi to their hosts. This problem remained unresolved until 1835 when Agostino Bassi, an Italian lawyer turned farmer, published the results of an investigation he had begun nearly thirty years before.

Agostino Bassi (Fig. 69) was born in Lombardy on 23 September 1773, at Mairago, four miles from Lodi, where he went to school and where three years before Bassi's birth Mozart had written his first string quartet. Then he studied law at the University of Pavia and after graduating in 1798 was appointed to the post of provincial administrator and assessor at Lodi, later becoming an official under the new Napoleonic regime. Like so

Fig. 69. Agostino Bassi (1773–1856).

DEL MAL DEL SEGNO

CALCINACCIO o MOSCARDINO

Malattia che affligge

I BACHI DA SETA

E SUL MODO

DI LIBERARNE LE BIGATTAJE

ANCHE LE PIÙ INFESTATE

Opera

DEL DOTTORE AGOSTINO BASSI

DI LODI

*la quale oltre a contenere molti utili precetti intorno al miglior governo
dei Filugelli, tratta altresì delle Malattie*

DEL NEGRONE E DEL GIALLUME

LODI

DALLA TIPOGRAFIA ORCESI

1835

Fig. 70. Title–page of Bassi's
Del mal del segno, 1835.

many men of genius, Bassi suffered from life-long ill health.
Defective eyesight forced him to relinquish office and to retire to
his father's farm at Mairago, and one result of this was the
publication in 1812 of his first book, *Il pastore bene istruito* ('The
well-educated shepherd'), a comprehensive work of 460 pages
on sheep farming. In 1815 an improvement in his health allowed
him to re-enter government service, but only for a year, after
which he returned to Mairago, where he devoted himself to
agricultural and scientific pursuits until his death in 1856. At first
financially embarrassed, Bassi in 1838 inherited a fortune from a
cousin, a poor tailor of Mairago who became a wealthy military
contractor and was created Count of Sommariva by Napoleon.

Bassi's interests were wide, and, while at the university, in
addition to his legal studies, he took courses on mathematics,
medicine and various aspects of natural science, his teachers

including the anatomist Scarpa, the physicist Volta, and Lazzaro Spallanzani, the versatile physiologist noted for his studies on spontaneous generation (see p. 20). Bassi was also of an inquiring turn of mind and in 1807 he decided to investigate the dread muscardine disease of silkworms which ravaged the silkworm industries of Italy and France. The notion prevailed that the cause of muscardine was 'environmental', that is, was due to the state of the atmosphere, the food, or the method of breeding, and it was to these aspects that Bassi first turned his attention. He 'bred silkworms in all ways, subjecting them even to the most barbarous treatment; the poor creatures died by thousands and in a thousand ways', but not a single case of muscardine disease resulted. He next tested the hypothesis that the disease was due to hyperacidity and studied the effects of phosphoric acid, but although the silkworms again died there was only putrefaction; no hardening of the corpse and no white bloom on its surface, the characteristic post-mortem signs of muscardine disease. Bassi was able to simulate death by muscardine disease by hanging up silkworms enclosed in paper bags in a chimney over a continuous fire and then placing the hardened corpses in a cellar or keeping them moistened with water. Silkworms so treated deceived the experts, but to Bassi's chagrin they lacked the one essential character of a silkworm which has died from muscardine disease. The 'contagious faculty' was missing. They did not induce the disease when placed among healthy silkworms.

Discouraged by his failure, Bassi was 'humiliated in the extreme, silent and idle' and 'oppressed by a terrible melancholy' for many months; but in 1816 his spirits revived and he began work again on a new line of attack suggested by his observations on the course of natural epidemic outbreaks. It occurred to him that the disease did not originate spontaneously but was caused by an 'extraneous germ'. He soon discovered that the infectious material was the white efflorescence which developed on the dead silkworms, and by subjecting this material to microscopic examination (Bassi possessed a compound microscope which after his death was acquired by the town of Lodi), he found it to be 'a plant of the cryptogamic kind, a parasitic fungus'. By a long series of comprehensive and often elegant experiments, Bassi was able to demonstrate that this fungus was the pathogen. He showed that silkworms of all ages were susceptible and that an infected worm became infectious only after 'calcination' – the sporulation of the fungus over the surface of the dead larva – and he concluded that every 'spontaneous outbreak of the disease could be traced to the introduction of infected silkworms or to the use of contaminated cages or utensils. He also developed measures by which the disease could be controlled.

By 1833 Bassi felt ready to submit his work for adjudication by the University of Pavia, and in 1834 he performed a series of the crucial experiments before a commission of nine professors from the Faculties of Medicine and Philosophy. The professors were convinced. They issued Bassi with a signed certificate which he printed in the preface of the first part of the monograph (*Del mal del segno, calcinaccio o moscardino*; Parte prima: 'Teoria', 1835; Parte seconda: 'Pratica', 1836) (Fig. 70) in which he gave a detailed account of his investigations and his conclusions. This work aroused much interest but was given a mixed reception. Guiseppe Balsamo-Crivelli, professor of natural history in the University of Milan, studied the causal fungus and named it *Botrytis bassiana* (now *Beauveria bassiana*), and confirmation of Bassi's findings was published in 1838 by the director of the Paris Natural History Museum, Victor Audouin, to whom Bassi had sent a copy of his monograph and an infected chrysalis (Fig. 71). There were opponents, to whose opinions Bassi was perhaps unduly sensitive, but his results were well founded and convinced the unprejudiced.

The study of muscardine disease was Bassi's major achievement. In the years that followed, he issued a second edition of *Del mal del segno* (1837) and published additional observations on this disease. He also wrote lengthily on pellagra and cholera, on diseases of the mulberry, vine and potato, and on other topics. He received a number of honours from both Italian and foreign academies, and to commemorate the centenary of Pasteur's birth a collected edition of Bassi's writings was published in Italy in 1925 under the auspices of a national committee headed by Mussolini.[14]

As with many other outstanding figures in the history of science, Bassi crystallised ideas which were becoming prevalent. The influence on Bassi of Spallanzani's views on spontaneous generation has already been implied, and Bassi must have been familiar with the frequent association of fungi with decay and

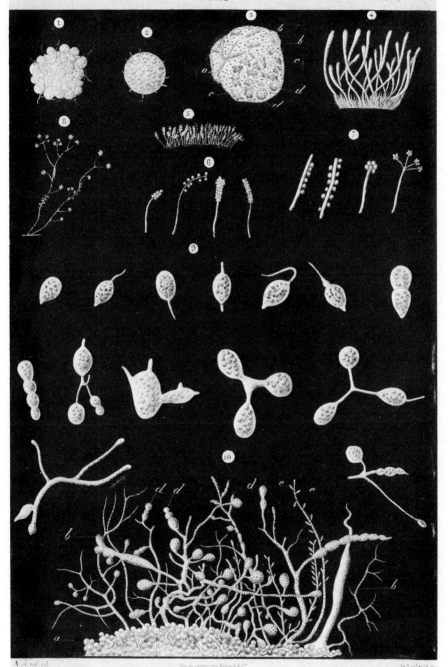

Fig. 71. *Beauveria bassiana*, the cause of muscardine disease of silkworms. (V. Audouin (1838): pl. 2.)

disease. In England an illustrated observation on salmon disease of freshwater fish had been made as early as 1748 (see p. 175), and in 1832 the zoologist, Richard Owen, when dissecting a flamingo which had died at the London Zoo, found the lungs lined with a vegetable mould and concluded that there were Entophyta as well as Entozoa. The trail became even warmer when, as cited by Bassi, Professors Configliachi and Brugnatelli, of Pavia, suggested that the muscardine disease had a mycotic origin because of the fungus smell associated with silkworms killed by this disease; but to Bassi must go the credit of being more perceptive and persevering than his contemporaries.

Bassi, if given to verbosity, was a modest and at times almost diffident writer. He was generous to others but he appears not to have suffered fools gladly, for in the preface to *Del mal del segno*, while appealing to savants and breeders for their observations, he warns them that 'to save time and trouble I shall reply only to those which seem to me sufficiently founded in fact to be able to further in some way the progress of science and art'. Bassi lived up to his motto: 'Quando il Fatto parla Ragion tace, perche la Ragione e figlia del Fatto, non il Fatto figlio della Ragione' ('When Fact speaks Reason is silent, because Reason is the child of Fact, not Fact the child of Reason').

Within a few years of the publication of Bassi's discovery, fungal diseases of man were recognised by Johannes Lukas Schönlein and Robert Remak in Berlin and David Gruby in Paris.

Remak,[15] a Pole, the son of a Jewish merchant, was born in 1815 at Posnan (then part of Prussia). At the age of eighteen he moved to Berlin where he spent the rest of his life, as student, medical practitioner, and in various posts. Finally, in 1859, six years before his death, he was made an 'extraordinary professor' (not a full professor as he had long hoped) at the university as the climax to a career hampered throughout by his faith and his nationality. Remak was apparently the first to recognise the fungal nature of favic crusts when in 1837 he noted that they were composed of spherical bodies and ramified fibres, an observation he allowed a fellow countryman to publish in a doctoral thesis. Remak did not recognise the causal role of the fungus until Professor Schönlein in 1839, stimulated by Bassi's publications, claimed favus as a mycotic disease when Remak (1842) showed the condition to be infectious by successfully inoculating himself. In 1843, unable to obtain a university appointment, Remak became an assistant in Schönlein's Clinic and two years later in a volume of reports of his investigations at the clinic and elsewhere Remak described the favus fungus which he designated in Schönlein's honour *Achorion schoenleinii* (Remak, 1845).

LE DOCTEUR GRUBY A SOIXANTE-QUINZE ANS

Fig. 72. David Gruby (1810–98).

In the meantime major contributions to the recognition of mycotic infections of man were being made in Paris by David Gruby (Fig. 72),[16] a Hungarian of humble Jewish parentage who, like Remak, was unwilling to pay the price of a certificate of baptism for a university chair. After the publication of his doctoral thesis, *Observationes microscopicae ad morphologiam pathologicam*, at Vienna in 1840, Gruby migrated to Paris where during 1841–4 he published the series of short papers in the

Compte Rendu of the Academy of Sciences on which his reputation as the founder of medical mycology rests.[17] Gruby first independently described the mycotic nature of favus and gave details of the causal fungus which he showed by inoculation experiments to be pathogenic for man and animals (Gruby, 1841). The next year (Gruby, 1842 *a*) he elucidated the aetiology of ringworm of the beard, an ectothrix trichophytosis. Again he did not name the pathogen but Charles Robin (1853) based *Microsporon* [*Trichophyton*] *mentagrophytes* on Gruby's description. In 1843 Gruby described classical human microsporosis and named the pathogen *Microsporum audouinii* in honour of Victor Audouin, director of the Paris Natural History Museum. Finally he completed the series (Gruby, 1844), with an account of endothrix trichophytosis caused by the *Trichophyton tonsurans* of Malmsten, 1848. In addition, in 1842, Gruby announced the mycotic nature of 'mouget' (thrush in infants; caused by the *Oidium* [*Candida*] *albicans* of Robin, 1853), a discovery made independently in the same year by F. T. Berg in Sweden. Gruby made no more mycological contributions and within a few years became a naturalised Frenchman and concentrated on his medical practice in Paris where he prospered as an eccentric but popular physician – Dumas *père* and *fils*, Liszt, Chopin and George Sand were among his many famous patients – until his death some fifty years later.

Ringworm had long been recognised. Favus was known to Celsus and there are records of the Tudors granting licences under the signet to enable loyal sufferers from ringworm of the head to remain covered in the king's presence. A famous painting by Murillo [1617–82] shows St Elizabeth of Hungary washing the heads of children suffering from favus and another interesting seventeenth-century record, possibly the first of a tropical mycosis, is that made by William Dampier who, when voyaging round the world, wrote in his journal after a visit to the Phillippines in 1686:

The *Mindanao* People are much troubled by a sort of Leprosie, the same as we observed in *Guam*. This Distemper runs with a dry scurf all over their Bodies, and causes great itching in those that have it, making them frequently scratch and scrub themselves which raiseth the outer Skin in small whitish flakes, like the scales of little Fish, when they are raised on edge by a Knife. This makes their Skin extraordinary rough.[18]

The same disease was subsequently observed in Polynesia and became known as 'Tokelau Itch' or 'Tokelau Disease', after the islands where the disorder was prevalent, but it was not until 1879 that one of the pioneers of tropical medicine, Patrick Manson, then Medical Officer to the Imperial Maritime Customs at Amoy, China, elucidated its aetiology, and named it tinea imbricata. Some twenty years later the French worker R. Blanchard described the causal fungus as *Trichophyton concentricum*. Others too confused ringworm with leprosy.[19]

In the early nineteenth century, one view on the status of ringworm was concisely stated by the English dermatologist F. Bateman in whose opinion ringworm of the scalp seemed 'to originate spontaneously in children of feeble and flabby habits, or in a state approaching marasmus, who are ill fed, uncleanly, and not sufficiently exercised'. That is to say he believed the disease to be constitutional, a view that proved difficult to eradicate. Finally, when the evidence for the implication of fungi as parasites became overwhelming, this hypothesis influenced opinion on the identity of the fungi themselves.

Gruby's findings though welcomed by some were not universally accepted, particularly by clinicians. This was in part because although Gruby's mycology was, for his time, outstanding, his association of 'phytoalopécie' (tinea capitis caused by *Microsporum audouinii*) with porrigo decalvans or alopecia (a non-parasitic baldness), did much to discredit his work and confuse his successors. In France A. P. E. Bazin advocated accepting the parasitic origin of ringworm, but in 1850 A. Cazenave was warning dermatologists against 'the illusions of micrography' and denying the presence of 'any pathogenic properties to the mysterious atoms of *Achorion schoenleini*', the only fungus which he recognised. In Britain likewise, opinion was divided. Sir William Jenner adopted the parasitic view of skin infections and thought that the simple way to effect a cure was to destroy the fungus by some topical application. 'But error soon betrayed itself to his mind and he candidly acknowledged that this mode of treatment ended in disappointment', according to Jabez Hogg, a constitutional diehard, who in 1873 reprinted in book form[20] a series of papers published some years earlier fulminating against the parasitic theory of disease or, when the involvement of fungi in skin disease became indisputable, advocating the view that the variations in these diseases were due to differences in the patients' constitutions modifying the soil for one essential fungus. W. Tilbury Fox in 1863, as noted in

Chapter 2 (p. 30), identified this one essential fungus as *Torula* but Thomas M'Call Anderson in 1861 believed each of the four cutaneous infections – favus, tinea tonsurans, pityriasis, and, following as he thought Gruby, alopecia areata – to be due to the presence of a distinct vegetable parasite. George Thin showed by careful cultural work on a variety of liquid media and on the 'meat-gelatine' recently introduced by Koch that *Trichophyton tonsurans* is a fungus in its own right 'totally distinct from the common fungi whose spores infest all the objects by which we are surrounded' (Thin, 1881, 1887). Others made similar observations and thus prepared the way for the suggestion by Furthmann and Neebe in 1891 that there was more than one *Trichophyton* fungus and for the publication in the *Annales de Dermatologie et Syphylographie* in November 1892 of 'Étude clinique, microscopique, et bactériologique sur la pluralité des trichophytons de l'homme' by the Parisian dermatologist Raymond Sabouraud who was a man of parts. He was an able sculptor and wrote a highly commended book on Montaigne. Although he became the world specialist on disorders of the scalp he was quite bald and always wore a kind of smoking cap.

Papers poured from Sabouraud's laboratory; the mycological findings of Gruby were vindicated, and the modern era opened. The climax of the renewed interest in the dermatophytes occurred at the Third International Congress of Dermatology held in London in August 1896 when an afternoon session, one of the most important events of the Congress according to the *British Journal of Dermatology*, was devoted to 'Ringworm and the Trichophytons'. The symposium was introduced by Dr Sabouraud who summarised his own investigations and exhibited a series of three hundred cultures, by Prof. Rosenbach of Göttingen, and by Mr Malcolm Morris of London. Subsequent speakers included Charles J. White (Boston, U.S.A.), Leslie Roberts, H. G. Adamson, Colcott Fox, and F. R. Blaxall (England), and Professor P. G. Unna of Hamburg, while other contributions by French, Italian, and Spanish workers were made or taken as read. The text of the discussion occupied more than a hundred pages of the *Official Transactions* (published in 1898) and Malcolm Morris later expanded his contribution into a book (Morris, 1898), illustrated by a notable series of photomicrographs. The plurality of the ringworm fungi was established as scientific fact and the main lines for the study of these fungi during the next fifty years laid down, even if then, as since, not all were willing to accede to 'the infinite series of species which M. Sabouraud asks us to accept'.

From the middle of the nineteenth century, in parallel with the investigations on ringworm in man, both medical men and veterinarians were recording ringworm in farm and domestic animals. J. H. Bennett in 1842 noted favus in the mouse. Tilbury Fox's (1871) record of seven cases of ringworm in man contracted from a white pony seems to be the first record of equine ringworm. Subsequently, particularly in France, there were increasingly frequent records of ringworm in horses, cats, dogs, cattle, and poultry and a growing realisation that it was from infected animals that many outbreaks of human ringworm originated.

The development of medical and veterinary mycology shows a sharp contrast to that of plant pathology. Professional mycologists have always moved freely in and out of the field of plant pathology where advances in general mycology have frequently originated. Very few mycologists ventured into the medical field and if at times plant pathology suffered from too much mycology, the study of mycotic disease in man and animals suffered from too little. The first excursions by mycologists into the medical field were not a conspicuous success while, though a number of medical men at the beginning of the present century became expert self-taught mycologists, the increasing complexities of both medicine and mycology led to much unsatisfactory work in medical mycology. The taxonomy of the dermatophytes (and other 'medical fungi') became very confused as can be seen from the compilation of medical fungi made by C. W. Dodge in 1935 in a book still useful because of the accuracy of the extensive documentation.

A new era began in 1936 with the appointment of Chester W. Emmons, a trained mycologist, by the United States Department of Health and in one of his early papers (Emmons, 1934) he set the taxonomy of the dermatophytes on a new and firm basis by defining the three genera *Microsporum*, *Trichophyton*, and *Epidermophyton* in mycological terms and absorbing the fourth widely used genus *Achorion* (up to then clinically defined as causing favus) in *Trichophyton*. A later breakthrough was the establishment of perfect states for a number of dermatophytes, especially by Arturo Nannizzi (1927), Phyllis M. Stockdale (1961), and Christine O. Dawson and J. C. Gentles in the same year.

Clinical topics are outside the scope of this chronicle but mention must be made of two techniques that have done more than any others to facilitate the control of head ringworm in children. The first of these is epilation by means of X-rays, the second the use of 'Wood's light' for diagnosis.

Microsporum audouinii parasitising the hair of children is notoriously difficult to eradicate. Such infections may linger on until puberty when for reasons not fully understood they spontaneously disappear. Treatment with efficient fungicides is often uncertain because of the difficulty of killing the fungus in the hair follicles and the insidious way in which the infection spreads to healthy hairs. Cropping or shaving off the hair, or the manual epilation of infected hairs, accompanied by the conscientious use of fungicidal preparations and precautions to prevent re-infection eventually succeed but at the expense of serious dislocation in the education of children of school age. In 1901 to obviate this educational disturbance, the Metropolitan Asylums Board (since absorbed in the London County Council) established a ringworm school, with Colcott Fox as the visiting dermatologist, where children from London could be both treated and instructed. Six hundred to a thousand children of, in 1910, an estimated total of three thousand infected, were treated annually.[21] The average stay of each child at the school was nineteen months. From 1905, this period was dramatically reduced to four months after the introduction of a technique developed by Sabouraud in 1904 as the result of a chance observation in 1897. A young woman who, three weeks or so after she had been used as a subject to demonstrate the effect of X-rays which revealed on a fluorescent screen a chain round her neck under her clothes, lost all the hair of the occipital region and consulted Sabouraud. The hair grew again normally and Sabouraud at once recognised the potential value of X-ray epilation for the elimination of infected hairs in the treatment of ringworm of the scalp. In collaboration with H. Noire he determined the conditions for the routine use of this technique in the clinic (see Sabouraud, 1910).

Twenty years later the French workers J. Margarot and P. Devéze[22] discovered that *Microsporum*-infected hairs showed a brilliant greenish fluorescence in light which had been filtered through soda glass containing nickel oxide (first described by the American R. W. Wood of Johns Hopkins University). *Trichophyton*-infected hairs do not fluoresce with the exception of favus hairs which show a similar but weaker fluorescence (Kinnear, 1931). The technique of diagnosis, the manual epilation of infected hairs, and the test of a successful cure were thus greatly facilitated and combined with X-ray epilation. *Microsporum audouinii* infection in children is now a disease of low incidence.

Panaceas for the therapy of ringworm have been legion. The most famous topical application, 'Whitfield's ointment' (a benzoic acid–salicylic acid formulation), was devised by the dermatologist Arthur Whitfield to treat an accidental infection in the clinic of his own forearm.[23] The one major recent novelty in ringworm treatment has been the introduction of the antibiotic griseofulvin (see p. 22) for oral administration.

Systemic mycoses

Ringworm has always been one of the main themes of medical mycology with which it has at times even been equated. As a disease it is ubiquitous – approximately three per cent of the male population of western Europe is infected by dermatophytes – troublesome among those who frequent swimming baths, and an occupational disease of soldiers and coal miners, but in general its effects are measured in degrees of minor inconvenience. There are, however, other mycoses which, if given less publicity, are medically more significant. In 1948 Emmons drew attention to the fact that of the 1 385 187 deaths recorded in the *Vital statistics of the United States for 1942*, 359 were attributed to fungi. Although this was less than 0.03 per cent of the total, it was nearly twice as many as all the deaths during the same period from paratyphoid fever, undulant fever, smallpox, rabies, leprosy, plague, cholera, yellow fever, and relapsing fever and more than half the number from typhoid fever, tetanus, and poliomyelitis. Some of these systemic mycoses have long been recognised. Others have only recently been characterised.

The earliest record of a mycotic disease in a vertebrate seems to have been that by William Arderon of what was clearly *Saprolegnia* infection in a roach, published in the *Transactions of the Royal Society* for 1748 (see Fig. 73). Bennett (1842) mentions a

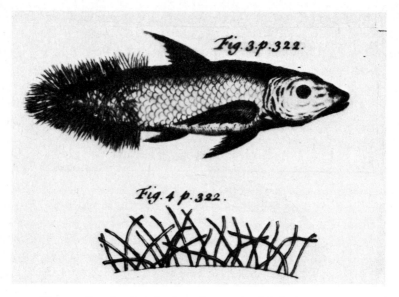

Fig. 73. *Saprolegnia* infection of a roach. (W. Arderon, 1748.)

similar infection in a goldfish and as 'salmon disease' this disorder gained notoriety in late Victorian times (Thomas H. Huxley was called in as a consultant in the epidemic in English and Scottish rivers of 1882) and the disease is still of sporadic economic importance.

In 1749 de Réaumur in France observed moulds (possibly *Aspergillus* species) in incubating eggs and since the first half of the nineteenth century, when Montagu in 1813 found a 'mould or blue mucor' in the thoracic air sac of a Scaup duck (*Aythya marila*) and Rousseau and Serrurier (1841) noted mycotic lung lesions in an axis deer (*Cervus axis*), there have been hundreds of records of aspergillosis in birds, particularly wild birds in captivity in zoological gardens, and other animals. Human pulmonary aspergillosis was first recognised by Virchow in 1856 and the commonest species responsible for aspergillosis, *Aspergillus fumigatus*, was described and named by Fresenius in 1863. Interest in aspergillosis reached a peak towards the close of the century with the appearance of the two monographs by Rénon and Lucet in France – both published in 1897 – and aspergillosis is still a significant medical problem of today.[24] During the eighteen-nineties and the first decade of the present century much attention was paid to sporotrichosis, a subcutaneous infection involving the lymph nodes and other tissues which occasionally becomes generalised. Sporotrichosis was first recorded by R. R. Schenck in 1898 in the United States, when the causal fungus was regarded by the plant pathologist Erwin F. Smith, to whom a culture had been submitted, as a species of *Sporotrichum*; an identification subsequently shown to be incorrect and the pathogen, following Hektoen & Perkins (1900), is now known as *Sporothrix schenckii*. During the early nineteen hundreds sporotrichosis was of frequent occurrence in France where extensive studies culminated in L. de Beurmann & H. Gougerot's massive monograph, *Les sporotrichoses*, 1912, which includes a comprehensive bibliography of the earlier literature. Sporotrichosis is an exogenous mycosis, infection being contracted from the environment, and its incidence though world wide is sporadic. The high incidence of the disease in France declined and between 1941 and 1944 the largest outbreak of sporotrichosis ever recorded occurred in the gold mines of the Witwatersrand, South Africa, where no less than 2825 miners were affected. Infection, following minor skin injuries, was from a reservoir of the pathogen growing saprobically on the mine timbers. The outbreak was brought under control by fungicidal treatment of the mine timbers and potassium iodide therapy (see p. 181) of the affected miners. There were no fatalities.[25]

The two systemic mycoses, coccidioidomycosis and histoplasmosis,[26] the recognition and investigation of which provided a major stimulus for the development of medical mycology since 1940, have a number of features in common. Both were first recognised in the Americas as rare and frequently fatal human infections and the causal agents were first thought to be protozoa. Later it became apparent that the common forms of these mycoses are benign and that millions of people living in endemic areas – which are semi-arid for coccidioidomycosis, humid for histoplasmosis – suffer infection which immunises them from future attack. It also became apparent that farm and domestic animals were similarly attacked and that the source of infection for both man and animals was the soil. Both pathogens are dimorphic (see p. 30) and both mycoses have since 1950 been the subjects for series of symposia in the United States each attended by several hundred research workers and clinicians.

Coccidioidomycosis[27] was first recorded in 1892 from the Argentine Pampa by Alejandro Posadas, a 21-year-old medical student working in the pathological laboratory of Robert Wernicke at Buenos Aires. Posadas and Wernicke identified the 'spherules' they found in affected tissue as a protozoan similar to *Coccidia*. This was also the opinion of E. Rixford and T. C. Gilchrist of the Johns Hopkins Hospital, Baltimore, in 1896 who named the organism *Coccidioides immitis*. When they tried to culture the pathogen the plates became overgrown by a mould, a 'contaminant' W. Ophuls & H. C. Moffitt (1900) showed to be the cultural form of the pathogen which *in vivo* occurs as spherical sporangium-like bodies (spherules) in the lung and other organs and in culture as a mycelial growth bearing vast numbers of dry, easily detached, arthrospores which on inhalation cause infection. Skin sensitivity (induced by infection) to a culture filtrate of the fungus (coccidioidin) was demonstrated by E. F. Hirsch and others in 1927. Coccidioidomycosis, because prevalent in the San Joaquin Vally of California, became known as 'valley fever'. The opening of the modern period began with the recognition of the primary form of the disease by two Californian physicians Ernest C. Dickson and Myrnie A. Gifford in 1936–7 (Dickson & Gifford, 1938). Study of coccidioidomycosis was given an additional impulse during the Second World War when many of the troops sent to the Arizona desert and elsewhere for training were attacked by the primary form of the disease. The incidence of infection was reduced by sixty-five per cent by grassing down dusty areas and surfacing roads and air strips, recommendations of a team headed by Professor Charles E. Smith of Stanford University, California, whose researches have added so much to knowledge of this mycosis. Elucidation of the ecology of the disease is, by mycologists, associated especially with the name of Chester Emmons who confirmed and extended earlier observations that small desert rodents were infected by *Coccidioides immitis* (and also he found by rather similar fungi now classified as *Emmonsia* species, the agents of adiaspiromycosis; see Jellison, 1969). At first he thought that the rodents constituted an animal reservoir of the pathogen but it later became clear that those rodents, like man, cattle, and other animals were all subject to the same hazard of infection, the air-borne arthrospores of a saprobic, soil-inhabiting fungus.

Histoplasmosis, an infection of the reticulo-endothelial system involving the lungs, spleen, kidneys, and many other organs of the body, was discovered by Samuel T. Darling who investigated three cases in Panama in 1906 and concluded that the causal micro-organism which he found to be plentiful within reticulo-endothelial cells was a protozoan. The next case was reported from the United States (Minnesota) where the first case diagnosed before death was noted in 1934 and W. A. DeMonbreun (1934) described the in-vitro cultural characteristics of *Histoplasma capsulatum* which is mycelial and exhibits large, warted macroconidia and smaller, smooth, spherical microconidia. The in-vivo form occurs as small yeast-like cells. As for coccidioidomycosis, at first only the fatal disseminated form was recognised. Later it was found that there was a benign or asymptomatic form induced by the inhalation of spores which immunises the subject who thereafter gives a positive skin reaction to the fungal antigen (histoplasmin). The endemic areas for histoplasmosis (which has a world-wide distribution) in the U.S.A., chiefly along the eastern seaboard, are more humid than those characteristic for coccidioidomycosis, but again, as Emmons (1949) showed, soil is the source of infection and millions of people living in such areas are histoplasmin sensitive. More recently *H. capsulatum* has been found to be particularly associated with the soil of old chicken runs and, in North and South America and South Africa, with the guano of cave-inhabiting bats from which speleologists have contracted infection.

A second species of *Histoplasma* with larger yeast-like cells in infected tissues (*H. duboisii* Vanbreuseghem, 1952), which causes a clinical variant of classical histoplasmosis, has been described from tropical Africa while *H. farciminosum* (the *Cryptococcus farciminosus* of Rivolta, 1873) is responsible for epizootic lymphangitis of horses and mules. This last disorder, characterised by subcutaneous and ulcerating lesions of the skin, particularly of the neck region, which run a very protracted course, is unlike human histoplasmosis in being contagious. It is endemic in countries bordering the Mediterranean and in Japan and southern Asia and was first recognised by French veterinarians about 1820. Epidemic outbreaks have often been associated with military campaigns and the disease was introduced into the United Kingdom in 1902 after the South African war but eradicated by a statutory slaughter policy as recounted in the little monograph by Captain W. A. Pallin (1904) of the Army Veterinary Department which may be consulted for additional historical details.

Studies of these and other mycoses have, particularly since 1950, set mycotic disease in man and higher animals in truer perspective. It has become abundantly clear that while there is no major fungal disease of man comparable to tuberculosis or cholera, which are the equivalents of such fungal diseases of plants as potato blight and black rust of cereals, mycoses of man are of sufficiently high incidence to merit special attention. It has also become clear that there are very few, if any, obligate fungal parasites of man and higher animals – *Rhinosporidium seeberi*, cause of the tropical rhinosporidiosis, and a pathogen of uncertain taxonomic position, is a notable exception; but even here infection seems to be contracted from contaminated water or soil. Most fungi which cause disease in man are 'opportunistic', to use a term that has come into common usage since 1962.[28] That is to say, they normally occur as saprobes, often common saprobes, in man's environment which possess the ability to act as pathogens when introduced by inhalation or via chance injury into a susceptible animal host. Sometimes it seems to be a constitutional change in the host that determines whether a fungus acts as a pathogen or not, as when *Mucor* or *Candida* infections develop in diabetic subjects or in patients undergoing steroid therapy but frequently, if the inoculum is large enough, the saprobe establishes itself as a pathogen.

A number of additional mycoses exhibit a pattern similar to that of the diseases already mentioned. The non-mycelial yeast *Cryptococcus neoformans* responsible for cryptococcosis of the central nervous system and the lungs is frequently found in pigeon droppings (although the pigeons themselves are not infected), in Madura foot (*Madurella mycetomi*) – one of the classical tropical mycoses; first noted in 1694[29] and the subject of H. V. Carter's monograph of 1874 – is associated with injury of the feet by thorns, while since the introduction of the 'hair-baiting technique' by Vanbreuseghem (1952) for the isolation of dermatophytes from the soil ringworm fungi and allied, but not always pathogenic, keratinophilic fungi have been isolated from soil in many parts of the world.

Therapy for the major systemic mycoses as satisfactory as the use against sporotrichosis of potassium iodide (which was first introduced against actinomycosis in cattle in 1885 and in 1903 suggested by Sabouraud to Beurmann and Gougerout for trial against sporotrichosis in man) is still lacking. During recent years the two most used commercially available antifungal drugs have been the polyene antibiotics produced by species of *Streptomyces*, nystatin (discovered in 1950)[30] and amphotericin B (1956),[31] although the use of the latter has to be balanced against irreversible liver damage to the patient.

The deployment of mycologists in the medical field was accompanied by an increase in educational facilities for medical mycology. The appointment of Norman F. Conant as medical mycologist to Duke University School of Medicine, North Carolina, and mycologist to Duke Hospital was particularly important in this connection. He developed a summer course in medical mycology which was an inspiration throughout the world and also, in collaboration with four of his medical colleagues, wrote the *Manual of clinical mycology*, published in 1944 as one of the Military Medical Manuals of the National Research Council of the United States. This raised the standard of medical mycology texts to a new level and proved itself a most influential book.

Finally, it may be noted that although there has never been a society restricted to students of fungus diseases of plants medical mycologists have shown a marked tendency to promote their interests by founding specialist societies. This first occurred – somewhat unusually – at the international level when a group of medical men and mycologists attending the sixth International Botanical Congress in Paris in 1954 inaugurated the International Society for Human and Animal Mycology (I.S.H.A.M.), which in 1961 started its own journal, *Sabouraudia*. Subsequently flourishing national societies or groups were organised in France, Japan, Germany, Great Britain, and other countries, most of which are affiliated with I.S.H.A.M. and participate in periodical international 'Assemblies'.

HYPERPARASITIC FUNGI

Hyperparasitic fungi – fungi parasitising other fungi – have long been known. Micheli (1729: 200; tab. 82, fig. 1) was the first to observe *Nyctalis asterophora* on *Lactarius piperatus* and other agarics which he illustrated very crudely. Bulliard (1791–1812: tab. 166) gave a good illustration of the same hyperparasite on *Collybia fusipes*, as did Bolton (1788–91:tab. 155) of *N. parasitica* on *Russula nigricans*. Zopf (1890:269–82) and Buller (1909–50, **3**:432–73) have compiled records of other fungal hyperparasites. Zopf also listed fungi parasitic on lichens which were later monographed by Keissler (1930).

In conclusion, this is perhaps an appropriate place to recall that fungi are also subject to disorders caused by both bacteria and viruses. A. G. Tolaas (1915) in the United States first recorded a bacterial disease of the cultivated mushroom and it was S. G. Paine (1919) who named the organism *Pseudomonas tolaasi* after studying an English outbreak of the disease which he designated 'brown blotch'. More recently a disorder of mushrooms, 'watery stripe', now generally known in England as 'die back', was shown to be viral in nature and Miss Gandy (1960) successfully transmitted the virus in in-vitro cultures. Since then several more viruses have been described from mushrooms and other agarics and viruses or virus-like particles have been demonstrated in species of *Penicillium, Aspergillus, Saccharomyces* and other microfungi. This is currently a rapidly expanding field.[32]

7. Poisonous, hallucinogenic, and allergenic fungi

Although in 1847 Badham correctly wrote of the larger fungi that 'the innocuous and esculent kinds are the *rule*, the poisonous the *exceptions* to it' fungi have since classical times been notorious for their poisonous properties. Man has always been impressed by calamities and the earliest mention of fungi in Greek writings is in connection with a multiple fatality. According to Eparchides, Euripides was on a visit to Icarus, in about 450 B.C., when a woman, her two full-grown sons, and an unmarried daughter died after eating fungi which they had gathered in the fields whereupon the poet composed an epigram which, in Houghton's translation, reads:

O Sun, that cleavest the undying vault of heaven, hast thou ever seen such a calamity as this? – a mother and maiden daughter and two sons destroyed by pitiless fate in one day?[1]

Poisonous fungi are a topic of perennial interest. Much has been written on how to recognise them and the first taxonomic division of fungi, by Clusius (1601), was into edible and poisonous. Battarra (1755) displayed on the title page of his monograph the motto (in Greek) 'We study fungi, we do not eat them' and in 1760 Scopoli's concluding and somewhat pessimistic diagnostic feature of the class Fungi was 'venenata saepius, suspecta semper'.

Much traditional advice on how to distinguish edible from poisonous fungi is unreliable. Horace wrote that 'Fungi which grow in meadows are the best; it is not well to trust others'.[2] Dioscorides believed that poisonous fungi 'grow amongst rusty nails or rotten rags, or near serpents' holes, or on trees producing noxious fruits; such have a thick coating of mucus, and when laid by after gathering quickly become putrid'.[3] Pliny stated that 'Some of the poisonous kinds are easily known by a dilute red colour ('diluto rubore'), a loathsome aspect, and internally by a livid hue; they have gaping cracks ('rimosa stria')

7-2

and a pale lip round the margin'.[4] These views were given renewed prominence by the herbalists together with such equally unreliable tests that the skin of the cap of a poisonous fungus does not peel; that poisonous, unlike edible, fungi discolour an onion cooked with them and tarnish a silver spoon. Many of these erroneous beliefs survived till recent times. The only way to escape trouble is to profit by the past misfortunes of others and correctly identify species known to be harmless.

Early writers offered much advice on the treatment of fungal poisoning. Nicander's was to

take the many-coated heads of the cabbage, or cut from around the twisting stems of the rue or old copper particles which have long accumulated, or pound clematis dust with vinegar, then bruise the roots of pyrethrum, adding a sprinkling of vinegar or soda, and the leaf of cress which grows in gardens, with the medic plant and pungent mustard, and burn wine lees into ashes or the dung of the domestic fowl; then putting your right finger in your throat to make you sick, vomit forth the baneful pest.[5]

Celsus recommended 'If any one shall have eaten noxious fungi let him eat radishes with vinegar and water ('posca') or with salt and vinegar'.[6] Galen too advocated 'raw radishes in quantities', also unmixed wine, lye ashes of the vine, a mixture of soda and vinegar, ashes of burnt lees of wine mixed with water, wormwood and vinegar, and rue either with vinegar or alone.[7] Dioscorides was more cautious and wrote that as a safeguard all fungi

should be eaten with a draught of olive-oil, or soda and lye-ashes with salt and vinegar, and a decoction of savory or marjoram, or they should be followed with a draught composed of bird's dung and vinegar, or with linctus of much honey; for even the edible sorts are difficult of digestion and generally pass whole with the excrement.[8]

Pliny said it was good to eat pears immediately after fungi.[9]

Another approach was to devise procedures for rendering poisonous fungi harmless. Celsus stated that unwholesome fungi 'can be rendered serviceable by a mode of cooking them; for if they have been boiled in oil or with the young twig of a pear-tree they become free from any bad quality'.[10] Pliny maintained that fungi are safer if cooked with meat or with pear-stalks. Paulet in his *Traité des champignons* (1790–3, **2**: 25) stated that if poisonous fungi are cut into pieces, steeped in water containing salt, vinegar, or alcohol they are then innocuous to animals. This procedure may have provided the inspiration for his fellow countryman Frédéric Gérard, an assistant at the Jardin des Plantes, Paris, who after cutting up the fruit-bodies of species believed to be poisonous soaked them for two hours in water to which vinegar or common salt had been added (or in several changes of water only), rinsed them well, and put them in a pan of cold water which was brought to the boil and then allowed to cool. After half-an-hour they were removed from the water, wiped, and used for cooking in the ordinary way. When describing his researches Gérard records that within one month he and his family of twelve had eaten seventy-five kilograms of poisonous fungi. In 1851 he convinced the Paris Conseil de Salubrité of the success of his method by practical demonstration before a commission. As it is not certain that Gérard experimented with *Amanita phalloides* (which is responsible for most fatalities in Western Europe) and as he is believed to have finally succumbed to his own experimentation, mycologists have remained sceptical of Gérard's claims and do not recommend his method.

Although the poisonous nature of some fungi was generally accepted this view was occasionally questioned – perhaps because since the time of Nero there have been cases of sudden death from poison added to dishes of edible species. For example, the Rev. Roger Pickering F.R.S., in 1746,[11] attributed the poisonous quality of mushrooms to 'the accidental *Ova* or *Animalcula*, which the Richness of their Nutriment has allured to them', a view promptly discounted by William Watson,[12] another Fellow of the Royal Society. It was, however, 120 years before a fungal toxin was isolated.

In Europe and North America the great majority of fatalities from fungal poisoning have been caused by *Amanita phalloides* and the related *A. pantherina*, *A. verna*, and *A. virosa* while the less poisonous but better known fly agaric (*A. muscaria*), which when broken up in milk kills flies, is sometimes involved. It was from this last that Schmiedeberg & Koppe (1869) isolated muscarine (*Muskarin*) which though toxic to man does not kill flies or cause the typical symptoms of *A. muscaria* poisoning. This was found to be due to a second toxin, muscaridine. It was similar for *A. phalloides* toxins. R. Kobert (1891) isolated a haemolytic toxin which he called phalline but from subsequent studies, by the American W. W. Ford in the early nineteen-hundreds and by H. and T. Wieland and their collaborators in Germany since 1940, *A. phalloides* is now known to contain ten or more peptide toxins of two main groups characterised by phalloidin and α-amanitine, respectively, and in 1969 an antitoxin, antamide, was isolated.[13] Currently, the most favoured treatment in cases of *Amanita* poisoning is antiserum therapy, first attempted by the French bacteriologist A. Calmette at the turn of the century and made generally available twenty-five years later as a result of work by Dujarric de la Rivière and others at the Pasteur Institute in Paris (see Dujarric & Heim, 1938).

Mycotoxicoses

Up to the middle of the present century almost all attention to poisonous fungi was concentrated on larger fungi which certainly have over the years caused much illness and many fatalities in man. Poisonous microfungi, with one major exception, received virtually no attention although it has recently become clear that microfungi or their toxic products consumed unwittingly by man and animals in food and feeding stuffs far outweigh in medical, veterinary, and economic importance poisoning by larger fungi. These poisonings, or 'mycotoxicoses' as they are now called, are of world-wide occurrence and since the nineteen-sixties have been a subject of world-wide investigation.

Ergotism

Ergotism caused by eating food, particularly rye bread, contaminated by ergot (the sclerotia of *Claviceps purpurea*) was the first mycotoxicosis to be recognised and because ergot is also used in medicine (see p. 216) the study of ergot and ergotism has generated a very large literature well reviewed by Barger (1931).

The first indubitable published reference to ergot is that by Adam Lonitzer (Lonicerus) in his *Kreuterbuch* of 1582 (Fig. 74) and ergot was first illustrated in Gaspar Bauhin's *Theatri botanici*, 1658, where ergoted rye is, following J. Bauhin, treated as a distinct species, '*Secale luxurians*' (Fig. 75). From the time of Johann Thal in the late sixteenth century (*Sylva hercynia*, 1588) to the early nineteenth century there were those who held the view that the sclerotia were abnormal rye grains resulting from insect or other damage or fungal infection. Hence on description of the ascostromata developed from germinating sclerotia they

SECALE LⱯXVRIANS.

Nota: Von den Kornzapffen / Latiné
Claui Siliginis: Man findet offtmals an de
ähren deß Rockens oder Korns lange schwar
tze harte schmale Zapffen / so beneben vnd zwi=
schen dem Korn / so in den ähren ist / herauß
wachsen/vnd sich lang herauß thun / wie lan=
ge Neglin anzusehen / seind inwendig weiß/
wie das Korn / vnd seind dem Korn gar vn=
schädlich.

Solche Kornzapffen werden von den wei=
ben für ein sonderliche hülffe vnnd bewerte
Artzney für dz auffsteigen vn wehthumb der
Mutter gehalten / so man derselbigen drey ce=
tlich mal einnimpt vnd jsset.

Fig. 74. First record of the use of ergot in
childbirth. (A. Lonitzer (1587): 286.)

were named *Sphaeria purpurea* by Fries. Others accepted the
suggestions made by Geoffroy in 1711 and Münchausen in 1764
that the sclerotia were fungal in nature and in 1815 de Candolle
codified this view by naming these bodies *Sclerotium clavus*.
Lastly, in 1827 Léveillé, in his monograph on ergot, described
the conidial phase (the 'honey dew' first noted by K. Schwenckfelt in 1600)[14] under the name *Sphacelia segetum* which he believed was responsible for inducing the abnormality of the rye
grains. The major contribution of L. R. Tulasne in his classic
paper in 1853 (Fig. 76) was to demonstrate that the sphacelial
phase, the sclerotia, and the ascocarps were all stages of one
fungus, *Claviceps purpurea*.

When the first outbreak of ergotism was recorded is a matter
for speculation but in Europe written records of what, under a
multitude of local names, was almost certainly ergotism go back
to the beginning of the present millennium. In man there are
two main types of symptom – the gangrenous and the
convulsive – the latter being associated with vitamin A deficiency, while in cattle the gangrenous type is the norm. Gangrenous ergotism takes the form of an intense burning sensation of

the limbs (hence the appellation St Antony's Fire) which may eventually mortify and drop off when death though frequent is not inevitable. In the convulsive type unbearable itching of the skin (hence the German name *Kriebelkrankheit*) is accompanied by spasms, convulsions, or other nervous symptoms. Ergotism was perhaps responsible for the dancing epidemics of the Middle Ages when hundreds of people were seized with the impulse to dance.[15] It was certainly prevalent during that epoch when the gangrenous type was most common west of the Rhine, the convulsive in Germany and eastern Europe, and also Russia where outbreaks were of frequent occurrence after they had become relatively rare elsewhere.

The Medical Faculty at Marburg in 1597, when reporting on a local outbreak of 1595, declared the cause to be the use of bread made from spurred rye[16] and gradually legal restrictions were introduced to prevent human consumption of ergot-contaminated rye grain and these, combined with improvements in agriculture, reduced ergotism to the status of a rare disease.

The chemistry of ergot proved to be complex. It gradually emerged from the studies of Barger, Stoll, and others that the toxicity and physiological properties were mainly due to a series of alkaloids, some of which are lysergic acid derivatives.

Mycotoxicoses of farm animals

In addition to ergotism, and the related 'paspalum staggers' (uncoordination of movement and other nervous symptoms caused by eating the sclerotia of *Claviceps paspali* on the fodder grass paspalum) which was recorded from the United States (H. B. Brown, 1916) and other parts of the world, occasional cases of other disorders of farm animals associated with mouldy feeding stuffs were noted. For example, the toxicity to pigs of barley blighted by *Fusarium graminearum*, the imperfect state of *Gibberella zeae* (Christensen & Kernkamp, 1936) and paralysis and other symptoms in cattle attributed to *Diplodia zeae* infected maize (see Theiler, 1927). It was, however, in the Soviet Union that disorders of this type were most prevalent, or as it would now seem, where most attention was paid to such disorders. A mycotoxicosis in horses due to *Dendrochium toxicum* was first described from the Ukraine in 1937 and a similar condition in the same host, attributed to *Stachybotrys alternans*, was recorded

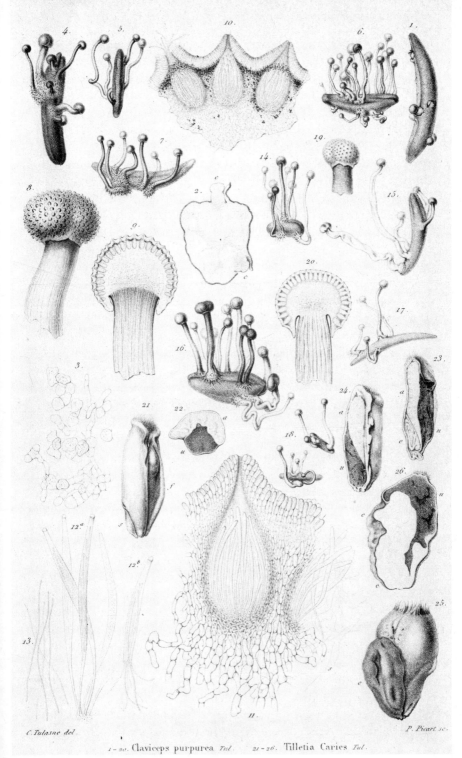

C. Tulasne del.

P. Picart sc.

1–20. Claviceps purpurea *Tul.*   21–26. Tilletia Caries *Tul.*

Fig. 76. *Claviceps purpurea* (ergot) and *Tilletia caries* (bunt of wheat).
(L.-R. Tulasne (1853): pl. 3.)

from the U.S.S.R. in 1938. These conditions, together with fusariotoxicosis in cattle and other farm animals caused by toxin-producing strains of *Fusarium sporotrichioides* were subject to much investigation in the U.S.S.R. and in 1954 A. K. Sarkisov published a general text on mycotoxicoses and ergotism. The Russian work made little impression elsewhere, no doubt largely because both the research results and Sarkisov's monograph were written in Russian, until two major mycotoxicoses were discovered in other parts of the world.

The first of these was facial eczema,[17] an important disease of sheep and cattle in New Zealand where it had been known since 1898 and was in some years responsible for disastrous economic losses. The disease could not be transmitted from an affected to a healthy animal and as early as 1908 the disease was attributed to 'dietetic errors'. It gradually became clear that liver dysfunction and the resulting photosensitisation of unprotected white or lightly pigmented skin areas, which accounted for the lesions on the face and ears of affected sheep, was associated with certain pastures at certain seasons, and after prolonged investigation the cause was finally proved to be the toxic spores of *Pithomyces chartarum*, saprobic on the pasture grass. Subsequently, the toxin was isolated and named sporidesmin, from *Sporidesmium*, the genus in which the fungus had previously been classified. More rational attempts to control the disorder were then possible.

The second mycotoxicosis was unsuspected until the summer of 1960 when in Great Britain some 100000 young turkeys and ducklings died after being fed with a product containing Brazilian groundnut meal. By the next year it had been established that the toxicity of the meal was caused by a hepatatoxin (subsequently named 'aflatoxin')[18] produced by strains of *Aspergillus flavus* (Sargeant *et al.*, 1961). Aflatoxin poisoning was also recorded for cattle, pigs, and guinea-pigs in all of which there was liver damage. Further, although rats fed with an aflatoxin-containing diet showed no acute symptoms they finally developed carcinomas of the liver and circumstantial evidence was adduced implicating aflatoxin in the aetiology of liver cancer in man.

This is not the place to review in detail the many developments resulting from these studies and the many additional mycotoxicoses which have now been recognised. Neither is it yet possible to set the topic of mycotoxicoses in true perspective although it is already certain that their investigation will be seen to have been a major feature of mycological research during the nineteen-sixties and beyond.

As is usual with any new development, the roots of mycotoxicoses are deeper than is commonly supposed and in conclusion a passage from the second volume (p. 337) of *The variation of plants and animals under domestication* by Charles Darwin, published in 1868, is relevant.

Three accounts have been published in Eastern Prussia, of white and white-spotted horses being greatly injured by eating mildewed and honey-dewed vetches; every spot of skin bearing white hairs becoming inflamed and gangrenous. The Rev. J. Rodwell informs me that his father turned out about fifteen cart-horses into a field of tares which in parts swarmed with black aphids, and which no doubt were honeydewed, and probably mildewed; the horses, with two exceptions, were chestnuts and bays with white marks on their faces and pasterns, and the white parts alone swelled and became angry scabs. The two bay horses with no white marks entirely escaped all injury.

These observations would be generally accepted today as characteristic references to a mycotoxicosis recalling facial eczema.

FUNGAL PHYTOTOXINS

The recognition of fungi poisonous to higher plants is a recent development although the concept that toxic metabolites (toxins) of pathogenic fungi play a part in the symptom expression of plant disease goes back to de Bary (1886) but the toxic principle he discovered proved (as previously noted, p. 159) to be an enzyme complex and it is not usual to treat the enzymes associated with soft rots or with the browning of vascular tissue in certain wilt diseases as toxins. Neither are growth-regulating substances such as the gibberellins toxins in the usual sense of the term which in phytopathology is reserved for toxic chemicals ('chemopathogens') able to cause effects similar to those of the disease symptoms induced by micro-organisms. One currently accepted terminology is to employ 'phytotoxin' for any product of a living organism toxic to plants and to categorise phytotoxins into 'vivotoxins', in the sense of Dimond & Waggoner (1953) who coined the term and defined a vivotoxin as 'a substance produced in the infected host by the pathogen and/or its host which functions in the production of disease, but is not itself the initial inciting agent of the disease', and 'pathotoxins' for toxins proved to play an important causal role in disease.

This last category, pathotoxins, by definition the simplest, is still a small group. The first to be recognised was the toxin noted by Tanaka in Japan in 1933 from *Alternaria kikuchiana* Tanaka, the cause of black spot of Japanese pears (*Pyrus serotina*) the symptoms of which Tanaka found could be reproduced by spraying fruits of a susceptible variety with a culture filtrate of the pathogen. Subsequently, the best known and most studied example of a pathotoxin, victorin – which is also host specific – from *Helminthosporium victoriae*, the pathogen responsible for blight of the South American oat variety Victoria, was discovered by Meehan & Murphy (1947). Later still the toxin from *Periconia circinata* was shown by Scheffer & Pringle (1961) to induce symptoms of the milo disease of grain sorghum (*Sorghum vulgare* var. *subglabrescens*).

Vivotoxins are of two main types. Toxins produced by the host in response to infection, e.g. ipomeamarone by the sweet potato on infection by either of the two taxonomically unrelated pathogens *Ceratostomella fimbriata* or *Helicobasidium mompa*, or toxins produced by the invading pathogen as are α-picolinic acid and piriculin, metabolites of the rice blast fungus *Piricularia oryzae* in its host which are not specific in their action and do not reproduce all of the disease symptoms. Further, piriculin is more toxic to *P. oryzae* than to the rice plant but there appears to be a 'piriculin-binding protein' which inactivates the toxic effects of piriculin on the pathogen but not on the host.

The status of most of the numerous other phytotoxins which have been investigated in greater or less detail, including alternaric acid (from *Alternaria solani*), colletotin (*Colletotrichum fuscum*), diaporthin (*Endothia parasitica*), and fusaric acid, lycomarasmin, and other toxins associated with fusaria causing wilt in tomato and other plants (work on which during the nineteen-forties and fifties was particularly associated with Gäumann and his school at Zurich; see Gäumann, 1957), is in varying degrees controversial. Further details of these and other phytotoxins may be obtained from the reviews by Wheeler & Luke (1963), Pringle & Scheffer (1964), and Wood (1967).

HALLUCINOGENIC FUNGI

Since earliest times man has made use of plants and plant products as stimulants and soporifics, intoxicants and hallucinogens, and some fungi, too, have been employed for such purposes. Until twenty years ago the most familiar mycological example was the use made by the men of certain tribes in north-eastern Siberia, particularly the Kamchadals and Koryaks, of dried fruit-bodies of *Amanita muscaria* (the fly agaric) to induce a violent erotic intoxication. The fly agaric is not endemic in the region and the dried basidiocarps were obtained by barter with Russian traders while the experience was so much sought after that poorer men induced it by drinking the urine of those fortunate enough to have eaten the fungus. This topic has been exhaustively reviewed and documented by Wasson (1968) who gives English translations of the accounts by travellers, explorers, and anthropologists, from the seventeenth century onwards, of the Siberian practice of eating the fly agaric. He assembled much evidence in support of the hypothesis that the enigmatic Vedic Soma, which 'was at the same time a god, a plant, and the juice of that plant', should be identified with the fly agaric; a theme popularised and given wider publicity a few years later by John Allegro in his more speculative *The Sacred Mushroom and the Cross*, 1971, where it is claimed that a mushroom cult underlay Christianity.

In China too the magic mushroom (*ling chih*) was a leading Taoist symbol for the better part of two thousand years and although the fungus fruit-bodies generally shown in later iconography have been identified as those of *Ganoderma lucidum*, *Amanita muscaria*, introduced from India, may have been the fungus employed in religious rites (see Wasson, 1967: 80–92[19]).

R. Gordon Wasson, a New York banker, in collaboration with his Russian-born wife Valentina P. Wasson, a pediatrician, and his mycological mentor Roger Heim, director of the Museum National d'Histoire Naturelle in Paris, may with some justification be claimed as the founders of ethnomycology – the study of fungi in folklore and ritual from prehistoric times to the present day.

Gordon Wasson is best known for his field work and literary studies on the use of hallucinogenic fungi in religious practices as described in the Wassons' two-volume *Mushrooms, Russia and history*, 1957, and his subsequent publications with Roger Heim and others (see Heim & Wasson, 1958; Heim *et al.*, 1967). It was in June 1955 that Gordon Wasson and Alan Richardson, a photographer, first shared with a family of Mexican Indians in a remote village in the Oaxaca district of Mexico a celebration of 'holy communion' at which 'divine' mushrooms were adored and then consumed in a night-long ceremony during which those participating experienced hallucinatory visions resulting from a heightened perception similar to that induced by lysergic acid diethylamine (LSD). Subsequently Wasson took part in further observances of the rite which is usually supervised by women curanderas or shamans and conducted in the Mixeteco language. These mushroom rites, which are taken with great seriousness and which are now overlaid with Christian symbolism, are of ancient origin. When the Spaniards conquered Mexico in the early sixteenth century they reported that the Aztecs were using in their religious ceremonies mushrooms which they called *teonanácatl* or 'god's flesh' and as recorded by Wasson and others (e.g. Lowy, 1972) the divine mushroom has been pictured in some of the few fifteenth-century manuscripts which survived the wholesale destruction by the Spaniards of 'idols' and pagan writings (see Fig. 77). Even earlier evidence of a mushroom cult in Central America is provided by frescoes in the Valley of Mexico (*c.* A.D. 400) and 'mushroom stones' (Fig. 78) carved by the highland Maya of Guatemala which are believed to date from the middle of the first millennium A.D.

Fig. 77. Representation in a Maya manuscript (Madrid Codex, p. xcv*b*) of the offering of the sacred mushroom. (B. Lowy (1972): fig. 3.)

Fig. 78. Maya mushroom stones (British Museum 5525 (*right*) and 1935-4 (*left*)). 5525 is possibly from Guatemala. 1935-4 probably dates from the Late Pre-Classic (500 B.C.–A.D. 200) or Early Classic (A.D. 200–600).

back to perhaps 100 B.C.[20] The main effect on mycology of these anthropological studies has been the description by Heim and others of the hallucinogenic agarics as a series of new species in the genera *Conocybe*, *Stropharia*, and, particularly, *Psilocybe*. Several of the species of *Stropharia* and *Psilocybe* have been cultured and the hallucinogenic principles characterised as simple substituted indole derivatives and designated 'psilocybin' and 'psilocin'. Both these metabolites have been synthesised and a tentative route for the biosynthesis of psilocybin in *P. cubensis* has been offered.[21] Psilocybin has found experimental use in the treatment of mental disorders.

More recently Wasson, Heim, and others (see Heim (1972) for a review) have investigated 'mushroom madness' in the very primitive, almost Stone Age, Kuma tribe of the Mt Hagen region

195

of Australian New Guinea where both the men and women at regular intervals eat the fruit-bodies of larger fungi and undergo 'mushroom madness' which in the opinion of the anthropologist Marie Reay has been institutionalised to serve as a social catharsis. The fungi involved in this instance have proved to be species of *Boletus* and *Russula*, again hitherto undescribed. Species of *Boletus* are eaten by men and women, the *Russula* species by women only.

FUNGAL ALLERGY

The classical signs of fungus allergy are respiratory effects such as bronchial asthma and rhinitis resulting from the inhalation of airborne spores although, like most foods, edible fungi sometimes cause digestive or other disorders, frequently of obscure aetiology, which are in some cases almost certainly allergic in nature.

Fungi are ubiquitous and fungal spores are always present in the atmosphere from the poles to the tropics though the concentration and composition of the 'airspora'[22] varies diurnally, seasonally, topographically, and geographically. Fungus spores are heavier than air and in spite of atmospheric turbulence there is continuous deposition which is compensated for by the constant liberation of fresh spores.

Spore liberation

Many fungi have special devices to ensure that spores become airborne. In such heavy fungal artillery as *Sphaerobolus* (the discharge of which was first described and illustrated by Micheli, 1729: tab. 101) and *Pilobolus*,[23] although the trajectories attained can be measured in metres and are thus relatively immense, the heavy projectiles (the spore masses and sporangia) never become passively airborne as do the light spores of many other fungi which are forcibly discharged to the distances of a few microns or at most a few centimetres, sufficient to allow the spores to clear the fruit-bodies and be carried away by air currents. In terrestrial ascomycetes violent discharge of the ascospores (in groups or singly) from turgid asci is of common occurrence. Micheli (1729: 204) was also the first to record and illustrate the visible 'puffing' of spores from the ascocarps of

Fig. 79. First illustration of spore discharge by a discomycete. (P. A. Micheli (1729): pl. 86, fig. 17.)

discomycetes (Fig. 79) and von Haller, who in 1740 made a similar observation, was apparently the first to hear the discharge[24]. Bulliard (1791–1812, **1**: 51–2) described and illustrated the same phenomenon but his explanation of the mechanism was, as pointed out by Tulasne (1861–5, **1**: 42; p. 44 of English transl.), fanciful. Persoon (1801: 103) noted puffing in *Rhytisma salicinum* (as *Xyloma salicinum*) and there are many later records. It was de Bary in his textbook (1866: 141–3) who gave the first satisfactory elucidation of the process. In basidiomycetes, Schmitz (1845) noted that the four spores of the basidium of *Stereum hirsutum* (as *Thelephora sericea*) were discharged in succession, Brefeld (1872–1912, **3**: 65, 66, 132) recorded that the basidiospores were violently discharged from the basidium, and Fayod (1889) was the first to describe the development of what was for long believed to be a drop of water at the hilum of the spore of *Galera tenera* shortly before discharge. Twenty-two years later this last observation was made independently by Buller who first illustrated (1915c) the 'water drop' (or 'bubble', claimed by Olive, 1964) (Fig. 80) and subsequently showed its development to be a general characteristic of spore discharge in hymenomycetes (see Buller (1909–50) **2**: chap. 1) but the explanation of the discharge mechanism is still controversial. Among hyphomycetes two general types of spores may be distinguished, dry spores and slime spores. Spores of the former category are usually carried away from the conidiophores by very weak currents of dry air, those of the latter are only liberated if the current of air is moist (Dobbs, 1942) or they become airborne by the splash dispersal of falling rain.

It was in connection with sporophore adaptation for spore

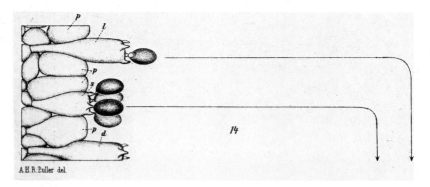

Fig. 80. First illustration of the 'water drop' (bubble) associated
with basidiospore discharge. (A. H. R. Buller (1915 c): pl. 2, fig. 14.)

liberation that tropic responses to gravity and light in fungi were
first noted. Attention was called to the negative geotropism of
the stipe of the agaric fruit-body by J. Schmitz in 1843[25] and by
1863[26] Sachs was noting the positive geotropism of the gills of
agarics and the teeth of *Hydnum* which provides the fine
adjustment to ensure the liberation of spores, topics later to
be investigated by Buller (Buller (1909–50) **1**). Phototropism
has been much studied in the Mucorales. Spallanzani found that
the sporangiophores of *Rhizopus stolonifer* were insensitive to
both gravity and light (see p. 20) but in many mucoraceous fungi
they are positively phototropic. Two much-studied examples are
Phycomyces nitens, in which the marked positive phototropism of
the sporangiophores was noted by Hofmeister in 1867 (*Die
Pflanzelle*: 289) and was the subject of the famous experiment by
Buder (1918) by which he reversed the response to one-
directional light by immersing the sporangiophores in liquid
paraffin, and *Pilobolus*, experimentally investigated by Brefeld in
1881 (Brefeld (1872–1912) **4**: 77) (Fig. 81) and later in depth by
Buller (1909–50, **6**) who offered an explanation of the mechanism involved.

The airspora

The 'floating matter of the air', as John Tyndall (1881) described
it, was for long a subject of interest in connection with the
problems of spontaneous generation and putrefaction. An early
study of the mycological component of the air was that by

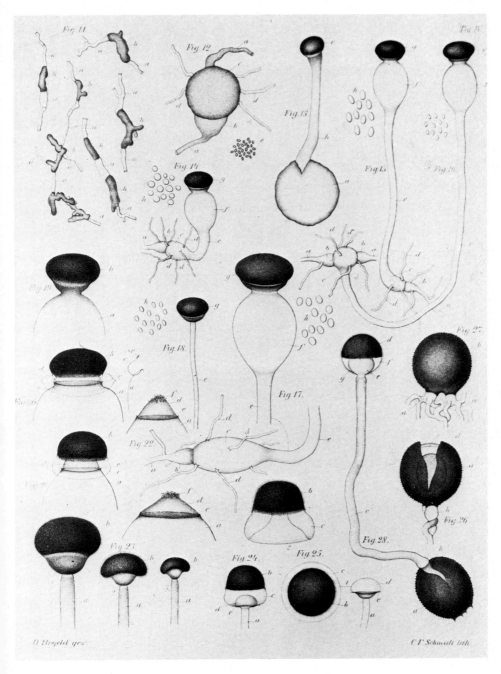

Fig. 81. *Pilobolus.* (O. Brefeld (1872–1912) **4**: pl. 4.)

J. G. Gleditsch who in the winter of 1748–9 took ten vessels sterilised by heat into each of which he placed pieces of ripe melon (also previously heated) and after covering the mouths of the vessels with muslin set them about in various situations in his garden and house. In due course species of *Byssus* and other fungi developed on the melon in most of the vessels and in concluding his report (Gleditsch, 1749) Gleditsch marvelled at the prodigious quantity of perfectly organised particles present in the air which attach themselves to all parts of animals and plants, both living and dead, and are present in the air we breathe and in all our food and drink.

In 1790, James Bolton speculated further:

the multitude of seeds produced by an Agaric or a Boletus is innumerable! is astonishing! yet not one in ten thousand answers the purpose of propagation. Is not the air we breathe charged with them in the declining part of the year? Do we not receive them into our Lungs with every breath we draw? Whence proceed the Quinsies, Coughs, and other complaints, which prevail in the Autumn?

I am not here to enquire into their noxious or salutary effect when received into our lungs; yet it cannot surely be amiss, to those whose province it is to make such inquiries.[27]

Such inquiries were made, but not for many years.

The main impetus for the detailed microbiological study of the air resulted from the demonstrations by Pasteur, Tyndall, and others that organisms responsible for putrefaction were airborne, from which it seemed reasonable to conclude that pathogenic micro-organisms were also present in the atmosphere. One of the first and most comprehensive attempts to correlate the incidence of disease in man with variations in the airspora was that undertaken in Calcutta during 1871–3 by D. D. Cunningham of the Indian Medical Service when attached to the Sanitary Commissioner. Cunningham exposed glycerine-coated microscope slides for 24-hour periods in an apparatus ('aeroconiscope'[28]) designed to keep the slide facing the wind and he attempted to correlate the daily catch as revealed by microscopic examination with the incidence of ague, dysentery, diarrhoea, dengue, and cholera in the Presidency jail. He found no correlation but his published findings (Cunningham, 1873), illustrated by drawings of typical catches (see Fig. 82), did for the first time reveal such features of the variation in the airspora as increase in the number of fungal spores trapped after rain. Shortly afterwards the French bacteriologist Pierre Miquel

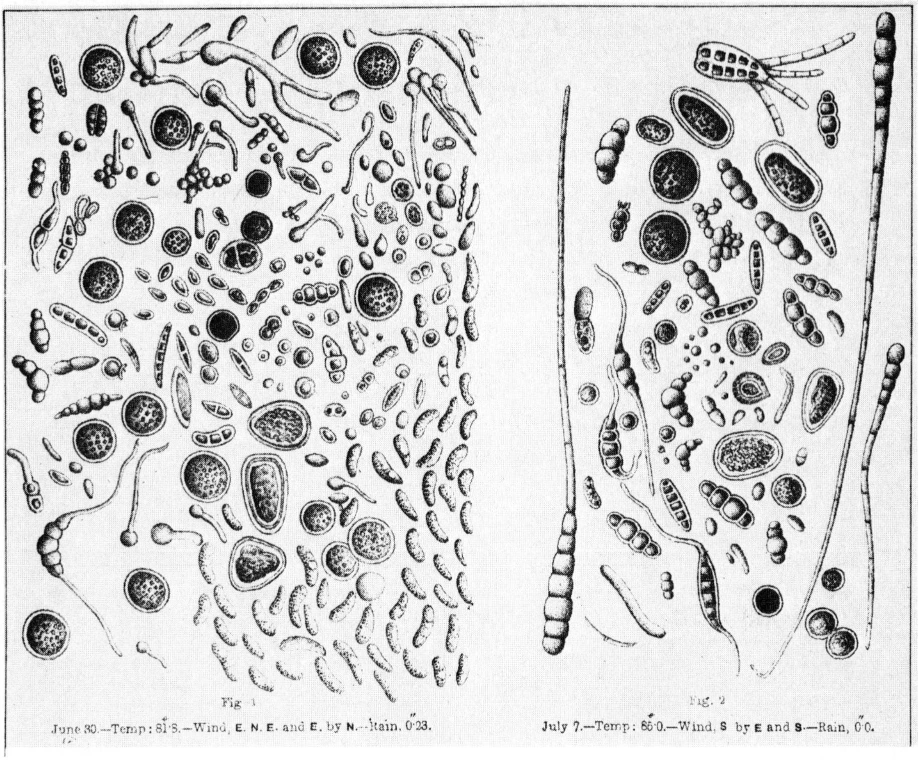
Fig 1
June 30.—Temp: 81·8.—Wind, E. N. E. and E. by N.—Rain, 0·23.
Fig. 2
July 7.—Temp: 86·0.—Wind, S by E and S.—Rain, 0·0.

Fig. 82. Examples of D. D. Cunningham's catches of airborne spores. (D. D. Cunningham (1873): part of pl. 9.)

working at the Observatoire Montsouris, Paris, where he became Chef du Service Micrographique, made detailed studies on the organisms present in the air not only of the Montsouris Park but also of dwelling houses, hospitals and sewers, the high Alpes and oceans, and of rain. These results were summarised in his doctoral thesis of 1883 and the annual reports of the observatory between 1878 and 1899.

During recent years much attention has been paid to aerobiology which will appear as a major trend when the microbiological history of the period comes to be written. There have been surveys of the fungal airspora of many regions and diverse habitats, not only in relation to allergic disorders but also in connection with the aetiology and epidemiology of fungal diseases of plants, animals, and man. Much of this work has been stimulated by the studies of P. H. Gregory (who was influenced in his choice of research by being asthmatic, cf. Charles Blackley below) and his colleagues at Rothamsted Experimental Station where the development in 1952 of the 'automatic volumetric spore trap' by J. M. Hirst (the 'Hirst trap') provided a most useful analytical tool which was a great advance over the much used 'gravity slide'. General developments in this field are summarised and documented in the two editions of P. H. Gregory's *Microbiology of the atmosphere*, 1961, 1973.

Mould allergy

The year of Cunningham's report also saw the publication of the famous, and still very readable, *Experimental researches on the causes and nature of* Catarrhus aestivus (*hay-fever or hay-asthma*), 1873, by the Manchester physician Charles H. Blackley. Blackley himself suffered from hay fever and from the evidence of twenty years' observation on his own case and on others met with in his practice, by spore trapping, and by self-experimentation he was able to prove that hay fever could be produced by the inhalation of pollen. In the course of these studies Blackley recorded the first case of an allergic reaction to the experimental inhalation of fungus spores. He wrote (p. 57):

I had noticed many years ago that the dust from straw sometimes brought on attacks of sneezing with me, and that this seemed to occur more frequently when we had had a long spell of wet weather. I determined to try what fungi could be generated on damp straw. For this purpose wheat straw, slightly moistened, was placed in a closed vessel, and was kept at a temperature of 100° Fahr. In about twenty-four hours a small quantity of white mycelium was seen; this increased slowly for three or four days, and in a short time after was followed by the appearance of minute greenish-black spots dotted here and there along the surface of the broken straw, apparently coming out more readily on the inner than the outer surface. This, I found on examination, was the *Penicillium glaucum*. After a few days another crop of dark-coloured spots were seen, but these became almost jet black, and had quite a different contour. These I found to be the *bristle mould* (*Chaetomium elatum*).

The spores of these two fungi were sown again separately on straw which had been placed in separate vessels after being subject to the action of boiling water for a short time. A separate crop of each fungus was thus obtained.

The *odour* of the *Penicillium* produced no perceptible effect upon me, but the odour of the *Chaetomium* brought on nausea, faintness and giddiness on two separate occasions. By inhaling spores of the Penicillium, in the involuntary experiment of which I have spoken, a severe attack of hoarseness, going on to complete aphonia, was brought on. This lasted for a couple of days, and ended in a sharp attack of bronchial catarrh, which almost unfitted me for duty for a day or two.

Blackley added: 'The sensations caused by these two agents were so unpleasant that I have never cared to reproduce them'.

Little attention was paid to fungal spores as allergens during the next fifty years. It was in 1924 that van Leeuwen noted in the island of Zuid-Beveland that 0.5 to 1.0 per cent of the inhabitants of some villages were asthmatic, an incidence greatly in excess of that prevailing in other parts of Holland. He failed to associate the condition with any meteorological pecularity and attributed the cause to 'miasma' or 'climate allergens'. Further, he showed that sufferers were relieved by air freed from 'miasmatic substances' by filtration through cotton wool or by passage through glycerin and he suggested that these airborne substances were mainly of fungal origin. Also in 1924, Cadham in Canada recorded three cases of agricultural workers who developed severe asthma on exposure to dust from rusted grain and infected straw. He considered stem rust (*Puccinia graminis*) spores to be the causal agent and he was able to induce a paroxysm by the inhalation of a few rust spores. According to Feinberg (1948), it is possible that the rust spores were contaminated with spores of *Alternaria* or *Cladosporium* because rust spores are ineffective as allergens and are obtained in pure culture with difficulty. Whether this was so or not, Cadham was able to afford some relief with a vaccine prepared from the rust. In 1930, in the U.S.A., Bernton described a case of asthma associated with *Aspergillus fumigatus* and Hopkins, Benham, & Kesten, in the same year described a similar case caused by *Alternaria* spores derived from vegetables stored in the basement of the patient's house. Since then many more instances have been recorded and Durham (1938) stated that after the storm of *Alternaria* spores during 6–7 October 1937 in the eastern United States when thousands of tons of mould spores were transported several hundred miles 'a brief inquiry has disclosed the correlation of clinical symptoms with the advent of the storm in some of the places affected'.

Interest in mould allergy has been greatest in the United States where many papers have been published and whilst the fact that mould allergy is of wide occurrence and of clinical significance appears to be beyond reasonable doubt much of the data has lacked precision. This was in part due to variation in the diagnostic procedures and in part due to the difficulty in obtaining proof of the specific allergens involved. In a summary by Feinberg (1948: 275, table 19) of the results of eleven investigations in the United States involving tests on more than 5000 individuals suffering or suspected of suffering from allergic disorders, the percentage giving a positive reaction to 'moulds' varied from less than 1 to 85 per cent and the reaction of one group diagnosed as suffering from 'mould allergy' (75 per cent) was probably not significantly different from that of other groups of 'respiratory' and 'general allergy' cases which gave 69 and 85 per cent positive mould reactions, respectively. Van Leeuwen in Holland found that half the asthmatics he tested gave a positive reaction to fungi. This compares with 15 per cent positive in Germany by Hansen in 1928 and 16 per cent in the same country by Fraenkel in 1938 who obtained 53 per cent of mould reactors in a series in England.

The fungus spores present in the air are normally derived from saprobic or parasitic growth of fungi on plants or plant remains. As Hyde & Williams (1946) pointed out, the straw of wheat and other cereals would appear to be an ideal medium for the growth of *Alternaria*, which would explain the predominance of *Alternaria* spores in the air of states in or adjacent to the wheat belt of central and eastern U.S.A., and the dominance in other parts of the world of *Cladosporium herbarum*, a ubiquitous and unspecialised saprobe, is likewise expected. In addition to allergic disorders caused by spores of such widespread fungi a number of instances of allergy to fungi of more localised distribution have been recorded.

Cobe, in 1932, described the first case of asthma due to tomato leaf-mould fungus (*Fulvia fulva*), the spores of which occur in high concentrations in glasshouses containing affected plants. The powdery mildew of oak (*Microsphaera alni*) has been associated with allergic disease in California and, according to Feinberg (1948), Sir John Floyer in his *Treatise on Asthma*, 1726, said; 'There is a Remarkable Instance in Bonetus of an Asmatic, who fell into a violent Fit by going into a wine-cellar where Must was fermenting', which is perhaps the first record of mould allergy.

8. Uses of fungi

Fungi have been put to many and diverse uses. For example, the fruit-bodies of *Fomes fomentarius* are the well-known source of amadou (German tinder) and they (and those of *Ganoderma applanatum*) are still used to produce a suede-like material from which hats, various articles of dress, handbags, and picture frames are made in Roumania and elsewhere.[1] Dried *Coriolus versicolor* brackets have been used for hat and costume decoration, bottle corks made from *Polyporus nidulans* fruit-bodies, and razor strops from those of *Piptoporus betulinus* and *Polyporus squamosus*. The green wood in Tunbridge ware (decorative wooden objects, a kind of treen) was originally wood infected by *Chlorosplenium aeruginascens*. *Coprinus* species have yielded ink, *Amanita muscaria* fruit-bodies have been used to kill flies, and pieces of wood rendered luminous by fungal attack have been employed as guide marks at night. These and other trivial uses, apart from maintaining popular interest in fungi, have done little to advance mycology. Very different has been the use of fungi in three important but overlapping fields – food, drink, and their preparation, medicine, and industry – which together with the use of lichens as dyes, has had an effect comparable to that of pathogenicity in widening and deepening mycological knowledge.

AS FOOD

Fungi are used as food in all parts of the world. In Europe fungi have been eaten since classical times. The Romans were familiar with agarics, boletes, and truffles (see p. 2) and following the introduction, during the last half of the second century B.C., of the Sumptuary Laws which were designed to curb the extravagant use of meat and fish at private banquets, edible fungi gained favour as gastronomic delicacies. Special silver vessels ('boletaria') were designed to cook them in and the delights of fungi were a subject for epigrams and verse. The nutritive status of edible fungi, however, remained in doubt (see p. 184) and in the seventeenth century van Sterbeeck in his *Theatrum fungorum* (1675: 23–4) considered both the gills ('plues') of agarics and the tubes ('mergh') of boletes to be inedible.

In the Far East two edible fungi have been popular for many centuries – *Lentinus edodes*, the *shii-take* of Japan (known as *hsiang hsin* and *hsiang ku* in China) and the Chinese *Auricularia polytricha* (*mu erh*) which is described in the first of the Chinese natural histories, the *Shen Nung Pên Tshao Ching* (Pharmacopoeia of the Heavenly Husbandman) of the second century B.C.[2] *Volvariella volvacea* (the padi straw mushroom) is also a well-established esculent in south-east Asia and Madagascar. Tuckahoes (the large sclerotia of *Poria cocos*, first recorded from Virginia in 1722) were at one time made into 'bread' by North American Indians and the comparable sclerotia of *Polyporus mylittae* (blackfellow's bread) were put to similar use by the Australian aborigines.

Among the many examples of travellers noting the use of edible fungi is the observation of Charles Darwin in 1834 during his voyage on the *Beagle* that in Tierra del Fuego 'a globular, bright-yellow fungus', which grew in vast numbers on the beech trees (*Nothofagus*), eaten uncooked, provided the Fuegians with an important article of diet. It was on the basis of Darwin's specimens and notes that M. J. Berkeley proposed a new genus for the fungus which he named *Cyttaria darwinii*. Again, the famous botanist Joseph Dalton Hooker recorded in his *Himalayan Journals*, 1854, that in the pine woods of Tibet a large mushroom known locally as *onglau* (and later described, also by Berkeley, as *Cortinarius emodensis*) was a favourite food of the natives.

During the past 150 years in both Europe and North America the cultivated mushroom (which was only in 1951 recognised as distinct from the field mushroom (*Agaricus campestris*) and named *Agaricus bisporus*) has been the most eaten species – in the United Kingdom and the U.S.A. to the virtual exclusion of others – but in western European countries, particularly France and Italy, a range of wild species is still regularly offered for sale in local markets. Originally all edible fungi must have been collected from the wild. Where the cultivation of edible fungi was first undertaken is unknown. It was possibly in the Far East but the practice probably originated independently and at different times for different species in different countries. The most primitive 'cultivation' of edible fungi of recent times is that of tuberasters and truffles. The former, the fruit-bodies of *Polyporus tuberaster* (the stone fungus, Pietra fungaia), were in the eighteenth century obtained by collecting the sclerotia (pseudosclerotia) which were then kept warm and moist until fructifications developed, and a similar procedure has been used for *Pleurotus tuber-regnum*, the '*olafata*' of African natives.

Truffle cultivation likewise has never been put on a scientific basis. It was about 1810, according to Chatin (1892: 176–7) that Joseph Talon, a peasant of Vauclause in France, sowed acorns in a piece of stony ground near his home and a few years later found that Perigord truffles (black or French truffles, *Tuber melanospermum*) had developed under the young oak trees. Recognising the value of this discovery he bought up more worthless land on which he planted oaks. His secret became known and his practice widely adopted. By 1868 the total French crop attained 1 500 000 kg and was valued at 150 million francs. Many truffles were exported from France and also from other European countries, especially Italy, Germany, Belgium, and Spain which had flourishing truffle industries.

Auricularia polytricha and the *shii-take* are similar in that they are both lignicolous and it is believed that the earliest method of 'cultivation' was to stack pieces of naturally infected wood in the forest and to harvest fruit-bodies that developed. The next stage in *shii-take* cultivation (which began in the mid-seventeenth century; the name *shii-take* dating from 1564[3]) was to mix healthy and naturally infected logs. Infection was later rendered more certain by inserting spore prints on paper into incisions in the healthy logs and later still an aqueous spore suspension was employed. It was not until the 1920s that pure culture spawn was introduced by K. Kitayima.

The padi straw mushroom is grown by methods similar to that of the cultivated mushroom but, typically, on damp rice straw. According to Singer (1961: 120) its cultivation was developed and disseminated by Cantonese farmers. Originally, the beds were spawned by inoculum from previous beds but now there is some pure culture spawn available (see Su & Seth, 1940).

Aspects of the cultivation of mushrooms were noted in Chapter 4 (p. 82) when considering the beginnings of the culture of fungi. In the West the practice started in France towards the end of the seventeenth century (Fig. 83) and the empirical

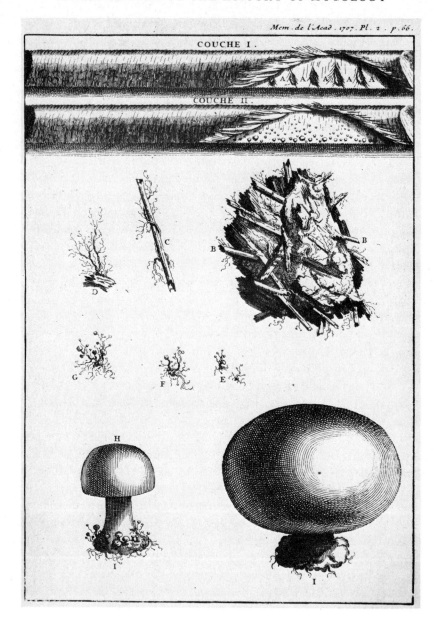

Fig. 83. Mushroom beds and stages in the development of the fruit-body. (J. P. de Tournefort (1707): pl. 2.)

Fig. 84. Mushroom-cave, 70 feet beneath the surface, at Mont-rouge, near Paris, July 1868. (W. Robinson [1880]: 58.)

method then evolved became, by the mid-nineteenth century, the basis of a large industry, particularly in the Paris region where the mushrooms were cultivated in disused underground limestone quarries (see Fig. 84). A major advance was the introduction of pure culture spawn which had been developed by J. Costantin by 1890 and in 1894, at the suggestion of Émile Duclaux, director of the Institut Pasteur, Costantin and Matruchot established in the rue Dutet, Paris, a commercial service for the supply of sterilised mushroom spawn derived from spores by Costantin's method which was patented. During the next decade France had a virtual monopoly of the production of pure culture spawn and this enabled the French to maintain their position as the leading mushroom producers until the nineteen-twenties when there was a great expansion of mushroom cultivation in both North America (as an outcome of the studies by B. M. Duggar (1905) on the production of pure culture spawn from tissue cultures and Margaret C. Ferguson's 1902 study on spore germination in mushrooms and other basidiomycetes) and in England where the recent developments in methods can be appreciated by consulting the successive editions of the Ministry of Agriculture's Bulletin (no. 34), *Mushroom-growing*, first published in 1931.

Fermentation

One of the most ancient uses of fungi in the preparation of food and drink has been the leavening of bread and the making of a diversity of alcoholic drinks. The introduction of these two practices must date from the dawn of civilisation and the latter has been attributed to divine intervention, the brewing of beer to Osiris by the Egyptians, wine making to Bacchus by the Greeks, while St Patrick is said to have taught the Irish the art of distilling spirits. Models of a bakery and a brewery on a wealthy estate were found during excavations of a XIth Dynasty site (*c.* 2000 B.C.) at Thebes and examination of sediments contained in beer urns from various sites revealed what appear to be yeast cells. It has even been claimed that there was a decided improvement in the purity of the yeast in such sediments between 3400 and 1440 B.C.[4] Perhaps the earliest treatise on the brewing of beer is that ascribed to Zosimos of Panopolis in Upper Egypt in the third century A.D.[5] It was not, however, until the nineteenth century that the nature of alcoholic fermentation was elucidated.

Over the centuries there has been much confusion regarding the concept of fermentation and its terminology. In the Middle Ages fermentation and digestion were considered synonymous and the latter was also applied by alchemists to any chemical reaction, ferments being the agents capable of inducing such reactions. Although early applied to alcoholic fermentation (which Johann Joachim Becher in 1682 noted to occur only in saccharine solutions,[6] from which alcohol was at first absent) the term fermentation was confused with effervescence. Georg Ernst Stahl in his *Zymotechnica fundamentalis*, 1697, while recognising alcoholic and acetic fermentation also included putrefaction and decomposition as related categories. This last view was discussed by Justus von Liebig as late as 1844 when he poured scorn on the opinion that fermentation or the decomposition of vegetable beings was caused by lower plants (and the analogous putrefaction of animal beings by microscopic animalculae). 'Can we imagine plants and animals to be the agents in destroying and annihilating the bodies of plants and animals, when they themselves and their own constituents suffer the same processes of destruction', he wrote. 'If the fungus be the cause of destruction of the oak...to what cause are we to ascribe the putrefaction of the fungus?'[7] The recognition of the biological nature of alcoholic fermentation and the leavening of bread, the discovery of diastase and pepsin (and later zymase) led to the differentiation of two main classes of ferments – *organised ferments*, such as yeasts and bacteria, and *unorganised* or *soluble ferments* for those for which the term 'Enzym' was coined by W. Kühne in 1877 and anglicised as *enzyme* by William Roberts in 1881. At the same time the circumscription of fermentation was broadened to include any decomposition or transformation brought about by the activity of living fungi or other micro-organisms. Thus the commercial production of antibiotics became a major branch of the fermentation industry.

Leeuwenhoek in 1680 observed the budding cells of yeast in beer (see p. 58) and in 1826 J. B. H. J. Desmazières in his study of the genus *Mycoderma* (proposed by Persoon in 1822) described five species one of which, 'Mycoderma cervisiae', he had isolated from a membrane growing on beer and from his illustrations it appears that he observed yeast cells. Desmazières, like Leeuwenhoek, did not associate these organisms, which he considered to be monads (*animalcula monadina*) with fermentation. Ten years later – in parallel with the elucidation of basidial structure by six independent workers (see p. 71) – three of Desmazières' contemporaries independently described the involvement of living cells in alcoholic fermentation: the physicist Charles Cagniard-Latour in France, in 1836, and Theodor Schwann (then assistant to Johannes Müller at Berlin) and the algologist Friedrich Traugott Kützing in Germany, both in 1837.

Although the last to publish (because on a visit to the Adriatic coast), Kützing was the first of the three to conclude that yeast was a living organism and the cause of fermentation; a view he communicated to Ehrenberg in 1834. Kützing (1837) described the nucleus of the yeast cell, but did not mention budding, and he believed yeast to be pleomorphic and to have a mould phase (cf. p. 30). Cagniard-Latour's account of his one mycological venture to the Société Philomathique on 18 June 1836 (Cagniard-Latour, 1837), which gives him priority of publication, and the later supplementary publications were more satisfactory. He described budding, estimated the size of the beer yeast cells as 1/150 mm, found similar cells in wines, showed dried yeast and yeast subjected to -5 °C in liquid carbon dioxide to be still active, and attributed the production of carbon dioxide and alcohol from sugar to the vital action of yeast. As Bulloch (1938: chap. 3) in his attractive treatment of historical aspects of fermentation emphasised, Schwann's contribution was the most significant of the three. He observed the growth of yeast by budding, noted differences between the yeasts of beer and wine, and that for the fermentation of saccharine solutions by yeast a source of nitrogen is also necessary. Schwann consulted the mycologist F. J. F. Meyen who agreed with him that the yeast was a living organism although Meyen was doubtful whether it was an alga or a fungus. However, the next year (1838) Meyen proposed a new genus *Saccharomyces* (the latinisation of Schwann's 'Zuckerpilz', sugar fungus) for brewers' yeast which he named *S. cerevisiae*.[8]

In 1838 these results were confirmed in France by T. A. Quevenne and accepted by Turpin but the idea that fermentation was biological in nature was received with much resistance by chemists led by J. J. Berzelius; particularly Liebig who, with Friedrich Wöhler (best remembered for having prepared urea from ammonium cyanate in 1828) published in 1839 in the *Annalen der Pharmacie* (of which they were the editors) an anonymous lampoon against the view that the role of yeast in fermentation was anything more than that of a catalyst. Liebig stubbornly maintained this view for the next quarter of a century, even after the researches of Pasteur, particularly the paper of 1860, had rendered such views untenable, while Oscar Brefeld, a believer in the biological nature of fermentation, was critical of Pasteur's conclusion that 'la fermentation est la consequence de la vie sans air'.[9]

During the next forty years there was much confusion in the taxonomy and status of yeasts associated with alcoholic fermentation and of yeasts in general. This was in part because they became involved in the final stages of the resolution of the problem of spontaneous generation and in the elucidation of the phenomenon of fungal pleomorphism, topics already covered in Chapter 2. It was also in part due to taxonomic confusions and uncertainties. Turpin (1838) was responsible for introducing one major and persistent taxonomic error by classifying the sporogenous *Saccharomyces cerevisiae* in the imperfect genus *Torula* properly used for hyphomycetes. This led many astray (including Pasteur, and Hansen who used *Torula* for certain asporogenous yeasts) and to a very broad and incorrect application of the generic name *Torula* to yeasts in general. Although

Schwann had observed endogenous spores in *Saccharomyces* as early as 1839[10] he failed to identify them with ascospores. De Bary in the first edition of his textbook (1866) was uncertain as to the affinities of yeasts and it was de Seynes (1868) and Max Rees (1870) who recognised the ascomycete relationship – the latter concluding *Saccharomyces cerevisiae* to be an ascomycete 'in weitesten Sinne des Wortes'. Shortly afterwards Engel in his doctoral thesis of 1872 introduced the gypsum block method for the induction of ascospore production in sporogenous yeasts. In the second (1887) edition of his textbook de Bary endorsed these findings and concluded that 'the only admissible assumption' was that members of the Saccharomycetes are 'greatly *reduced* Ascomycetes..., with extensively interrupted homology which is only restored with the appearance of the asci'.

During the last two decades of the nineteenth century and the first of the twentieth, the taxonomy of yeasts was put on a much more secure basis by the researches of a number of workers, especially E. C. Hansen who between 1879 and 1906 was the author of more than ninety scientific papers on yeasts and fermentation. Emil Christian Hansen, born in 1842 and trained as a school teacher, was in 1876 awarded a gold medal by the University of Copenhagen for his study of the coprophilous fungi of Denmark and three years later gained a Ph.D. for a thesis on yeasts. He subsequently became director of the Carlsberg Physiological Laboratory in Copenhagen and a university professor. Hansen was the first to obtain single-spore cultures of yeasts which he achieved by a dilution method (Hansen, 1883: 23–6). He made microscopic determinations of the concentrations of dilute suspensions of yeast cells and then inoculated each of a long series of flasks with a single drop of inoculum calculated to contain a single spore. Cultures were made from flasks in which not more than one colony developed. By this means he was able to initiate brewing fermentations with pure cultures and to differentiate a number of the commoner genera and species of yeasts on morphological and physiological characters. Later developments are summarised by Alexandre Guilliermond (1912) while the 1952 monograph by the two Dutch workers Johanna Lodder and Nellie Kreger-van Rij raised the taxonomy of the yeasts to a new level.

Another fundamental advance in the understanding of alcoholic fermentation was the isolation of the enzyme complex

zymase by Hans Buchner in 1897, thus confirming the suggestion made by Moritz Traube as early as 1858. This discovery
stimulated as much controversy as had the announcement of the
role of yeast sixty years before.[11] Disbelief in Buchner's findings
was particularly prevalent among those engaged in brewing
research whose views were endorsed by such weighty names as
those of Alfred Fischer the bacteriologist, Martius Beijerinck,
professor of microbiology at Delft, and Carl Wehmer. Support
for zymase came from Émile Roux and Émile Duclaux of the
Pasteur Institute and many of those first critical soon became
ardent Buchner supporters. For example, Reynolds Green in his
book *The soluble ferments*, 1899, treated zymase with the
enthusiasm of a convert. This acceptance did not close the
account. From 1904 A. Harden, of the Lister Institute of
Preventive Medicine, London, and others showed that zymase
action depends on the presence of co-enzyme and phosphate
(see Harden, 1911–32).

Yeasts are not the only fungi able to effect alcoholic fermentation as had been demonstrated by Pasteur. Fitz (1873) showed
alcohol to be produced from sugar by *Mucor racemosus* (as *M.
mucedo*) and in 1897 a German patent was taken out for the
production of alcohol by the so-called 'Amylo' process involving
the saccharification and fermentation of maize or other starchy
materials by *M. rouxii*. This process is still in use today with the
mucor replaced by *Rhizopus* species.

The many and diverse alcoholic drinks of the world are too
numerous even to catalogue. Most countries have their own
national variants of beer or wine or other traditional fermented
beverages. There are also a number of other fermented food
products, particularly in the Far East where one famous series
which includes *miso* (soybean paste), *shoyu* (soy sauce), *sake* (rice
wine), and *shocho* (a distilled spirit) of Japan and China is
associated with *Aspergillus oryzae* fermentation (see Hesseltine,
1965 for a review). The fermentation necessary for these foods
and drinks is initiated by different types of starters or *koji*, which
consist of soybeans or various cereals infected by *A. oryzae* and it
was the German H. Ahlburg, while a teacher at the Tokyo Kaisei
School (the forerunner of the University of Tokyo) who first
isolated the Japanese koji-fungus in 1878.

Food yeast

In addition to brewing and bakery, yeast has been found to have other nutritional applications. It is a valuable source of vitamins of the B group and various autolysed yeast preparations have been put on the market. Yeast is also rich in protein and a notable attempt was made in Jamaica to produce food yeast for human consumption by culturing *Candida utilis* (syn. *Torulopsis utilis*) in a medium of molasses or surplus raw sugar and ammonium nitrate. Food yeast has been most used as an addition to animal feeding stuffs.

Cheese

Another food of ancient origin is cheese in the ripening of which fungi play an important role. Penicillia are particularly impor-tant in this connection when their implication, as in the camembert and brie types ripened by strains of *Penicillium camemberti* and *P. caseicola*, is not always as obvious as in blue-veined cheeses such as Roquefort in which the mould is *P. roqueforti*. Originally, the ripening was effected by natural contamination by local moulds. It was in an effort to put cheese making on a more scientific basis that Charles Thom was appointed in 1904 as a mycologist at the Storrs (Connecticut) Experiment Station to study the ripening of cheese. The result was Thom's description of both *P. camemberti* and *P. roqueforti* in 1906 and his famous *U.S.D.A. Bulletin* no. 118 of 1910 which was the forerunner of a series of monographs on aspergilli and penicillia (see 1926, 1929).

IN MEDICINE

Since earliest times many different fungi have been advocated in medicine and folklore as prophylactics or cures for a wide range of human disorders. Cauterisation by burning fungi, typically tinder (*Fomes fomentarius* fruit-bodies, frequently soaked in saltpetre ($NaNO_3$) and dried), on the skin of the affected part, as practised by the Greeks in the fifth century B.C., survived into the nineteenth century in Lapland and Nepal. The dried fruit-bodies of polypores and puff balls and the spores of the latter have been used to staunch bleeding. *Daldinia concentrica* ascocarps were carried about on the person to prevent cramp (hence the common name 'cramp balls'). Pliny attributes to Glaucias the view that suilli (*Boletus*) are good for fluxes of the bowels, removing freckles, sore eyes, and bites inflicted by dogs. *Phallus impudicus* and other phallales and *Elaphomyces granulatus* were eaten for supposed aphrodisiac properties, preparations of the Jew's ear (*Hirneola auricula-judae*) used as a specific for throat infections, and *Cordyceps sinensis*, still attached to the parasitised caterpillar, was a famous Chinese drug (see p. 25). Lichens too have been employed in medicine. For example, as their scientific names commemorate, *Lobaria pulmonaria* was used to treat lung disease and *Peltigera canina*, mixed with black pepper, against rabies.

The most famous of all fungi formerly used in medicine was the classical agaricum or agarick (*Fomes officinalis*), a universal remedy, the scope of which may be gauged by the claims made by Dioscorides who wrote (Houghton, 1885: 26–7):

Its properties are styptic and heat-producing, efficacious against colic (στρόφους) and sores, fractured limbs, and bruises from falls; the dose is two obols weight with wine and honey to those who have no fever; in fever cases with honeyed water; it is given in liver complaints, asthma, jaundice, dysentry, kidney diseases, where there is difficulty in passing water, in cases of hysteria, and to those of a sallow complexion in doses of one drachma; in cases of phthisis it is administrated in raisin-wine, in affections of the spleen with honey and vinegar. By persons troubled by pains in the stomach and by those who suffer from acid eructations, the root is chewed and swallowed by itself without any liquid; it stops bleeding when taken in three-obol doses; it is good for pains in the loins and joints, in epilepsy when taken with an equal quantity of honey and vinegar; it assists menstruation and relieves flatulence in women when taken with equal proportions of honey and vinegar. It prevents *rigor* if taken before the attack; in one- or two-drachma doses it acts as a purgative when taken with honeyed water; it is an antidote in poisons in one-drachm[a] doses with dilute wine. In three-obol doses with wine it is a relief in cases of bites and wounds caused by serpents. On the whole it is serviceable in all internal complaints when taken according to the age and strength of the patient; some should take it with water, others with wine, and others with vinegar and honey or with water and honey.

Few fungi are now used as drugs. Agaricum and Iceland moss (the lichen *Cetraria islandica*), the latter still sold in apothecaries' shops in Sweden, survived in early twentieth-century pharmacopoeias and *Amanita muscaria* finds application in homeopathic medicine but the only fungus drug officially recognised today is ergot (sclerotia of *Claviceps purpurea*) (see p. 186), preparations of which are used in obstetrics.

Ergot

The Chinese were using ergot (but from wheat rather than rye) in medicine in the eighth century A.D.[12] In the West it was Lonitzer (1582) who first recorded the use of ergot to accelerate delivery at childbirth (Fig. 74) but official recognition of ergot as a drug was delayed until 1820 when it was included in the first edition of the *Pharmacopea of the United States of America*. In Britain ergot was introduced into the 1836 issue of the *Pharmacopoeia of the Royal College of Physicians of London* and today, though no longer included in the *British Pharmacopoeia*, it is still specified in the *British Pharmacopoeia Codex*. Ergot of commerce was originally obtained mostly from naturally occurring infections of rye in eastern Europe and the Spanish peninsula (which produced ergot with the highest alkaloidal content). More recently there has been successful economic culture of ergot in Switzerland and elsewhere by spraying flowering rye with an aqueous suspension of conidia, and less successful attempts at laboratory culture.

Antibiotics

Today the main interest in the use of fungi in medical practice centres around the deployment of a number of fungal metabolites possessing valuable chemotherapeutic properties. The best known of these fungal antibiotics is penicillin, the discovery and development of which resulted in the evolution of a vast new branch of the pharmaceutical industry. The story of the discovery of penicillin by Alexander Fleming at the Inoculation Department of St Mary's Hospital in London has frequently been told but many details regarding this event remained somewhat mysterious until 1970 when Professor Ronald Hare, who as a young man was one of Fleming's colleagues at St Mary's Hospital at the time of the discovery, gave a most perceptive and convincing account of what in all probability really happened.[13]

The traditional account of the discovery – that a penicillium spore came in through the window and contaminated a plate of staphylococci – is based on somewhat equivocal remarks made by Fleming in lectures and conversation and on the opening paragraph of Fleming's 1929 paper which reads:

While working with staphylococcal variants, a number of culture plates were set aside on the laboratory bench and examined from time to time. In the examinations, these plates were necessarily exposed to the air and they became contaminated with various micro-organisms. It was noticed that around a large colony of a contaminating mould, the staphylococcal colonies became transparent and were obviously undergoing lysis.

If this account is correct and the contamination occurred when the plates were being examined there would have been no observable lysis of the staphylococcal colonies because, as was proved in 1942, penicillin only acts on bacterial cells which are young and multiplying. Additional evidence that Fleming's explanation was incorrect is that apparently Fleming himself did not understand what had happened and made numerous unsuccessful attempts to 're-discover' penicillin, the discovery of which was due to a most extraordinary series of coincidences.

Alexander Fleming was a bacteriologist noted among other things for his discovery of lysozyme (the enzyme present in tears, nasal mucus, and saliva which dissolves many saprobic bacteria) and for his interest in staphylococci on which he was invited to contribute a chapter in the Medical Research Council's nine-volume *System of bacteriology*. When engaged on that task a paper was published by a group of workers in Dublin on the differentiation of staphylococci by their colony characters on solid media, one of the methods of incubation being at laboratory temperature. Fleming repeated these experiments and presumably also incubated inoculated plates in the laboratory. What happened next, according to Hare, was that before going away for his summer holiday towards the end of July or the beginning of August 1928 Fleming must have inoculated a series of plates which he left piled up on his bench in a corner out of the sun (his room was being used by a research student in his absence). One of these plates was contaminated by a spore of a penicillin-producing penicillium. Weather records at Kew for 1928 show that after a very hot spell in mid-July there was a cold spell of nine days from 28 July when the maximum daily temperature only exceeded 20 °C on two occasions, and there is little doubt that the temperature in Fleming's laboratory in central London was sufficiently low during the cool period to allow growth of the penicillium to take place (with the resulting production of penicillin which diffused into the medium) and to inhibit development of the staphylococci. Then higher daily temperatures returned when growth of the staphylococci occurred and colonies developing near the penicillium were lysed. On his return from holiday Fleming looked over his plates and discarded so many that only the bottom layers of those placed in a tray of lysol came in contact with the disinfectant. Shortly after Fleming's return D. M. Pryce, one of his former research students, called to see him and Fleming while explaining the staphylococcal work to Pryce, to demonstrate a point, picked up one of the top layer of the discarded plates when he noticed the lysis of the staphylococci (his cry of *Eureka* was 'That's funny!') and then and there made a subculture of the mould. If Fleming had been incubating his cultures at 37 °C – the optimum temperature for the growth of staphylococci – contamination by the penicillium would not have revealed penicillin production and during the three months he was working on the staphylococcal variants the laboratory temperature was only low enough for the discovery of penicillin in the way suggested for about nine days. Further, the shortage of technicians in the Inoculation Department allowed the discarded plates to accumulate and a chance visitor led to the retrieval of a plate which excited the trained eye of a worker long interested in chemotherapeutants, but not of his colleagues and others to whom he later showed the historic plate.

But there was another set of most suggestive coincidences. That the spore of the contaminating mould came in through a window is most unlikely as the windows of Fleming's room were never open. The door usually was. Microfungi were being cultured in a makeshift laboratory directly below Fleming's and both rooms opened on to the same stairs surrounding a lift. Some time before the events related, the Dutch allergist Storm van Leeuwen (see p. 203) had given a series of lectures at St Thomas's Hospital which John Freeman of St Mary's had attended when he was so impressed by van Leeuwen's suggestion that mould spores were able to cause asthma that he was able to arrange for the appointment of a young Irish mycologist, Charles J. La Touche, as an assistant. La Touche isolated moulds from the homes of asthmatic patients and prepared fungal extracts used by Freeman for desensitisation purposes. Fleming referred his mould to La Touche, who identified it as *Penicillium rubrum* (an identification subsequently changed to *P. notatum* by Charles Thom, a specialist on the group). He also tested the antibacterial properties of a number of cultures of different fungi kept in culture by La Touche. All proved negative except for one penicillium, of eight strains examined, which 'possessed exactly the same cultural characters as the original one from the contaminated plate'. As Hare points out, Fleming's culture was an unusual one because when during the war the Americans undertook a world-wide search for penicillin-producing strains of penicillia, Fleming's strain was found to be one of the three best of the many examined. If La Touche's strain produced penicillin as freely as Fleming's it is difficult to avoid the conclusion that they had a common source.

The dozen years of intermittent investigation which finally resulted in the isolation of penicillin and the clinical trials which established its value as a therapeutant being largely matters of biochemistry and medicine rather than mycology need not be reviewed here. This climax, commemorated by the award of the 1945 Nobel Prize for Physiology and Medicine to Fleming, the biochemist Ernst B. Chain, and Howard W. Florey, professor of pathology at the University of Oxford, ushered in a period of unsurpassed activity in the search for further antibiotics and techniques for their large scale manufacture.[14]

The novelty in the discovery of penicillin lay in the high toxicity of penicillin to a range of important disease-inducing bacteria and its stability and very low toxicity *in vivo*, which enabled it to be safely used as a therapeutant. The deleterious effect of one organism on another (antagonism) and also the restricting effect shown by the metabolic products of an organism on its own growth (the staling of cultures) had long been known. During 1872–3 William Roberts, a Manchester physician and a good Darwinian, who introduced the term antagonism into microbiology, was one of the first to note antibiotic action, possibly even that of penicillin, of which he wrote (Roberts, 1874):

the growth of fungi appeared to me to be antagonistic to that of *Bacteria* and *vice versa*. I have repeatedly observed that liquids in which *Penicillium glaucum* was growing luxuriantly could with difficulty be artificially infected with *Bacteria*; it seemed, in fact, as if the fungus played the part of the plants in an aquarium, and held in check the growth of *Bacteria*, with their attendant putrefactive changes. On the other hand, the *Penicillium glaucum* seldom grows vigorously, if it grows at all, in liquids that are full of *Bacteria*. It has further seemed to me that there is an antagonism between the growth of certain races of *Bacteria* and certain other races of *Bacteria*...when an organic liquid is exposed to the contamination of air or water...There is probably...a struggle for existence and a survival of the fittest. It would be hazardous to conclude that because a particular organism was not found growing in a fertile infusion, that the germs of that organism were really absent from the contaminating media.

Rather similar observations were made by Joseph Lister in 1871 and by John Tyndall and T. H. Huxley in 1875.[15]

In the years following 1940 there was a feverish testing of thousands of microfungi for antibiotic properties but then and since none of the many new and interesting antibiotics discovered equalled penicillin, usually because of higher toxicity to man. The first major success to follow penicillin was streptomycin,[16] produced by the soil actinomycete *Streptomyces griseus* and subsequently actinomycetes (and other bacteria) were found to yield a great variety of antibiotics. So too were fungi and in a review in 1951 P. W. Brian was able to list no less than 96 named and apparently distinct antibiotics produced by fungi and of these 57 were well characterised. During the same period hundreds of other fungi representing many taxonomic groups (particularly Hyphomycetes and Agaricales *sensu lato*)[17] had been observed to show antagonistic effects against bacteria, fungi, or other organisms.

Most of the antibiotics produced by fungi are of mainly academic interest, very few have found wider application. The effect of adding *Gliocladium virens* to soil in controlling damping-off of citrus seedlings had been noticed by Weindling in 1932 and subsequently Weindling & Emerson (1936) isolated an anti-bacterial and antifungal antibiotic, gliotoxin, from the culture filtrate of this fungus. Patulin (from *Penicillium patulum*) was later advocated by George Smith and others for the control of *Pythium* infection in plants (and a claim was made for success against the common cold in man). Griseofulvin (from *Penicillium griseofulvum* and other *Penicillium* species) was first characterised by A. E. Oxford *et al.* in 1939 and later independently discovered by Brian as the 'curling factor' which distorted the growth of chitinised fungal hyphae. This antifungal antibiotic has found limited application as a systemic fungicide for the control of grey mould (*Botrytis cinerea*) in lettuce but it is now commercially available under many proprietary names as an effective therapeutant when taken orally against ringworm infections in man and animals. Another group of important fungal antibiotics is the antibacterial cephalosporins (from *Cephalosporium* spp.) which have come into prominence since 1952 and are now manufactured in large quantities.[18]

The commercial production of edible fungi has already been touched on. So too have the major industrial uses of fungi in brewing, wine-making, and the manufacture of various foods and in the production of antibiotics, but a number of examples of the commercial exploitation of fungi remain to be considered. The historically most famous of these is citric acid manufacture.

Organic acids

It was in 1893 that Carl Wehmer, professor at Hannover, demonstrated citric acid production by penicillium-like moulds for which he proposed the new genus *Citromyces*, a name later synonymised with *Penicillium* by Charles Thom. Two years before, Wehmer (1891) had shown oxalic acid to be produced by *Aspergillus niger* but he did not discover that this mould too produces citric acid, and in larger quantities than penicillia. The commercial supply of citric acid remained dependent on the natural product extracted from citrus fruit until the foundation for the production of citric acid by *A. niger* on the large scale, was laid by the American J. N. Currie, in 1917, who showed that strains of *A. niger* would grow at an initial pH of 2.5–3.5 and produce sufficient citric acid to lower the pH to below 2.0, no oxalic acid being formed as at pH 4–5 when oxalic acid replaces citric acid. Within a decade virtually all commercial citric acid was manufactured by the *A. niger* fermentation of sugar, a process covered by many patents and still subject to a certain secrecy.

One essential factor determining the success of all such fermentation processes is the use of strains of micro-organisms which give high yields of the required products. This criterion though frequently met by empirical selection has also initiated research which has deepened knowledge of fungal genetics and introduced the concept of strain specificity. An equally important factor is pure culture maintenance. This can constitute a major problem and the difficulties associated with the preparation and handling of very large volumes of sterile media and its aeration are frequently considerable. They were a major cause of delay in the introduction of many microbiological processes on a commercial scale. The very low pH of the medium in citric acid production is unfavourable for the growth of most contaminating micro-organisms but for many antibiotics the best medium is frequently favourable for the growth of a wide range of fungi and bacteria. It was unfortunate experiences in the early days of the antibiotic industry that led to the recruitment of mycologists who in collaboration with chemical engineers devised and perfected methods for the sterilisation, inoculation, and incubation of large quantities of media and its subsequent processing. These developments enabled the at one time universal method of surface culture to be replaced by submerged culture in very large tanks. Submerged culture has found many applications and has been used to supplement the conventional cultivation of mushrooms by the production in bulk of mushroom flavoured mycelium for incorporation in soup.

Other organic acids derived from the activities of fungi have included gallic, gluconic, itaconic, and D-lactic acids. The first was the subject of van Tieghem's monograph of 1867 in which he described his investigations on the production of gallic acid from tannin by *A. niger* and it was Calmette's German patent 129164 of 1902 that put the process on a commercial basis. Gluconic acid, widely used as the calcium salt to correct deficiencies in calcium nutrition, was first discovered in culture filtrates of *A. niger* (as *Sterigmatocystis nigra*) by Molliard in France in 1922. Large-scale production was developed in the United States at first by the surface culture of *Penicillium purpurogenum* var. *rubri-sclerotium* and subsequently by *P. chrysogenum* and *A. niger* and a method of submerged culture devised by Herrick, Hellbach & May (1935) was a forerunner of the major developments in submerged culture technique made by the antibiotic industry.

Itaconic acid, an unsaturated dicarboxylic acid, was first characterised by Kinoshita in Japan in 1929 as a metabolic product of *Aspergillus itaconicus* but the commercial product is obtained from *A. terreus* following studies in England and the U.S.A. while D-lactic acid is produced from *Rhizopus oryzae* (Ward *et al.*, 1936).

Fats

In the same way that pressures imposed by the Second World War accelerated the development of penicillin production, shortages of essential materials during the 1914–18 war led to the microbiological production of glycerol for explosives. As long ago as 1859 Pasteur had shown that small amounts of glycerol were formed during alcoholic fermentation of sugars by yeast (*Saccharomyces cerevisiae*). Details of the German process which yielded a million tons of glycerol a month towards the end of the war, were published by Connstein & Ludecke in 1919. Similarly, lack of fats led to their commercial production by another yeast – *Endomyces vernalis* (Haehn & Kintoff, 1924) but the process proved too costly for peacetime use.

Enzymes

Since 1894 when J. Takamine in the United States took out a patent for the production from aspergilli of the *Aspergillus flavus-oryzae* series of enzyme mixtures, mainly diastase (marketed as Takadiastase and other trade names for use in medicine and for desizing textile fibres; see Takamine, 1898) this and other enzymes, including amylases from strains of *A. niger* (see Le Mense *et al.*, 1947), produced by fungi have been put on the market.

Steroid transformations

Chemists have made use of the ability of some fungi to distinguish closely related compounds or to bring about chemical changes that are difficult to accomplish *in vitro*. The classical example of the former was the use by Pasteur, to obtain the D-form of ammonium tartrate from a racemic mixture, of the preference shown by strains of *Penicillium* for the L-form. An important example of the latter is the discovery made by Peterson & Murray (1952) that *Rhizopus arhizus* was able to hydroxylate the sterol progesterone at carbon-11. Since then a number of other fungi have been found capable of modifying the structure of steroids, a property that has proved important in the development of physiologically active compounds for steroid therapy.

9. Distribution of fungi

GEOGRAPHICAL DISTRIBUTION

Fungi are ubiquitous and the distribution of many common moulds is world wide. The basis of the knowledge of the geographical distribution of fungi has been derived from the long tradition of naturalists and mycologists to publish lists, and often at the same time to describe, the fungi found in the localities where they live and regions which they visit. Such studies when carefully made, even when repetitive, do much to maintain the interest in fungi of successive generations of workers who by their current listing update classifications, correct past errors, and add many hitherto undescribed forms to the primary mycological census which is still so very incomplete. Also, many research workers have received their initial mycological stimulus by collecting fungi in the field. It would be impossible, except as a catalogue, to notice even the more important of the vast number of publications on this aspect of mycological endeavour and in what follows only representative examples are given to illustrate the development of this branch of mycology.

The first regional account of fungi was that by Clusius in 1601 in his *Rariorum plantarum historia* of the fungi of Pannonia (formerly the Roman province comprising parts of the modern Austria, Hungary, Czechoslovakia, and Yugoslavia) made while residing in Vienna (see p. 41). Little was published during the rest of the century but 1675 saw the appearance of van Sterbeeck's *Theatrum fungorum* in Antwerp and in 1690 John Ray's *Synopsis methodica stirpium Britannicarum* (and later editions during the next 34 years), included a list of fungi. In the early years of the eighteenth century J. Loeselius (1703) catalogued the fungi of Borussia (East Prussia, now Poland) as did Dillenius for the Geissen region, J. C. Buxbaum (1721) for Halle, and S. Vaillant for the environs of Paris in his posthumous *Botanicon parisiense*, 1727. Micheli's *Nova plantarum genera*, 1729, began the census for Italy to which A. J. A. Battarra's attractively illustrated account of the fungi of Rimini, 1755, was a notable supplement. The next 150 years saw the publication of a flood of regional works on the fungi of Europe, especially of Germany, France, and Italy. In Germany the eighteenth century works by Gleditsch (1753), Schaeffer (1762–74), Batsch (1783–9), and Tode (1790–1) prepared the way for the major floras of Fuckel (*Symbolae mycologicae*, 1869–75) on Rhenish fungi, Schroeter on the fungi of Silesia in Cohn's *Kryptogamenflora* (1885–1908), and the monumental *Kryptogamenflora von Deutschsland, Oesterreich und der Schweiz* by Rabenhorst and others (1884–1960) which proved itself internationally an outstanding taxonomic achievement. In France during the last decade of the eighteenth century, Bulliard's *Histoire des champignons* (1791–1812) and Paulet's *Traité* (1790–3) laid a foundation for the notable publications associated with the names of Quélet (*Les champignons du Jura et les Voges*, 1872–6), Gillet (*Les champignons (Fungi, Hyménomycètes) qui croissent en France*, 1878–90) and Patouillard (1883–9) in the last quarter of the nineteenth century and Boudier (1905–10) and others in the twentieth. Viviani (1834–8), Vittadinni (1835), and Venturi (1845–60) were among the forerunners of P. A. Saccardo (Fig. 85) (see p. 280) and the Abbé Bresadola (known especially for his *Fungi Tridentini*, 1881–1900 and *Icones*, 1927–60) in Italy where seventeen parts of the *Flora Italica cryptogama* issued between 1906 and 1938 have dealt with fungi.

At the same time studies were being initiated, if on a smaller scale, in many other European countries, and with most the names of one or more pioneer mycologists are permanently associated. The classical instance is that of Elias Fries with Sweden. Others include the name of P. A. Karsten with Finland and T. Holmskiold (for his superbly illustrated *Beata ruris otis fungis Danicis impensa*, 1790–9) with Denmark where later E. Rostrup made a major contribution to the knowledge of the regional fungi (see Lind, 1913). Albrecht von Haller (1768), Secretan (1833), and Trog (1845–50) all gave accounts of Swiss fungi and did much to initiate a strong local mycological tradition while Krombholz (1831–46) and Kalchbrenner (1873–7) were notable early students of fungi in Czechoslovakia and Hungary, respectively. Jean Kickx laid a foundation for the mycology of Belgium in the two-volume posthumous cryptogamic flora brought out by his son (Jean Kickx *fils*) in 1867 and

n. 1845 P. A. Saccardo ✝. 1906

Fig. 85. Pier Andrea Saccardo (1845–1920).

C. A. J. A. Oudemans did the same for the Netherlands in a series of contributions extending from 1873 to 1903 which he consolidated in his *Catalogue* of 1904. Weinmann (1836) gave an account of the larger fungi of Russia where mycological development was slow.

The first comprehensive list of English fungi was that by the Rev. M. J. Berkeley (1836), which contains much original observation and thought, and M. C. Cooke's two-volume *Handbook of British fungi*, 1871, is still the most recent attempt to compile all British fungi, only four volumes of the *British fungus flora* by George Massee (1892–5) having been completed. Robert Greville made important contributions to Scottish mycology and in 1879 the Rev. John Stevenson compiled *Mycologia Scotica*, a list of fungi indicating their distribution both within Scotland and world wide.

Biologists have frequently shown a tendency (or a desire) to travel. For example, Clusius was a much travelled man for his time. John Ray, after a series of journeys through much of Great Britain, made an extended European tour during 1663–6, while in the early eighteen-twenties what Elias Fries 'wished most of all...was to visit more distant countries, that however the straightness of my circumstances did not allow, and in those days – how much more fortunate you, who study nowadays! – there was no hope of obtaining public aid. Therefore I could not get about further than my feet would carry me. But at that time, if you were tall and a good walker and in good health, it was not impossible to walk 75 km in twelve hours.'[1] Corda made a collecting trip to Texas in the autumn of 1848 (his last letter to Berkeley was written from Cowes, Isle of Wight, where his boat from Bremen had put in en route for New Orleans) but he found Houston (where he 'collected a lot') – then a settlement of three to four thousand inhabitants – 'extremely tasteless', its geographical location unfavourable, and surmised that 'it will disintegrate just as quickly as it rose'![2] Unfortunately, Corda and his Texan collections perished in the Carribean on the homeward voyage in 1849.

In spite of such activities there was relatively little mycological work undertaken in other countries up to the end of the nineteenth century. European mycologists did, however, examine increasing numbers of fungi from other parts of the world collected by travellers or the results of voyages of exploration. For example, the Hookers, both father and son, who were successively director of the Royal Botanic Gardens, Kew, routinely submitted fungal specimens sent to Kew to the Rev. Berkeley who from this and other sources, such as the fungi collected by Darwin on the voyage of the *Beagle*, published accounts of fungi from North America, the West Indies, South America (including British Guiana, Surinam, Brazil, Chile, and Tierra del Fuego), Pacific Islands, Kerguelen Islands, Australia, Tasmania, the Philippines, India, Ceylon, Cape of Good Hope, Greenland, and the Arctic. Saccardo in Padua and Paul Hennings in Berlin also published on many exotic fungi and in reverse, Fidalgo (1968) when reviewing the history of mycology in Brazil noted some thirty-five European authors who had given accounts of Brazilian fungi up to the opening years of the twentieth century and before the serious cataloguing of Brazilian fungi began locally, mainly as a post-1940 development by plant pathologists such as Bitancourt and Viégas in the state of São Paulo and the late A. Chaves Batista at the University of Recife.

Fig. 86. Lewis David von Schweinitz (1780–1834). (*Ex* C. L. Shear, *Plant World*, **5**:45 (1902).)

Schweinitz (1834) (Fig. 86) initiated the study of fungi in North America and later outstanding contributions were made by Charles Peck in his long series of annual reports from 1867 to 1915 as botanist to the State Cabinet of Natural History (later the New York State Museum) at Albany. In South America the versatile Carlos Spegazzini, an Italian who emigrated to Argentina where he became professor at the University of La Plata, laid the foundation for studies of the fungi of Argentina, Chile, and Uruguay. All aspects of the study of fungi in Canada owe much to the work and inspiring teaching of A. H. R. Buller (Fig. 57) who, like that other eccentric batchelor Samuel Butler, left himself excessively well documented.[3] Buller was born in Birmingham, England, where he was a student at Mason College, and after post-graduate training in Germany and a short period on the staff of Birmingham University he was appointed in 1904 to the chair of botany at the University of Manitoba, Winnipeg, where he remained for the next thirty-two years. He spent most of each summer in England working at

Birmingham or Kew and crossed the Atlantic by sea no less than sixty-five times. He had hoped to end his crossings on an even number but died at Winnipeg in 1944.

During the twentieth century, major additions have been made to the knowledge of the geographical distribution of both pathogenic and saprobic fungi by plant pathologists. An understanding of the distribution of pathogens is essential for the drawing up of appropriate import and export regulations and quarantine procedures designed to prevent the spread of plant disease but, as indicated in Chapter 6 (p. 143), during the first half of the present century there was much emphasis on the pathogen. Plant pathologists were frequently able mycologists and when posted to what are now known as the developing countries – usually tropical – they made general collections of fungi, particularly microfungi, and major contributions to the mycology of the region.

This was particularly true for India, British Africa, and other parts of the British Empire.[4] In 1901 E. J. Butler (Fig. 104), after a medical training, went to India where in 1906 he became the first Imperial Mycologist. By the time he left in 1919 to become the first director of the Imperial Mycological Bureau (later the Commonwealth Mycological Institute) at Kew he had, in collaboration with H. Sydow of Berlin and others, made major additions to the knowledge of Indian fungi resulting from the earlier pioneering work of D. D. Cunningham (see p. 200) and A. Barclay, two Indian army medical officers, published between 1871 and 1897. These results were summarised in a list, *The fungi of India*, compiled with the help of G. R. Bisby and published in 1931. Tom Petch as botanist and mycologist to the Department of Agriculture, Ceylon, did similar work for that island and two years after his death in 1948 a complementary volume to Butler's on the fungi of India appeared (Petch & Bisby, 1950).

M. C. Cooke (1892) compiled a somewhat unsatisfactory account of the fungi of Australia where the fungus census was initiated locally by Daniel McAlpine who, after a training under T. H. Huxley and the botanist Thistleton-Dyer at the Royal College of Science in London and seven years experience as a lecturer at the Heriot-Watt College in Edinburgh, emigrated to Australia to take up a lectureship in the University of Melbourne. In 1890 he became Vegetable Pathologist to the Department of Agriculture, Victoria and in 1895 published the first

Australian list of Australian fungi. Subsequently he published volumes on the rusts (1906) and the smuts (1910) of Australia as well as phytopathological topics. The outstanding contributor to the mycology of New Zealand was G. H. Cunningham, the first government mycologist in New Zealand, who as a side line to his plant pathological duties monographed the rusts, smuts, gasteromycetes, polypores (1965), and thelephoraceous (1963) fungi of the country.

In the Far East the beginnings of mycology were delayed. The isolation of Japan from contacts with the Western world from the end of the sixteenth century to the arrival of the American trading expedition, led by Commodore Matthew Perry, in the Bay of Yedo on 8 July 1853 was virtually complete. Manuscript accounts of Japanese fungi did appear during this period – the most famous being *Kimpu* (or *Kinbu*) by Konen Sakamoto in 1834 which included descriptions of nearly a hundred species accompanied by watercolour drawings – and a few fungi had been included in various European publications on Japanese plants. It was not until 1886 when M. Shirai became a professor at the University of Tokyo, teaching plant pathology, that the study of mycology in Japan began in earnest and by 1905 Shirai was able to publish a list of 1200 species of Japanese fungi. Less is known about the fungi of China,[5] but there have been several recent lists of the fungal pathogens of plants. Penzig & Saccardo (1897–1904) described and illustrated fungi from Java.

A minor geographical specialisation has been the study of polar fungi - particularly those of the Arctic. Much of this work has been done in the herbarium by the examination for microfungi of plants collected on polar expeditions. Examples of such investigations are provided by the two papers by Fuckel (1874) and by the more comprehensive monograph by Lind (1934) who listed 422 species of microfungi found on specimens of arctic plants in the herbarium of the Botanical Museum at Copenhagen.

ECOLOGICAL DISTRIBUTION

In addition to general listing mycologists have always paid special attention to the fungi of particular habitats. As detailed in Chapter 6, fungi parasitic for plants, animals, and man have attracted much interest and in Chapter 7 aspects of the fungi found in air were touched on.

Soil fungi

The ecological fungal group first noted, and one which has received perennial attention since, was subterranean soil fungi. Truffles were the subject of a little monograph by Alfonso Ciccarelli in 1564. Ray (1686–1704, **1**) dealt with 'De Fungis subterraneis' and van Sterbeeck (1675) devoted a chapter to 'Aerd-buylen' which included not only truffles but potato tubers and the root nodules of legumes. Vittadini (1831) in Italy monographed truffles and later (1842) dealt with the Hymenogastrales while hypogeous fungi were the subject of the Tulasnes' classic account of 1851. Since then there have been numerous treatments of the hypogeous fungi found in many European countries, in North America, and elsewhere.[6]

The investigation of the microfungi of the soil had to await the development of pure culture methods and a number of special techniques. The dilution plate technique proved itself the most popular, and interest in this line of approach attained a peak in the nineteen-twenties when a standardisation of the method by W. B. Brierley and his colleagues at Rothamsted Experimental Station (Brierley, Jewson & Brierley, 1927) attracted attention. Direct microscopic examination of soil was first undertaken by H. J. Conn (1922) who observed mycelium in soil. Subsequently, Rossi (1928) sampled soil fungi by staining the micro-organisms adhering to a clean microscope slide after it had been pressed against a recently exposed soil surface, and two years later Cholodny (1930) modified this technique by leaving the slide in contact with the soil for varying periods before laboratory examination. H. L. Jenson (1935), in Australia, used the Rossi–Cholodny technique to make quantitative assessments. Among other techniques employed (the review by Chesters (1949) may be consulted for details) mention need be made only of the valuable 'soil plate' method, devised by J. H. Warcup (1950), by which small fragments of the soil to be examined are placed on the surface of agar medium in a petri dish and subcultures made from the fungal growth developing from the inoculum. One result of the deployment of this last technique was the isolation of basidiomycetes from soil.

Apparently, the first intentional isolations of soil fungi were made by Adametz in 1886, who obtained a number of different species by inoculating flasks of sterile media in the field with different types of soil, and although others confirmed and extended these findings little progress in the understanding of the role of fungi in soil was made before the second decade of the present century. It was the American microbiologist Selman A. Waksman of Rutgers University in the course of studies which culminated in the two editions of his famous *Principles of soil microbiology* (1927, 1931) who, in 1916, established that fungi did live in the soil and were not merely contaminants derived from the deposition of airborne spores, and did play an important part in the essential biological processes of the soil. The next year he distinguished two classes of soil fungi – the 'soil inhabitants' and the 'soil invaders' (Waksman, 1917). This concept was endorsed and elaborated by others, particularly S. D. Garrett (see Garrett, 1944) who has contributed so much to the development of studies on the ecology of soil fungi, a topic of much current mycological interest.

Two related topics are fungi of sand dunes (the simplest of soils) and coprophilous fungi. Although larger fungi had been recorded on sand dunes from the time of Linnaeus it is only since the publication by Olof Andersson in 1950 of his minor classic on the mycology of sand dunes in Scandinavia that the subject has come into prominence.

Coprophilous fungi have for long received attention. The first major treatment of these forms was that of the coprophilous fungi of Denmark by E. C. Hansen in 1876–7 and twenty years later, according to the monograph of Massee & Salmon (1901–2), Saccardo compiled in the *Sylloge* (1897, **12**(1):873–902) no less than 757 species of coprophilous fungi distributed in 187 genera. There is now a large literature covering many aspects of their taxonomy and biology.[7]

Myxomycetes

It is perhaps permissible to treat myxomycetes as an ecological group. They are of world wide occurrence on rotting wood and moist decaying vegetation. Many mycologists have specialised in their study since the time of Micheli who in *Nova plantarum genera*, 1729, described and illustrated a number of myxomycetes for which he proposed several new generic names, including *Lycogala* and *Mucilago*. Bulliard (1791–1812) illustrated additional species but for many years myxomycetes were confused with gasteromycetes and mucoraceous fungi and it was de Bary (1866) in the first edition of his textbook who first treated Myxomycetes (or Mycetozoa as he called them) as an independent taxon. In 1874–5 one of de Bary's students, J. T. Rostafinski, a Pole, published the first comprehensive monograph on myxomycetes, the *Śluzowce monografia*. This stimulated M. C. Cook in England to teach himself Polish so that he could translate descriptions for British species which he published, together with twenty-four plates (twenty-three of them reproductions of Rostafinski's figures), as *The myxomycetes of Great Britain* in 1877. A few years later Arthur Lister and his daughter Gulielma, two English Quakers, began sixty years of study which established them as world authorities on the myxomycetes. Arthur Lister, a retired wine merchant, took up the study of myxomycetes in 1887 (his fifty-seventh year) and after working over the collections in the British Museum (Natural History) and at Paris and the de Bary collection at Strasbourg, with the help of his daughter as amanuensis, his *Monograph of the Mycetozoa* appeared in 1894. The subsequent editions of 1911 and 1925 were brought out by Miss Lister. In a series of seventy-five notebooks extending from January 1887 to August 1947[8] the Listers left a unique day-to-day record of their researches and their voluminous correspondence with almost all the leading contemporary students of the group. These included authors of the well-known North American texts – T. H. MacBride, R. Hagelstein, and G. W. Martin – and also the Emperor of Japan in whose honour Miss Lister named *Hemitrichia imperialis*, based on a specimen found in the palace gardens.

Fungi of caves, mines, and cellars

The abnormal fruit-bodies produced by larger fungi in cellars and mines have for long attracted attention. The earliest record seems to be the illustrated account, given by John Martyn, professor of botany at Cambridge, to the Royal Society in January 1745, of an abnormal fructification of *Polyporus squamosus* found growing on an elm log in a London cellar. Later, a similar English record was made by Bolton (1789: tab. 138; *Polyporus rangiferinus*) from Leeds. Both these authors treated the teratological forms as distinct species. J. A. Scopoli (then professor of chemistry at the mining academy at Schem-

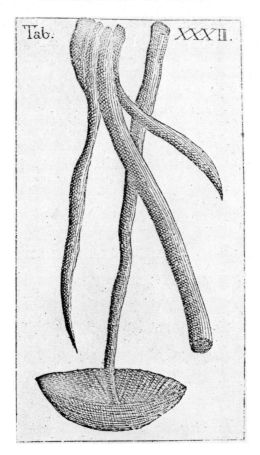

Fig. 87. *Lentinus lepideus* from a cave. (J. A. Scopoli (1772): pl. 32.)

nitz, Hungary) in his *Plantae subterraneae* of 1772 noted seventy-five fungi from caves including the horn-like apileate sporophores of what must have been *Lentinus lepideus* (see Fig. 87). Other eighteenth-century records were made by Alexander von Humboldt[9] in his *Flora Fribergensis*, 1793, which included observations made while he was a student at the Freiberg Academy of Mines and later as the 23-year-old chief inspector of mines for the Prussian government.

Rhinocladiella cellaris, the mould commonly known as the 'cellar fungus' and up to 1971 usually under the binomial *Racodium cellare*, was the first microfungus to be noted in cellars, particularly wine cellars, where it forms large mycelial masses. It was recorded by Ray in the second (1696) and third (1724) editions of his *Synopsis methodica stirpium Britannicarum* (as *Byssus*) and also by Micheli (1729: 211, tab. 89, fig. 9) who gave an illustration. Later it was treated in some detail by Guéguen (1906). During 1913 to 1922, J. Lagarde published a series of papers on fungi found in the caves of southern France and the Pyrénées and added a number of miscellaneous records. There are probably no fungi confined to caves. All are invaders.

An interesting recent association of fungi with caves was, as already noted in Chapter 6, that of the pathogenic *Histoplasma capsulatum* with the large deposits of bat guano found in certain caves in Africa, Central America, and elsewhere which have provided the source from which speleologists have contracted histoplasmosis (see p. 179). Another unwelcome invader was *Sporothrix schenckii* which in the early nineteen-forties was responsible for a major outbreak of sporotrichosis among workers in the gold mines of the Rand in South Africa (see p. 177).

Aquatic fungi

Aquatic fungi, particularly those of fresh water, have been studied for many years. The most important taxonomically were the phycomycetes which, as the name implies, were believed to be phylogenetically related to the algae with which they were at first frequently confused. These aquatic forms are typically zoosporic – but as noted in Chapter 3 (p. 62) zoospores were first observed in fungi in 1807 by Prévost in the terrestrial phycomycete *Albugo*. Nineteenth-century students of these forms included such well-known workers as A. Braun, de Bary, M. Cornu, W. Zopf, and P. A. Dangeard whose taxonomic findings were consolidated by M. von Minden, together with his own, in his contributions between 1911 and 1915 to the *Kryptogamenflora der Mark Brandenburg*; an account which was the forerunner to the later comprehensive monograph by F. K. Sparrow (1943), which excluded the Saprolegniaceae and Pythiaceae because they had already been treated by W. C. Coker (1923) and Velma D. Matthews (1931), respectively. These studies revealed many complicated life histories and much detail was added to knowledge of the sexuality and cytology of fungi.

A major new development began in 1942 when C. T. Ingold

described a series of aquatic hyphomycetes growing on decaying leaves in an English stream. These fungi, which proved to have a world-wide distribution in both temperate regions and the tropics, have been studied intermittently by the discoverer ever since and have provided the basis of a minor mycological industry for others.

Studies on marine fungi came later. A. D. Cotton in 1909 summarised the few records up to then. Soon afterwards Sutherland described additional marine pyrenomycetes.[10] E. S. Barghoorn and D. H. Linder's paper of 1944 on the marine fungi of California marks the beginning of the modern interest, an interest which both stimulated and was stimulated by T. W. Johnson and F. K. Sparrow's *Fungi in oceans and estuaries* of 1961.

Associations with insects and other invertebrates

The study of entomogenous fungi – that is, of fungal parasites, symbionts, and saprobes of insects – is one of the better known by-ways of mycology. Relationships of insects to fungi, especially that of the insect host to its fungal pathogen, had to be solved in establishing the status of fungi (see p. 25) and Agostino Bassi's elucidation of the muscardine disease of silkworms in the 1830s (see p. 163) is a landmark in the history of microbiology. The large literature generated in this field has never been the subject of a book-length review but the interest during recent years in attempts to effect the biological control of insect pests of plants by pathogenic fungi (particularly by species of *Entomophthora*) provided a stimulus for the publication of *Pilzkrankheiten bei Insekten* by E. Müller-Kögler in 1965.

Knowledge of several of the major associations of fungi with insects and other invertebrates has frequently been derived from the work of individual specialists. An outstanding example is the Laboulbeniomycetes. Members of this group of ectopara-sites were first noted in 1840 by the French entomologists Alex Laboulbène and Auguste Rouget whom Montagne and Robin commemorated in the names for the genus *Laboulbenia* and its type species *L. rougetii* (Robin, 1853:622). There were subse-quent investigations by Peyritsch and Berlese but contributions to a knowledge of the laboulbeniomycetes are dominated by those made by Roland Thaxter from Harvard University over a

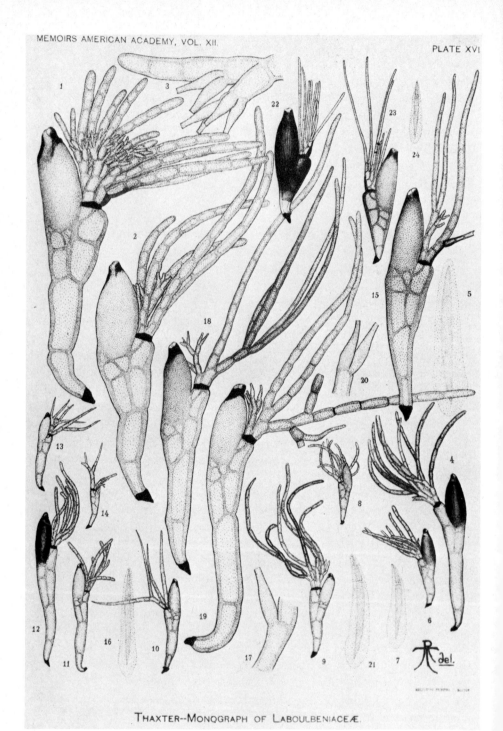

THAXTER--MONOGRAPH OF LABOULBENIACEÆ.

Fig. 88. *Laboulbenia* species (*L. elongata*, 1–14; *L. europaea*, 15–7; *L. pterostichi*, 18–21; *L. parvula*, 22–4). (R. A. Thaxter (1896–1931) **1**: pl. 16.)

period of more than forty years and especially by his monumental account published in five parts between 1896 and 1931 (see Fig. 88). Similarly knowledge of the complex parasitic relationship of *Septobasidium* with scale insects in the tropics and subtropics is based on J. Couch's beautiful monograph of 1938, while the initiation of modern developments on studies of predacious fungi (both zoopagales and hyphomycetes) (Fig. 89) will always be linked with the name of Charles Drechsler for the many papers published by him since 1933 (from 1934 mostly in *Mycologia*). Another long-overlooked but ubiquitous group of fungi, associated as parasites or commensals with the cuticle or the gut lining of many small arthropods, is the Trichomycetes, studies on which began in France with the monograph published by J. F. Manier in 1951.

Two other well-known entomogenous associations of fungi are those with ambrosia beetles and termites. It was in 1836 that J. Schmidberger[11] designated the white substance lining the burrows of certain wood-boring beetles (on which the larvae feed) 'ambrosia' which T. Hartig in 1844 recognised to be of fungal nature. Since then the involvement of a number of different genera of monilioid fungi has been established; mostly Endomycetales (see L. R. Batra, 1967), which not infrequently show a specificity for particular species of beetles. The association of the fungi with the beetles is very intimate and the adults, especially females, of certain Scolytidae have in their exoskeletons special cavities or *mycangia*[12] in which there is a growth of the appropriate fungus from which new tunnels can become infected. It has been observed since the eighteenth century[13] that similar fungal growths occur in the tunnels of the large nests (termitaria) of white ants but the modern view is that these growths do not provide food for the larvae but are those of a series of agarics of the genus *Termitomyces* (proposed by Heim in 1942) which have become adapted for this ecological niche and are tolerated by the termites.

Lichens

Lichens, too, are cosmopolitan and lists of those of many geographical regions and habitats have been compiled.[14] They are characteristically opportunistic and are able to colonise sites, such as polar and alpine regions, rock faces, and other exposed surfaces, which are inhospitable for most other organisms. They are also, as recognised for more than a hundred years,[15] sensitive to air pollution and during the last decade a large literature has developed on this aspect of lichenology[16] from attempts to correlate the occurrence and composition of lichen communities with the degree of atmospheric pollution.

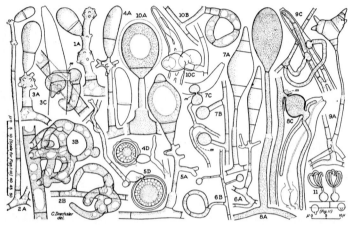

Figs. 1–10.—Various nema-capturing fungi, each numeral denoting a separate species, and all species drawn with the aid of the camera lucida at the same magnification; ×500. A, Conidiophore (shown completely only in Fig. 2) with attached conidium of approximately average size, shape and condition with respect to septation. B, Organs of capture, either adhesive hyphal loops or adhesive knob-cells. C, Internal disruptive development of fungus, or its external constrictive swelling, Figs. 3, 7 and 8 showing condition at the time the animal's movements ceased. D, Intramatrical resting reproductive structures. m, Adhesive mucous substance. Fig. 11.—Portion of fertile branch of *Harposporium anguillulae*, ×1000.

Fig. 89. Drechsler's first illustrations of predacious fungi. (C. Drechsler (1933): figs. 1–11.)

10. Classification

Although among the oldest, and certainly the most popular, aspect of mycological endeavour, classification has been relegated to the penultimate chapter of this survey because, as all characters of fungi merit consideration as taxonomic criteria, classification is best seen in perspective when treated as a climax. But before considering the development of fungal classification in detail a digression on the relationship of taxonomy to classification may be helpful.

Taxonomy, according to Cowan who equates the discipline with systematics, 'cannot be considered a science...but rather a scientific art intended to bring order into what otherwise would be untidy or disorderly'.[11] If few taxonomists would give unqualified assent to this stricture, all would agree that there is much looseness in the current usage of the term. In other words, there are differences of opinion on what taxonomy is and, in particular, its relationship to systematics and classification. Heslop-Harrison's definition of *taxonomy*, 'the study of the principles and practice of classification'[2] reflects the most widely held modern view. Taxonomy is thus only a part of *systematics* which (*fide* Heywood) is 'the scientific study of diversity and differentiation of organisms and the relatedness which exists between them';[3] biological *classification* being the ordering of organisms into hierarchically arranged groups (or *taxa*, sing. *taxon*[4]). It is, however, usual to designate as a taxonomist one who classifies after scientific investigation of the material under study as well as considers the principles underlying the practice of his profession. In addition, a taxonomist, because of his special knowledge, is frequently called upon to make identifications. He is also obliged to deal with questions of nomenclature, a topic dealt with in the next chapter.

Classifications are as many and diverse as the needs they are designed to meet. Augustin Pyramus de Candolle (*Théorie élémentaire de la botanique*, 1813: 26 *et seq.*) recognised two main types of classification as *empirical* – such as the alphabetical

Fig. 90. Jean-Paul Vuillemin (1861–1932). *Aet.* 63. (G. Percebois, *Ann. med. Nancy,* **9** February 1972.)

classification used in dictionaries and indexes – and *rational*. The latter he subdivided into 'practical classifications' based on properties in relation to their value to man, e.g. the classification of fungi into edible and poisonous; 'artificial classifications', such as Saccardo's 'Spore Groups' (see p. 64) designed to facilitate identification; and 'natural classifications' designed to express natural affinities. De Candolle considered the ultimate aim of biological classification to be the development of a natural classification, one that may be considered to epitomise knowledge of the organisms classified, a view subsequently re-enforced by the acceptance of the theory of evolution and one generally held today as the ultimate, if unattainable, goal. Buffon's dictum 'Se proposer de faire une méthod parfaite, c'est se proposer un travail impossible' is still true.

The most comprehensive historical account of the classification of fungi is that by Paul Vuillemin (1912) (Fig. 90) in which he reviewed systems based on morphology, ontogeny, phylogeny, and biological features, respectively. He differentiated classifications of the first category into 'morphographic' – classifications based on superficial external form (which are usually highly artificial) or on organography (which tend to be more natural) – and 'morphologic' – morphological in the narrower sense of based on anatomical and histological criteria. Biological systems included those which take into account the interaction between one fungus and another or between a parasitic fungus and its host and such biochemical evidence as serological findings. The summary of the evolution of fungal classification which follows is illustrated by only a selection of representative examples of the many attempts to classify fungi. Others are referred to by Vuillemin (1907, 1912) and, especially for larger fungi, by Richon & Roze (1885-9).

The first difficulty in classifying fungi was to recognise that they constitute a special group. For example, as noted in Chapter 3 (p. 38), in the herbal of Matthioli (1560), 'Tubera' (truffles), 'Agaricum' (*Fomes officinalis*) and 'Fungi' (agarics) are widely scattered and it was l'Obel in his *Kruydtboeck* of 1581 who first treated fungi as a single group. Following classical tradition, Clusius (1601) divided fungi into 'Fungi esculenti' (twenty-one genera) and 'Fungi noxii et perniciosi' (twenty-six genera) and the separation of edible from poisonous also underlay the compilation by G. Bauhin in his *Pinax*, 1623, and van Sterbeeck's classification in the *Theatrum fungorum*, 1675, the 'Eerste Deel' of which dealt with 'goede Campernoelien', the 'Tweede Deel' with 'quaede Fungi', while truffles were included with potatoes and other tubers in the 'Derde Deel' devoted to 'goede' and 'quaede' 'Aerd-buylen'.

The first major advance towards the classification of fungi on a more rational basis was made by the English naturalist John Ray in the first volume of his *Historia plantarum*, 1686-1704, in which the criteria for the separation into four sections of the ninety-three kinds of fungi he recognised were primarily ecological although he did introduce one morphological criterion. The first two sections comprised terrestrial fungi, the third arboreal, and the fourth hypogeous. Sections one and two were distinguished according to whether the fruit-bodies were lamellate or not, respectively, and tradition was again respected by subdividing the first section into edible and poisonous series.

Three years later, Pierre Magnol, who was born in Montpellier, where he became professor of botany, and in whose honour the genus *Magnolia* was named, made a more complete break with tradition when in *Prodromus historiae generalis plantarum in quo familiae plantarum per tabulis disponuntur*, 1689, he classified fungi together with mosses, ferns, seaweeds, and corals in his third family which was characterised by the absence of leaves. Magnol included sponges and Actinozoa (polyps) as fungi and in his primary subdivisions of the fungi proper into terrestrial, arboreal, and hypogeous he followed Ray. However, his secondary divisions into pileate, cancellate, and digitate forms, and the pileate into those lamellate below or porate, were based on structural criteria and for the first time the main morphological categories of the larger fungi began to be reflected in their classification.

Up to this point students of fungi and of plants in general had usually been content to delimit families and other suprageneric

Genre V.
Lycoperdon.

L A Veſſe de Loup eſt un genre de plante dont le caracte- Pl. 331. re peut être établi dans la figure de ſes eſpeces. Ces ſortes de plantes ſont des veſſies membraneuſes A B C D E G, qui en ſe crevant répandent une pouſſiere tres-fine. Il y en à qui ſont ſoûtenues par un pedicule aſſez long, comme l'eſpece E qui eſt ſoûtenue par le pedicule F. On en trouve quelques autres qui ſont envelopées d'une capſule aſſez forte qui en ſe crevant devient un baſſin recoupé en pluſieurs parties, comme on le voit en la figure H, & laiſſe voir la Veſſe de Loup G.

Les eſpeces de Veſſe de Loup ſont,
Lycoperdon Alpinum maximum.
Lycoperdon vulgare. *Fungus rotundus orbicularis* C. B. pin. 374.
Lycoperdon minus, & multiplex.
Lycoperdon Pariſienſe minimum pediculo donatum. *Fig.* E F.
Lycoperdon veſicarium ſtellatum. *Fig.* G H.

Fig. 91. 'Lycoperdon'. (J. P. de Tournefort (1694): 441, pl. 331.)

Pl. 331.

Lycoperdon. *Vesse de Loup.*

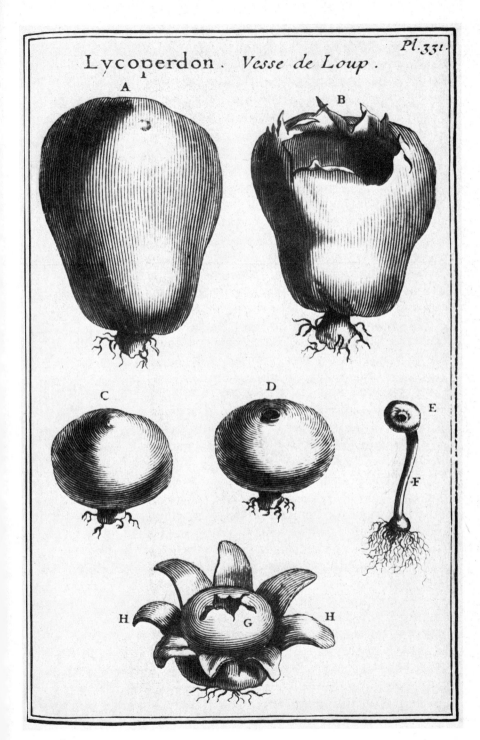

A

B

C

D

E

F

H   G   H

taxa and to employ few generic names. Clusius (1601), for example, divided the fungi he treated into genera which he differentiated merely by numbers. J. Bauhin (1651) used but two generic names, *Fungus* and *Tuber*, as did Ray (1686–1704), and G. Bauhin in the *Pinax* only used one more, *Agaricus*. One of the main contributions of the renowned Joseph Pitton de Tournefort (from 1683 professor of botany at the Jardin Royal des Plantes, Paris; he died in 1708 at the age of 52 from injuries received from a passing cart) was his emphasis on genera to which he gave names accompanied by concise descriptions, frequently illustrated, of their diagnostic characteristics. Tournefort is best known for his *Élémens de botanique, ou méthode pour connoîtres les plantes* published, in three octavo volumes illustrated by 489 copper plates, in 1694. A revised edition, in Latin, with rewritten generic diagnoses appeared in 1700 as *Institutiones rei herbariae*, issued in quarto and illustrated by basically the same series of plates. In the *Élémens* fungi (together with mosses, covered by the single genus *Muscus*) comprise Class XVII, characterised as 'Des Herbes dont on ne connôit ordinairement ni les fleurs, ni les grains', and they are treated under six generic names – *Fungus* (agarics and boletes), *Boletus* (morels, *Clathrus*, and *Phallus*), *Agaricus* (polypores and other bracket fungi, *Lycoperdon* (puff-balls), *Coralloides* (clavarioid forms), and *Tubera* (truffles) – and accompanied by seven plates (nos. 327–333). In the *Institutiones* a seventh generic name – *Fungoides* (cup-fungi) – was introduced without an illustration. In Fig. 91 Tournefort's illustrations of 'Lycoperdon', which clearly covered not only *Lycoperdon* of today but also *Bovista*, *Tulostoma*, and *Geastrum*, and his description of the genus are reproduced. The latter also provides an example of Tournefort's use of phrase-names for species, a practice in which he showed no advance on his predecessors whose names he frequently adopted.

The next important modifications in the general classification of larger fungi were made by Johann Jakob Dillenius who was born in Darmstadt in 1687 and, largely on the strength of his *Catalogus plantarum sponte circa Gissam nascentium* (published at Frankfurt in 1718 and re-issued the following year with an appendix), was, in 1721, invited to England by William Sherard to help with the preparation of Sherard's intended index of plants. After Sherard's death in 1728 he became the first

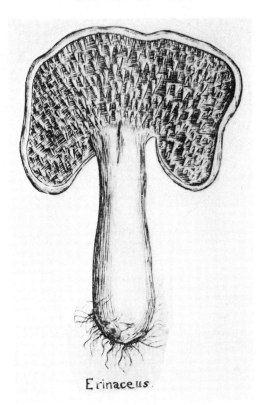

Erinaceus.

Fig. 92. First illustration of *Hydnum*. (J. J. Dillenius (1719): appendix p. 84.)

Sherardian Professor of Botany in the University of Oxford. The major contribution by Dillenius to mycological taxonomy is in his *Catalogus* where, as the arrangement of the plants is by their months of appearance, the 160 fungi included are dealt with (in eight genera) under October and November; lichens (and mosses) being assigned to December and January. Dillenius's primary division of fungi was into stipitate pileate forms and those lacking a pileus. The first of these groups he subdivided according to whether gills were present (*Amanita*; agarics) or absent, the last being subdivided into fungi with fruit-bodies characterised by spines (*Erinaceus*; Latin for hedgehog; the name introduced by Dillenius for *Hydnum* (Fig. 92)), pits (*Morchella*; a replacement of the *Merulius* of J. Bauhin and the *Boletus* of Tournefort for morels), or pores (*Boletus*; boletes). The non-pileate forms were divided into those with

247

(*Fungoides* (*Clavaria*, *Xylaria*)) or without a stalk, the three genera of the latter series being *Agaricus* (polypores, *Lenzites*, etc.), *Peziza* (cup fungi), and *Bovista* (puff-balls, truffles, etc.). Dillenius only used edible and poisonous, arboreal and terrestrial, for his lowest infraspecific categories. Subsequently, as the anonymous editor – his sponsors thought it unwise that the name of a foreigner should appear on the title-page – Dillenius updated the classification in the third (1724) edition of Ray's

Table 2. *Micheli's classification of lichens and fungi in the* Nova plantarum genera, *1729 (p. 73 et seq.)*

I
Plantae plerumque crustaceae, substantia vel coriacea, vel farinacea, vel gelatinosa, vel tartarea, floribus apetalis nudis a semine sejunctis.
  Lichen; *Lichen-agaricus* Mich. [*Xylaria*]; *Lichenoides* Mich. [*Pertusaria*].

II
Plantae simplicissimae, plerumque carnosae, anomalae, seu irregulares, flores apetalo, monostemone, scilicet unico filamento constante, a semine sejuncto.
  Agaricum [hymenomycetes with dimidiate fruit-bodies and pores, gills, or teeth]; *Ceratospermum* Mich. [*Sphaeria*, etc.]; *Linckia* Mich. [*Nostoc* (blue-green alga)]

III
Plantae simplicissimae, carnosae, regulares, floribus apetalis monostemonibus, seu unico filamento constantibus, a semine separatis.
  Suillus [*Boletus*]; *Polyporus* [polypores with central stripe]; *Erinaceus* [Hydnaceae with central stipe]; *Fungus* [agarics with central stripe]; *Fungoidaster* [*Craterellus*, *Leotia*, etc.]; *Phallus*; *Phallo-boletus* Mich. [*Morchella*]; *Boletus* [*Morchella*, *Mitrophora*, etc.]

IV
Plantae simplicissimae, plerumque non capitatae, seminibus in superficie ornatae.
  Fungoides [*Helvella*, various cup fungi]; *Clavaria* [*Clavaria* (unbranched), *Geoglossum*]; *Coralloides* [*Clavaria* (branched)]; *Byssus* [mycelium, filamentous algae]; *Botrytis* Mich. [*Botrytis*, etc.]; *Aspergillus* Mich. [*Aspergillus*, *Penicillium*, etc.]; *Puccinia* Mich. [*Gymnosporangium*, *Ceratiomyxa*]

V
Plantae simplicissimae, seminibus interna parte donatae.
  Clathrus Mich.; *Clathroides* Mich. [*Arcyria*]; *Clathroidastrum* Mich. [*Stemonitis*]; *Mucor* Mich.; *Lycogala* Mich. [*Lycogala*, *Reticularia*]; *Mucilago* Mich.; *Lycoperdon* [*Lycoperdon*, *Tulostoma*]; *Lycoperdoides* Mich. [*Polysaccum*]; *Lycoperdastrum* Mich. [*Scleroderma*]; *Geaster* [*Geastrum*]; *Carpobolus* Mich. [*Sphaerobolus*]; *Tuber*; *Cyathoides* [*Crucibulum*, *Cyathus*]

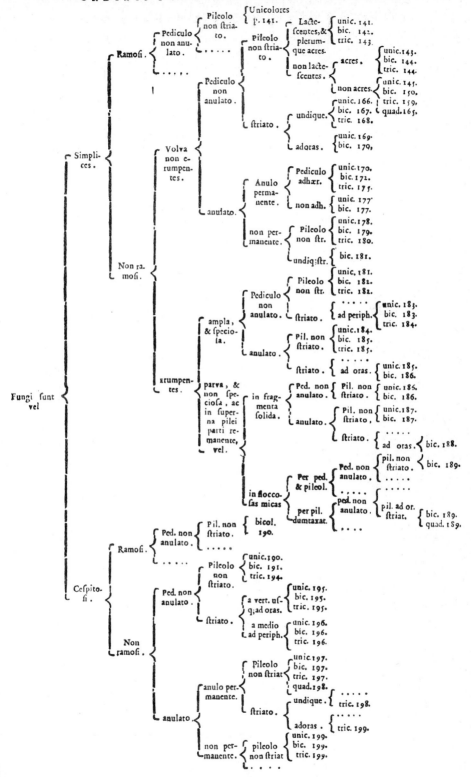

Fig. 93. Micheli's key to *Fungus* (agarics). (P. A. Micheli (1729): 140.)

Synopsis (1690) and by introducing additional fungi increased the British list to 161 species (from less than a hundred in the second (1696) edition), including an illustrated record of the teliosori of the rust *Tranzschelia anemones* on *Anemone nemorosa* as a 'fern' gathered by the Conjurer of Chalgrave. He also compiled ninety-one lichens.

Vaillant when cataloguing the flora of the Paris region in his *Botanicon Parisiense* (1727) listed the plants in alphabetical generic order so that both the descriptions and the fine illustrations of the fungi recorded are widely scattered. Eight genera were recognised: *Agaricus, Boletus, Fungoides, Fungus, Lycoperdon,* and *Tubera* in the sense of Tournefort but Tournefort's *Coralloides* was reserved for branched clavarias, etc., and supplemented by *Clavaria* for unbranched forms (*Geoglossum, Cordyceps, Clavaria fragilis*). No attempt was made to integrate the genera into a system but it may be noted that in subdividing the 108 species of the genus *Fungus* into 'families' and subordinate categories Vaillant introduced for the first time solid versus hollow stipe and the presence or absence of an annulus as criteria for differentiating agarics.

Although Micheli did so much to advance the taxonomy of fungi at the generic level his general classification of the group was vitiated by his belief that he had observed 'flowers' in some fungi. As can be seen from Table 2 he divided fungi into two main groups: those with 'flowers' and seeds and those with seeds only. the first group was subdivided into those with irregular fruit-bodies and those in which the fruit-bodies are radially symmetrical and the second group into those in which the seeds are borne externally or internally, respectively. He did, however, like Vaillant, have to subdivide the genus *Fungus* (agarics with a central stipe), and in the detailed key (see Fig. 93) with which he prefaced his account of the species he used a variety of morphological criteria including presence or absence of a ring or volva, whether caespitose or not, and, in the final differentiation, colour.

LINNAEUS

It is impossible in a few words to asses the influence of Carl Linnaeus[5] who, although a complex and controversial figure during much of his life, came to exert an unsurpassed authority throughout the biology of his period and beyond, which assured him a permanent place in the history of science and of taxonomy in particular. 'Deus creavit, Linnaeus disposuit.' In spite of this hard-earned reputation Linnaeus merits scant mention in a history of mycology. Throughout his working life his treatment of fungi was unsatisfactory as Linnaeus himself recognised for in the *Philosophia botanica* (1751: 240, and later editions) he wrote: 'Fungorum ordo in opprobium artis etiamnum Chaos est, nescientibus Botanicis in his, quid Species, quid Varietas sit' ('The order of Fungi, a scandal to art, is still chaos with botanists not knowing what is a species, what a variety').

In the *Systema naturae*, 1735, he listed nine genera of fungi and eighteen years later when he codified his views on the taxonomy of the plant kingdom in the *Species plantarum*, 1753, only ten (*Agaricus*, *Boletus*, *Hydnum*, *Phallus*, *Clathrus*, *Elvela*, *Peziza*, *Clavaria*, *Lycoperdon*, and *Mucor*; together with *Byssus* and *Tremella* among the algae). Ninety species were recognised, fewer than in the *Flora suecia*, 1748 – the best of his accounts of fungi[6] – and less than a tenth of the number recognised by Micheli (1729). Lichens, classified as algae, were given a more extended treatment, eighty species (half the number in Haller, 1742) of *Lichen* being compiled in seven sections (Leprosi tuberculati, Leprosi scutellati, Imbricati, Foliacei, Coriacei, Scyphiferi, Filamentosi); an ordered sequence which, as Annie Lorrain Smith (1921: 7) pointed out, showed his appreciation of the transition from the leprose crustaceous thallus to the most highly organised filamentous forms.

One debt that mycology, like other branches of biology, does owe to Linnaeus is binomial nomenclature which was derived from a paper-saving device first used in the text and index of *Öländska och Gothländska Resa*, 1745,[7] and subsequently systematised in the *Species plantarum*. For Linnaeus every species was properly designated by a generic name (*nomen genericorum*) plus a diagnostic phrase (*nomen specificum legitimum*) not more than twelve words long by which one 'can distinguish this species from all others of the same genus speedily, safely and pleasantly' but he also provided for each species a 'catch word', epithet, or trivial name (*nomen triviale*) which in association with the generic name provided a convenient designation for everyday use. It was this latter practice, introduced incidentally by Linnaeus, that was so widely adopted by others and became the binomial system for naming each species by a *generic name* plus a *specific epithet*. Thus for example, 'Fungus capestris, albus superne, inferior rubens', the name used by G. Bauhin, John Ray and others for the field mushroom became *Agaricus campestris* 'Erinaceus parvus, hirsutus, ex fusco fulvus, pileo semi-orbiculari, pediculo tenuiore', of Micheli, *Auriscalpium vulgare* (syn. *Hydnum auriscalpium*).

Linnaeus affected taxonomic mycology in two other ways. He introduced the name Cryptogamia for the twenty-fourth class of the Regnum Vegetabile, the class in which fungi have been traditionally classified ever since, and it was his authority which stabilised the application of several common generic names of fungi in senses that did not conform to classical usage. *Agaricus* was used by G. Bauhin, Ray, Tournefort, Dillenius, and Micheli (as *Agaricum*) in the classical sense of agaricum (see p. 216) for polypores (and some lamellate bracket fungi) but Linnaeus took the name up in the modern sense for gill-bearing hymenomycetes. Again, boletus for the Romans was *Amanita caesarea* but as a generic name it was used by both Tournefort and Micheli for morels, etc. and for boletes, as understood today, by Dillenius whose interpretation was followed by Linnaeus, as was that of *Peziza* (the pezicae of Pliny; = puff balls, *fide* Houghton (1885:48)) for cup-fungi; but for Dillenius's *Erinaceus* Linnaeus substituted *Hydnum*. Only two of the genera proposed by Micheli (*Clathrus* and *Mucor*) were accepted by Linnaeus; his opinion of Micheli was low.

AFTER LINNAEUS

Although in the previous section the date for Linnaeus was taken as 1753, the year of the publication of the first edition of the *Species plantarum*, during his life time he published works containing references to fungi from the first edition of the *Systema naturae* in 1735 to the thirteenth edition of the vegetable part of that work (*Systema vegetabilum*) in 1774 (four years before his death) so that the decision as to which contemporary authors came before or after Linnaeus is somewhat arbitrary.

The mycological reputation of Johann Gottleid Gleditsch (director of the botanic garden of the Academy of Sciences, Berlin, and lecturer in the medical school) mainly rests on his *Methodus fungorum* published in 1753, the same year as the *Species*

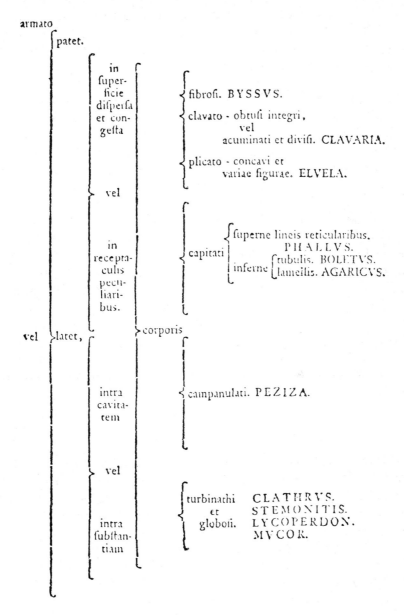

OMNIS
FRVCTIFICATIO FVNGORVM,
PARVITATE OCVLOS NVDOS SVPTERFVGIENS

armato

patet.

in fuperficie difperfa et congefta

fibrofi. BYSSVS.

clavato - obtufi integri, vel acuminati et divifi. CLAVARIA.

plicato - concavi et variae figurae. ELVELA.

vel

in receptaculis peculiaribus.

capitati

fuperne lineis reticularibus. PHALLVS.

inferne tubulis. BOLETVS. lamellis. AGARICVS.

vel latet, corporis

intra cavitatem

campanulati. PEZIZA.

vel

intra fubftantiam

turbinathi et globoii.

CLATHRVS. STEMONITIS. LYCOPERDON. MVCOR.

Fig. 94. Gleditsch's classification of fungi. (J. G. Gleditsch (1753): 16.)

253

plantarum and notable as being the first book exclusively devoted
to fungi. Although based on ten of the genera recognised by
Linnaeus, their grouping (see Fig. 94) shows an advance on that
of earlier and contemporary authors. The primary subdivision is
into open and closed fruit-bodies, the first category being
divided into those having the fructification superficially displayed or on a receptacle, the second category into those with the
fructification in cavities or within the substance of the fungus;
the final differentiation being based on texture, morphology, or
topography of the fruit-body. During the next forty years the
Linnean fungal genera were the only ones used by a number of
authors including Scopoli (1760, 1772), Schaeffer (1762–74), and
Batsch (1783–9) and little was done to modify the general
classification of fungi. Michel Adanson in his *Familles des plantes*
(1763), as could be expected from a man of unorthodox outlook,
made a serious attempt to arrange the fungi according to the
increasing complexity of their fruit-bodies in the first two
families of his system. Family 1, 'Les Byssus. *Byssi*', excluded
from the fungi, was used for powdery or filamentous forms and
included *Aspergillus* and *Botrytis*. Family 2, 'Les Champignons.
Fungi', the fungi proper, was divided into seven groups characterised by having spores powdering the whole surface of the
fruit-body, localised on the ridges of a network, on the walls of
sacs or tubes, on irregular furrows, or, in the final section, on
lamellae. Adanson was the first to unite lichens, hitherto treated
as algae, with the fungi.

One notable advance at this time was the gradual adoption of
binomial nomenclature. Gleditsch used phrase-names, as did
Scopoli in the first edition of the *Flora carniolica* (1760) but he
adopted binomials in the second edition (1772) and so did
Schaeffer (1762–74) and Batsch (1783–9). Among the conserva-
tives (including Adanson) who resisted binomial nomenclature
was Albrecht von Haller, the Swiss naturalist, poet, and novelist
who was for a time professor of medicine and botany at
Göttingen. Von Haller was an infant prodigy who by the age of
ten had a thorough knowledge of Greek and Hebrew, by fifteen
had written an epic and some tragedies, and at nineteen
qualified in medicine. He is most generally remembered as a
physiologist but he made important contributions to botanical
taxonomy as a rival to Linnaeus; especially in the two versions of
his account of the plants of Switzerland, published in 1742 and

1768, both of which include fungi. Von Haller's most notable innovation was to divide up the agarics (his genus *Fungus* in 1742; changed to *Amanita* in 1768) by the colour of the gills, a criterion taken over by both Scopoli and Batsch and one for which Albertini & Schweinitz in 1805 substituted the more precise character of spore colour.

PERSOON

What Linnaeus and Jussieu did to integrate the systematics of the flowering plants in the mid-eighteenth century Persoon and Fries did for fungi some fifty years later.

Christiaan Hendrik Persoon (Fig. 95) was born at the Cape of Good Hope on, it is conventionally held, 31 December 1761 (but his birth may well have been a year later, or on 1 January 1763). His father, Christiaan Daniel Persoon, who came from Pomerania, emigrated in the service of the Dutch East India Co. to the Cape, where he eventually settled. There he married a well-to-do Dutch farmer's daughter from Stellenbosch (and not a Hottentot as was once supposed), who died a few weeks after Persoon's birth, when the care of Christiaan and his two sisters became a charge of the Orphan Chamber. Their father remarried, prospered as a general dealer in anything from beer to stockings and slaves, and after his death in 1776 the Orphan Chamber embezzled funds held on behalf of the young Persoon. The Amsterdam Chamber, however, came to his rescue, if parsimoniously.

In 1775, Persoon had been sent to Europe, never to return to the Cape. He went to school at Lingen-on-Ems, Germany, and in 1783 proceeded to Halle as a theological student. Three years later having inherited a small income he was studying medicine at Leiden, a study he continued at Göttingen, then in the van of botanical research. Later still, at the instigation of Prof. J. C. D. von Schreber, a noted cryptogamist, he obtained his doctorate from the Academy of Natural Sciences, Erlangen. The next four years seem to have been spent travelling, and in 1803 Persoon settled in Paris where for thirty years he lived the life of a recluse on the sixth floor of 2 Rue des Charbonniers, a mean street near the present Gare de Lyon.

In later life at least, Persoon was of unprepossessing appearance and clearly eccentric. He confessed that he loved the study

Fig. 95. Christiaan Hendrick Persoon (1761–1836).
(*Z. f. Pilzkunde* N.F. **12**, pl. 8 (1933).)

of fungi because it could be carried out alone in a room. But
Persoon, in his room surrounded by his specimens and books,
was far from isolated. He corresponded regularly with many
leading scientists, particularly botanists, from all over Europe,
and received much material from many parts of the world. He
held no official appointments, and a price he paid for his
freedom was life-long financial worries, which are reflected in
his correspondence. In later life he was reduced to poverty, but
being unwilling to accept money as a gift he made over his
herbarium in 1825 to the Dutch Government in return for a
pension of 800 florins (1760 fr.). Since 1829 this herbarium has
been in the safe keeping of the Rijksherbarium, Leiden,
together with the greater part of his library, while the University
Library at Leiden has a valuable collection of letters written to
Persoon.

Persoon sought membership of scientific societies (in 1799 he
was elected a foreign member of the Linnean Society of London)
and he was honoured by others. In 1794 the twelfth issue of the
Annalen der Botanik was dedicated to him and he has been

commemorated by a number of generic names (including *Persoonia* J. E. Smith, 1798; Proteaceae) and specific epithets. Most recently, the mycological journal of the Rijksherbarium, Leiden, begun in 1959, has been named *Persoonia*.

Persoon died in November 1836, but the exact date, like that of his birth, is not quite certain. It is given as November 14 in the Archives du Departement de la Seine et la Ville de Paris, as November 16 in the obituary notice in the *Government Gazette*, but as November 15 in the register of Le Père Lachaise cemetery, where he was buried, and also on the gravestone which, according to Persoon's testamentary instructions (and for which he composed the inscription), was 'simple and vertical'.[8]

Little is known as to how Persoon's interests in natural history were aroused and developed or what induced him to concentrate his attention on fungi, but his talent became widely recognised. As a botanist his most important work is the two-volume *Synopsis plantarum*, 1805–7, which, in 1200 small, closely printed pages, lists according to the Linnean system all the phanerogams then known. It is, however, as a mycologist that he will always be best remembered and his reputation rests mainly on four works: *Observationes mycologicae*, 1796–9; *Tentamen dispositionis methodicae fungorum in classes, ordines, genera et familias*, 1797; *Synopsis methodica fungorum*, 1801; and the unfinished *Mycologia Europaea*, 1822–8.

Peroon's germinal work on classification was his 1794 paper in Römer's *Neues Magazin für die Botanik*. The scheme there set out was only slightly modified in the later *Tentamen* and the *Synopsis* and it substantiates Vuillemin's claim that Persoon's broad groupings of the fungi are comparable with the first natural arrangement of flowering plants by Antoine-Laurent de Jussieu in his attempt to replace the empirical approach of Linnaeus.[9] Persoon based his scheme on the gross form of the fruiting structures and paid little attention to microscopical detail. In the *Synopsis* (see Table 3), his primary division of the six orders (seven in the 1794 version) was into two classes – Angiocarpi and Gymnocarpi – in which the fruit-bodies were closed and open, respectively; a distinction made by both Scopoli (1760) and Batsch (1783–9). Persoon introduced the concept of the hymenium and although he believed that basidiomycetes bore asci it can be seen that he kept the Helvelloidei (discomycetes) apart from the hymenomycetes. He brought the rusts and smuts

Table 3. *Persoon's classification* (Synopsis methodica fungorum, *1801*)

Classis I. ANGIOCARPI. Fungi clausi, seu semina ut plurimum copiosa interne gerentes.

    Ordo I. Sclerocarpi. Fungi duriusculi substantia interna molli. (*Sphaeria, Tubercularia,* etc.)

    Ordo II. Sarcocarpi. Fungi carnosi farcti. (*Sclerotium, Tuber, Pilobolus, Sphaerobolus,* etc.)

    Ordo III. Dermatocarpi. Fungi membranacei, coriacei aut villosi intus pulvere farcti.

        1. Trichospermi, Pulvere seminali filis intertexto. (*Lycoperdon* and other gasteromycetes; myxomycetes)

        2. Gymnospermi, Pulvere nudo s. filis non reticulato. (*Mucor, Aecidium, Uredo, Puccinia*)

        3. Sarcospermi. Fructibus luculentis carnosis. (*Cyathus*)

Classis II. GYMNOCARPI. Fungi carnosi, semina (parca) in receptaculo (Hymenio) aperto gerentes.

    Ordo IV. Lytothecii. Membrana fructicans s. hymenium in laticem (gelatinam) demum solutum. (*Clathrus, Phallus*)

    Ordo V. Hymenothecii. Hymenium membranaceum indissolubile, sporulis pulverulentum.

        1. Agaricoidei. Hymenio lamelloso aut venoso. (*Amanita, Agaricus, Merulius*)

        2. Boletoidei. Hymenium in tubos varios prominens. (*Daedalea, Boletus*)

        3. Hydnoidei. Hymenium in aculeos aut dentes prominens. (*Sistotrema, Hydnum*)

        4. Gymnodermata. Hymenium laeve aut papillosum. (*Thelephora, Merisma*)

        5. Clavaeformes. Fungi carnosi, elongati, pileo cum stipite confluente. (*Clavaria, Geoglossum*)

        6. Helvelloidei. Pileus stipitatus, membranaceus, a stipite distinctus. (*Helvella, Peziza* and other discomycetes; also *Stilbum, Aegerita*)

    Ordo VI. Naematothecii. Fungi byssoidei. (*Aspergillus, Botrytis,* and other hyphomycetes)

---

together in the Dermatocarpi and if he did also include mucors and united various gasteromycetes with the myxomycetes (in the Trichospermi) almost all the hundred genera and subgenera he recognised are universally accepted genera of today. Although he made more use of a hand lens than a microscope Persoon accurately described many common microfungi – both saprobes and parasites – and what has given his descriptions and nomenclature particular value is that the specimens on which they are based are still available for study.[10] A firm foundation was thus laid on which others could build.

FRIES

Elias Magnus Fries, who was seven years old when Persoon's *Synopsis* (1801) was published, has frequently been compared with his fellow countryman Linnaeus for both were sons of country clergymen, both were born in the Småland district of southern Sweden, both went to school in Växjö and from there to the University of Lund, and each in his turn became professor of botany at the University of Uppsala. Also, both were voluminous writers.

According to his autobiography (published in *Monographia hymenomycetum sueciae*, 1857),[11] Fries had by the age of twelve become acquainted with the more notable plants of his neighbourhood and more than half a century later he remembered the day in 1806 when he went with his mother to a burnt-down forest to pick strawberries and found an unusually large specimen of *Hydnum coralloides*. This find stimulated his interest in fungi and by the time he left school in 1811 he had taught himself to distinguish three to four hundred species of fungi to which he was unable to give scientific names until he had access to the university library and received encouragement from two of his teachers, the professor of natural history A. J. Retzius (who gave Fries a copy of Albertini & Schweinitz, *Conspectus fungorum*, 1805) and the algologist C. A. Agardh (who gave him Persoon's *Synopsis*). Fries's long and hardworking life was uneventful. Apart from a brief visit to North Germany and Berlin in 1828 he did not travel outside Scandanavia but he received much material for examination from all parts of the world. When he died in 1878, in his eighty-fourth year, he was respected everywhere. Subsequently, the name Fries has been kept alive by distinguished mycologists and botanists of three succeeding generations[12] including Elias Fries's eldest son Theodor Magnus Fries, the lichenologist; his grandson Robert Elias Fries who published on myxomycetes; and the fungal physiologist Nils Fries whose researches as a professor at Uppsala are so ably complementing the work of his great-grandfather (see Fig. 96).

Fries would have preferred to have been a student at Uppsala, the oldest seat of botanical learning in Sweden, but Lund was chosen because of the warning he and his fellow school-leavers received from the masters in the upper school of the dangers of

Fig. 96. Elias Magnus Fries (1794–1878) (*top left*); his son Theodore
Magnus Fries (1832–1913) (*top right*); grandson Robert Elias Fries
(1876–1966) (son of T.M.F.) (*bottom left*); and greatgrandson Nils
Thorsten Elias Fries (nephew of R.E.F.) (aet. 59) (*bottom right*).

German romantic philosophy (*Naturphilosophie*) which had recently infiltrated Uppsala. Nature, as understood by this philosophy, was a spiritual whole, built up of polar forces in constant development, in which every part mirrored the whole. Because both the world and man were spiritual, study of a single part could lead to knowing the whole. Romantic philosophy was all embracing and so provided a theoretical basis for the construction of 'natural' arrangements of both animals and plants as was done by the German romantic biologist L. Oken in his *Lehrbuch der Naturphilosophie* (1809–11). Fries did not, however, escape infection although it was not until his early postgraduate years that his adherence to romantic philosophy became evident.[13] He was possibly influenced by Agardh (who succeeded Retzius as professor at Lund in 1812) and certainly by the romantic approach to fungi of *Das System der Pilze und Schwämme* (1816) by Christian Gottfried Nees von Esenbeck who, the year after the publication of his mycological classic, became the first professor of botany at the new University of Bonn. He subsequently moved to a similar post at Breslau where in old age he became an extreme radical, advocating the reform of Christianity and free-marriage without State intervention; views which led to his dismissal.

Fries's romanticism was at its height at the time of the publication of the first volume of the *Systema mycologicum* (1821) in the introduction to which he set out the speculative theoretical basis for his classification of fungi. Water and earth are the cosmic forces which determine the growth of a vegetable. There is a special force (*nissus reproductivus*) which when acting alone gives rise to fungi; their subsequent development being determined by air and the two components of fire, heat and light. Hence the four classes of fungi (Coniomycetes, Hyphomycetes, Gasteromycetes, Hymenomycetes), which are mainly determined by the reproductive force acting alone, air, heat, or light, respectively, and which may be distinguished by characteristic fruit-bodies: Coniomycetes by spores only ('sporidia nuda'); Hyphomycetes by spores on a floccose thallus; Gasteromycetes and Hymenomycetes by fruit-bodies bearing spores internally or externally, respectively. Each class was subdivided into four orders and each order in its turn into four tribes or main genera; each group being composed of a 'centrum', typical of the group, and three 'radii', the constituents of which border forms they resemble in the other groups. (For an example of this treatment, see Table 4). Species within one group show 'affinity' to one another, those of different groups which resemble one another 'analogy'. By 1830 his romanticism had weakened and his approach to taxonomy become more realistic but he retained a belief in vitalism and a conservatism which was unsympathetic to Darwinism.

Table 4. *Summary of Fries's subdivisions of the class Hymenomycetes* (Systema mycologicum, *1821, 1: liii–lvi*)

Class IV. HYMENOMYCETES (H)
  Order I. Sclerotiacei (HE) [radius to Coniomycetes]
    Genera: 1. *Erysiphe* (HEE)   2. *Rhizoctonia* (HEM)
            3. *Sclerotium* (HEX)   4. *Tuber* (HEG)
  Order II. Tremellini (HM) [radius to Hyphomycetes]
    Genera: 1. *Agyrium* (HME)   2. *Dacromyces* (HMM)
            3. *Tremella* (HMX)   4. *Hygromitra* (HMG)
  Order III. Uterini (HU) [radius to Gasteromycetes]
    Genera: 1. *Cyphella* (HUE)   2. *Solenia* (HUM)
            3. *Peziza* (HUU)   4. Mitrati (HUH)
  Order IV. Hymenini [centrum]
  Suborder I. Clavati (HH^1)
    Genera: 1. *Pistillaria* (HH^{1}E)   2. *Typhula* (HH^{1}E)
            3. Spathularia (HH^{1}U)   4. Clavaria (HH^{1}G)
  Suborder II. Pileati (HH^1)
    Genera: 1. *Thelephora* (HH^{2}E)   2. *Hydnum* (HH^{2}M)
            3. *Polyporus* (HH^{2}X)   4. *Agaricus* (HH^{2}G)

The capital letters in parentheses indicate supposed relationships with other groups. E, Entophytae (Coniomycetes, Ord. I); G, Geogenii; H, Hymenomycetes; M, Mucedines (Hyphomycetes, Subord. II (I)); U, Uterini veri (Gasteromycetes, Ord, III); X, Xylomacei (Gasteromycetes, Subord. II (II)*).

[*Note*: asterisks and superscript numbers denote complex subdivisions.]

In the event, the general classification outlined in the introduction to the *Systema* was modified as the work progressed. Later, the six classes (Hymenomycetes, Discomycetes, Pyrenomycetes, Hyphomycetes, Coniomycetes) advocated in the *Epicrisis systematis mycologia* (1836–8) marked a major advance. In the separation of Discomycetes from the Hymenomycetes (a distinction made, it may be noted, before the elucidation of basidial structure, see p. 68) he followed Persoon and he distinguished Pyrenomycetes from Gasteromycetes.

Fries's theories and general classifications are now of minor interest. His fame rests on his wide and intimate knowledge of fungal species and his taxonomic treatment of the hymenomycetes on which he left a permanent mark. Nearly five thousand species of fungi are concisely described in the three volumes of the *Systema* (1821–32) and the supplement (*Elenchus fungorum*, 1828) – the last including many fungi from outside Europe – while in the *Hymenomycetes Europaei* (1874), the preface of which was written on Fries's eightieth birthday, he described 1860 species of 'agarics', in twenty genera. As already mentioned, Haller (1768) used gill colour as a taxonomic criterion. Albertini & Schweinitz (1805) realised that spore colour was a more reliable taxonomic character but they were unable to employ this criterion themselves because of lack of data and it was left to Fries, first in the *Systema*, to apply spore colour as a differential by which the large genus *Agaricus* was divided into five series according to whether the spores were white (Leucospori), rosy red (Hyporhodii), rusty-red, brown, etc. (Dermini), purplish black or dark in colour (Pratellae), or black (Coprinarii); the twenty-eight subgenera (increased to thirty-five in *Hymenomycetes Europaei*) into which the genus was divided being common genera of today.

AFTER FRIES

The many books and papers by Fries published during his life time covered a span of more than sixty years (1814–77) during which the general classification he had developed by 1830, and to which he subsequently made few significant amendments, was surpassed by schemes devised by contemporaries who, in varying degrees, had been influenced by his writings. Attention need be drawn to only two examples.

J.-H. Léveillé, who in 1837 had elucidated the structure of the basidium (see p. 71), was a microscopist (he is said when on active service as a military surgeon to have used his microscope in the front line) and in 1846 he offered a novel general classification in which for the first time many histological criteria were employed; as can be seen in Table 5. Léveillé's primary subdivisions were based on spore characters and many of the lower categories were also differentiated on microscopical features, the final grouping of genera frequently being according to whether the spores were septate or not. This approach was not altogether satisfactory but if very dissimilar forms were at times brought together, for example, the inclusion of myxomycetes in the 'Basidiés', the scattering of hyphomycetes through several classes, and his treatment of rusts and smuts (which according to Vuillemin (1912) he intended to modify), his separation of the basidiomycetes and ascomycetes has stood the test of time and his use of microstructure as a taxonomic criterion was an example to others.

Table 5. *Léveillé's classes of fungi* (Considérations mycologique, *1846: 105–33*)

Class	Circumspection
I. Basidiosporés (spores on basidia)	Hymenomycetes, Gasteromycetes also Myxomycetes
II. Thécasporés (Spores in thecae (asci))	Ascomycetes
III. Clinosporés (Spores in clinodes – fascicles of filaments bearing terminal spores, which ma y be enclosed in horny receptacles)	Hyphomycetes (Tuberculariaceae, Stilbaceae), Ustilaginales, Uredinales, various Sphaeropsidales
IV. Cystosporés (Spores in terminal sporangia)	Mucorales
V. Trichosporés (Spores borne externally on simple or branched filaments)	Hyphomycetes
VI. Arthrosporés (Spores in chains)	*Alternaria, Aspergillus, Cladosporium, Penicillium, Torula*, etc.

The publication of Charles Darwin's *Origin of species* on 24 November 1859 affected every branch of biology. Evolution became an accepted fact and among the first to attempt an evolutionary approach to the taxonomy of fungi was Anton de Bary who in the preface to the first (1866) edition of his textbook offered a general classification of fungi which closely resembles that widely accepted today (see Table 6). The classes were arranged in an evolutionary series beginning with forms possibly

Table 6. *De Bary's classification of fungi* (Morphologie und Physiologie der Pilze, Flechten und Myxomyceten, *1866: vi*)

I. Phycomycetes
    a. Saprolegnieae
    b. Peronosporeae
    c. Mucorini
II. Hypodermii
    a. Uredinei
    b. Ustilaginei
III. Basidiomycetes
    a. Tremellini
    b. Hymenomycetes
    c. Gasteromycetes
IV. Ascomycetes
    a. Protomycetes
    b. Tuberacei
    c. Onygenei
    d. Pyrenomycetes
    e, Discomycetes
Flechten
Myxomyceten

evolved from algae,[14] and hence designated by de Bary as the Phycomycetes, through basidiomycetes via the rusts and smuts to the climax in the ascomycetes. De Bary was not primarily a taxonomist but his wide knowledge of many fungi gave him a sensitive approach to classification and the taxonomic treatment adopted in the second (1884) edition of his textbook corresponded even more closely to that most generally followed during the next hundred years. He tentatively restored the chytrids to the Phycomycetes from which he had excluded them, placed the Ascomycetes next in the series, and gave the rusts ordinal rank (Uredineae) as a bridge to his climax in the Basidiomycetes composed of Hymenomycetes and Gasteromycetes. He was unable to decide the phylogenetic relationships (still unsettled) of the Laboulbenieae, *Taphrina* and *Saccharomyces* to the Ascomycetes, and of *Protomyces* and the smuts (Ustilagineae) which he appended to his account of the Phycomycetes. Lichens and myxomycetes he retained in separate groups.

A competing classification was that of Brefeld, which was given wide circulation by von Tavel in his text-book of 1892, although it never achieved unqualified support because of the underlying assumption that only phycomycetes exhibited sexuality and that ascomycetes and basidiomycetes (both considered to be derived from zygomycetes) were asexual. It had certain merits and it was variously adapted by a number of authors including George Massee as late as 1910. Another contemporary proposal was that of L. Marchand (1896) who used a new nomenclature for what was basically the de Bary system. The Mycophytes (Fungi) were subdivided into the Mycomycophytes (fungi proper) and Mycophycophytes (lichens) and a number of new class names were introduced, e.g. Siphonomycetes (for Phycomycetes), Thecamycetes (Ascomycetes).

While these developments were taking place fundamental subdivisions were being made in some of the larger classes. Patouillard (1887) differentiated homobasidiomycetes from heterobasidiomycetes on basidium structure. Two years earlier, Boudier (1885) had divided the discomycetes into operculate and inoperculate series according to the methods of ascus dehiscence but the current emphasis on the taxonomic significance of the unitunicate and bitunicate structure of the ascus (the latter noted as early as 1838 (b) by Berkeley[15]), which led to the erection of Loculoascomycetes for the bitunicate series is mainly a development from the monographic study on the pyrenomycetes by Luttrell (1951). E. A. Bessey (1935) made a separate class, Teliomycetes (as Teliosporeae) for the rusts and smuts. The phycomycetes were always a heterogenous assemblage. Gäumann (1926) separated off the Archimycetes for certain chytrids and other forms which he considered primitive, and in this he was widely followed. It was Cejp (1957–8, 1) in Czechoslovakia and Sparrow (1958) who first introduced a fundamental re-structuring of the phycomycetes as the series of classes (Plasmodiophoromycetes, Chytridiomycetes, Oomycetes, etc.) now very generally accepted. It is interesting to note that the wheel has in one respect come full circle and an increasing number of modern authors (e.g. Kreisel, 1969) are excluding oomycetes from the Fungi and classifying them as algae.

Speciation

In recent years attention has frequently been drawn to the fact that there are few objective criteria for determining speciation in

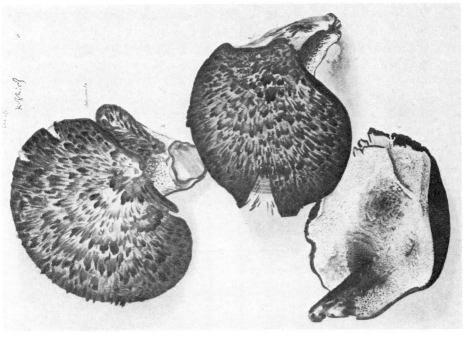

Fig. 98. A page of the Code de L'Escluse from which van Sterbeeck copied (cf. Fig. 97). (G. Istvánffi (1900): pl. 6.)

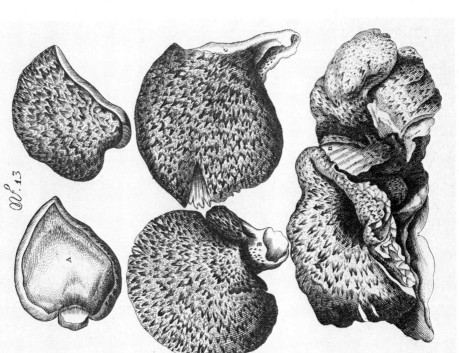

Fig. 97. Three of van Sterbeeck's 'species' of *Polyporus squamosus* (F. van Sterbeeck (1675): pl. 13). (Cf. Fig. 98.)

267

fungi which is largely based on tradition, convenience, and personal opinion. It seems to be a general rule that a taxonomist's judgement on first approaching a new group is biased in the direction of small differences; for example, van Sterbeeck (1675) distinguished no less than six 'species' of *Polyporus squamosus* based on striking individual specimens, three of which are illustrated in Fig. 97. Thus species and genera are multiplied to a greater extent than is subsequently considered justified in the light of wider experience and the suggestion has been made that any major fungal taxon in which less than half the names of the included taxa are in synonymy is in urgent need of critical revision. Another generalisation is that the average number of species per genus is higher for parasitic than saprobic fungi. This is partly subjective, due to the special interest shown in these forms. It is also an objective reality due to intimate host–parasite relationships (see p. 158) and today agriculturalists and horticulturalists are certainly inducing variation in many fungi by exposing them to a range of cultivars of novel genetic constitutions.

For both animals and plants, and also fungi (with the major exception of yeasts), morphology has been traditionally the most important specific criterion. This has been reflected in the choice of specific epithets. A random sampling of Persoon's *Synopsis* of 1801 showed, in round numbers, 40 per cent of the specific epithets to be based on morphological features, 25 per cent on colour terminology, and 10 per cent on the names of higher plants. A marked change is reflected by both the 1897 index volume of Saccardo's *Sylloge* (vol. **12**) and the first volume of the *Index of fungi* covering new names published during the period 1940–9 where between one-quarter and one-third of the epithets are based on morphological characters, approximately 10 per cent on colour, and 30 per cent on the names of host plants.

It is now customary to distinguish varieties on morphological grounds but many other sub- or intraspecific taxa are based on non-morphological characters. Of these last, formae speciales and physiologic races were discussed in Chapter 6 (p. 157) because of the urgent and practical need of plant pathologists to distinguish fine taxonomic gradations in fungi pathogenic for economic crops.

Modern trends

The main line of taxonomic development during the foresee-able future is likely to continue to be the traditional description (but with increasing precision) and naming of fungi because the census of fungi is so far from complete. In 1954 attention was drawn to the constancy of the ratio of new species to new genera described annually during recent years.[16] This is what would be expected statistically if a random sample of forms is described each year from a large population of still undescribed fungi. It was also noted that the ratio of new combinations (transfers) to new species, which also appears to be a relative constant, rose markedly during the years following the 1939–45 war, presum-ably because taxonomists being confined to their laboratories during hostilities devoted their energies to revisionary studies which were the first post-war publications. These traditional descriptive studies are, however, being increasingly sup-plemented by new and usually more experimental approaches to taxonomy such as the assessment of affinities by developmen-tal studies, numerical methods made possible by the use of computers, and a range of biochemical approaches including those of serology, cell wall chemistry, and the constitution of cytochrome *c*. This should establish new groupings and confirm many of the old while among the hymenomycetes (where an outstanding contribution to the ordering of agarics has been made by the successive editions of Rolf Singer's *The Agaricales in modern taxonomy*, 1951 (edn 3, 1975)) major reorganisa-tions – many stemming from the recent extensive taxonomic and bibliographic studies of the late M. A. Donk of Leiden – will follow when the significance of convergent evolution within the group has been assessed.

11. Organisation for mycology

In every branch of research, proper organization is essential for satisfactory progress. [S. P. Wiltshire, *Trans. Br. mycol. Soc.* **27**: 1 (1944).]

Communication is essential for the advance of mycological knowledge, as it is for science in general. Research is incomplete until the results have been made available to others for scientists build on the foundations laid by their predecessors. This communication has always been effected by personal contacts between individuals of like interests (the so-called 'invisible college'), including that between teacher and pupil, and by associations or societies of varying degrees of formality but for the past 500 years the traditional method of publicising scientific findings has been by the printed word which generates what is known as 'literature'. Some aspects of the development, characteristics, and problems of mycological literature will first be considered.

MYCOLOGICAL LITERATURE

Up to the year 1600, apart from casual references to fungi in a dozen or so herbals, there were only two specifically mycological printed publications, the little monographs by Adriaen Jonghe on *Phallus hadrianus*, 1562, and by Alfonso Ciccarelli on truffles, 1564. The seventeenth century was hardly more productive. There were additional, and more comprehensive, compilations by herbalists such as those by Clusius in his *Rariorum plantarum historia*, 1601, and the posthumous edition of Jean Bauhin's *Historia plantarum universalis* (1651) which culminated towards the end of the century in the treatment of fungi by John Ray in his *Historia plantarum*, 1686, and J. P. de Tournefort in *Élémens de botanique*, 1694. The century did, however, see the publication of the first book on fungi, van Sterbeeck's *Theatrum fungorum*, 1675. Also, van Leeuwenhoek's first letter to the Royal Society of London in which he describes some observations on mould (see p. 58) was published in the *Philosophical Transactions* and is the earliest journal article referred to in this history. The increase in mycological literature then became exponential. The number of mycological publications in the eighteenth century exceeded the total up to the year 1700 and this pattern was repeated during the nineteenth century and the first half of the twentieth. Aspects of this growth in the size of the literature are illustrated graphically in Fig. 99. The current (1974) output exceeds 8000 items a year.

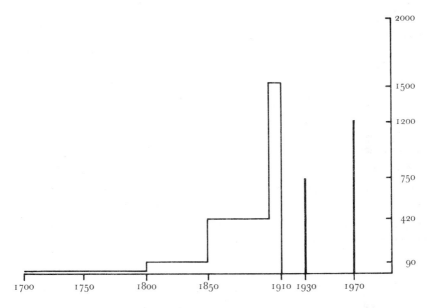

Fig. 99. Annual rate of publication of mycological literature up to 1910, and the rates for taxonomic literature only for 1930 and 1970.

Until the beginning of the nineteenth century almost all publications on fungi originated in Europe – especially France and Germany. A number of exotic specimens collected by travellers in the tropics and elsewhere were described from Europe but it was not until 1834, which marked the appearance of Schweinitz's *Synopsis fungorum in America boreali media digentium*, that a major mycological contribution was published elsewhere.

Until the mid-nineteenth century many publications, particularly taxonomic manuals such as those of Persoon and Fries, were written in Latin and the practice still survives in the use of 'botanical Latin'[1] for the descriptions of new species. Van Sterbeeck's *Theatrum fungorum*, 1675, was the first book to be written in the vernacular. By the closing years of the nineteenth century, French and German were the dominant languages but during the first half of the twentieth century with the shift of the van of mycological endeavour from Europe to North America they were replaced by English in which currently more than two-thirds of the mycological literature is written. Mycological authors have always wanted to be as widely read as possible. To write in Latin formerly achieved this end but with the decline in the classical education of mycologists and, throughout the English-speaking world, in their facility with foreign languages not only those belonging to minority language groups, such as the Dutch and Scandinavians, but also those whose native language is French, Italian, or German tend to write their more important publications, or at least to publish their main conclusions, in English. At the same time increased nationalism is leading to a much wider scatter of the mycological literature through the many languages of the world.[2]

Textbooks

The first attempts to organise the literature and spread mycological knowledge were by textbooks written by individual authors. Van Sterbeeck wished to popularise knowledge of edible and poisonous fungi by a well-illustrated text written in Flemish. Many similar books followed and in the late eighteenth and early nineteenth centuries the publication of accounts of the larger fungi of many European countries illustrated by hand-coloured engravings, frequently of considerable artistic merit, reached a peak. Many of these works were by botanical artists or naturalists of high artistic abilities; e.g. Bolton and Sowerby in England, Bulliard in France, Schaeffer in Austria, and Sturm in Germany. They were mostly issued in parts as material came to hand but some were later re-issued with the plates systematically arranged as, for example, the 1809 edition of Bulliard's *Histoire des champignons*. Although appreciated by mycologists these works were also designed for the libraries of the wealthy among whom natural history was then in fashion. The first textbooks available to working mycologists were a development from the taxonomic accounts – mainly lists with varying amounts of description – found in the many regional floras published before 1800, and took the form of taxonomic handbooks in which attempts were made to give comprehensive summaries of all the fungi then known. The two classical examples are the *Synopsis methodica fungorum* by C. H. Persoon (1801) and the three-volume *Systema mycologicum* by E. M. Fries (1821–32) which were subsequently recognised as the main starting points for the nomenclature of fungi (see p. 294). Both were written in Latin and were without illustrations. They provided the pattern for many subsequent compilations or 'floras', mostly of German origin, such as the seventeen volumes on fungi and lichens of Rabenhorst's *Kryptogamen-Flora von Deutschsland, Oesterreich und der Schweiz*, 1884–1960, compiled by a number of authors and the fungal and lichen volumes of A. Engler & K. Prantl's *Die natürlichen Pflanzenfamilien*, 1897–1928. Microfungi were given extended treatment by Corda in his impressive six-volume *Icones fungorum hucusque congitorum*, 1837–54, but there was no general textbook on fungi before 1866 which saw the publication of de Bary's *Morphologie und Physiologie der Pilze, Flechten und Myxomyceten* (Fig. 100).

HANDBUCH

DER

HYSIOLOGISCHEN BOTANIK

IN VERBINDUNG MIT

A. DE BARY, TH. IRMISCH, N. PRINGSHEIM UND J. SACHS

HERAUSGEGEBEN VON

WILH. HOFMEISTER.

ZWEITER BAND.

ERSTE ABTHEILUNG.

MORPHOLOGIE UND PHYSIOLOGIE DER PILZE, FLECHTEN UND MYXOMYCETEN.

VON

A. DE BARY.

LEIPZIG

VERLAG VON WILHELM ENGELMANN.

1866.

MORPHOLOGIE UND PHYSIOLOGIE

DER

PILZE, FLECHTEN UND MYXOMYCETEN.

VON

Dᴿ· A. ᴅᴇ BARY,

PROFESSOR AN DER UNIVERSITÄT FREIBURG I. B.

MIT 101 HOLZSCHNITTEN UND EINER KUPFERTAFEL.

LEIPZIG

VERLAG VON WILHELM ENGELMANN.

1866.

Fig. 100. Title-pages of the first edition of Anton de Bary's textbook, 1866.

Anton de Bary, son of a practising physician of Walloon descent, was born in 1831 at Frankfurt where he went to school. After attending the universities of Heidelberg and Marburg, in 1850 he qualified in medicine at Berlin where his teachers included A. Braun, Ehrenberg, and Johannes Müller. Then, after a short period of medical practice in Frankfurt, de Bary became in 1853 'Privat-docent' at the University of Tübingen under Hugo von Mohl. Two years later he was appointed Naegeli's successor at Freiberg and in 1867 took the chair of botany at Halle in succession to Schlechtendahl. In 1872, following the annexation of Alsace–Lorraine by Germany after the Franco-Prussian war, he made his final move to Strasbourg where he served a term as Rektor and died of cancer sixteen years later at the age of 57, still at the height of his powers.

This is not the place to catalogue the many and diverse contributions to mycology made by de Bary. His name has been referred to in every chapter of this history.

De Bary was a most influential teacher, although as Rees[3] recalled 'He was not a brilliant lecturer. His words were simple and his presentation without ornament' but his lectures were carefully prepared and the material well arranged. De Bary was, however, socially inclined and had a fund of sparkling wit. He travelled relatively little but during vacations made collecting trips to the Black Forest, the Harz mountains, or the Alps. In appearance, again according to Rees, 'He was of medium height, slender, active, and quick in his movements'. 'He had sparkling blue eyes'. 'None of his portraits do him justice' (Fig. 101).

De Bary's textbook, which includes the results of so many of his own observations and discoveries, was an outstanding mycological achievement, perhaps the most outstanding of the nineteenth century. The impact of the book can now only be fully appreciated by coming to it after scanning the texts available up to the time of its publication when the comprehensiveness of its cover and the modernity of its approach is most impressive. It was well received at the time and the enlarged second edition of 1884 (which appeared in English translation in 1887) is still much more than a historical record. A few years later Zopf's *Die Pilze*, 1890, was another broadly based and comprehensive account of the fungi as were M. C. Cooke's and George Massee's popularisations of 1895 and 1906 but most

Fig. 101. Anton de Bary (1831–88), aet. 36.

mycological textbooks since de Bary's have been confined to a more or less detailed taxonomic outline spiced with phylogenetic speculation. Von Tavel's text of 1892 was of this category as was the crop of textbooks published between the two World Wars which included E. A. Gäumann's *Vergleichende Morphologie der Pilze*, 1926; H. C. I. Gwynne-Vaughan & B. Barnes, *The structure and development of the fungi*, 1927; and the first North American text, E. A. Bessey's *A text-book of mycology*, 1935. In France, Maurice Langeron in his *Précis de mycologie*, 1945, made a more individual approach and in so doing, by reproducing many of Buller's illustrations, did for Buller what von Tavel had done for Brefeld. In spite of its neglect in the standard textbooks, rapid advances were concurrently being made in the knowledge of the physiology of fungi, and the first student text on fungal physiology – by Lilian E. Hawker – appeared in 1950 and the next year a similar volume by V. G. Lilly & H. L. Barnett was published in North America, as was J. W. Foster's *Chemical activities of fungi*, 1949, a most valuable and influential book. The first book on fungal genetics was that by J. R. S. Fincham & P. R. Day (1963). Lichens received their first general comprehensive treatment by Annie Lorrain Smith in her textbook of 1921.

Journals

After discussion between themselves, officers of the Royal Society of London (founded in 1660 and incorporated by Royal Charter of Charles II in 1662) concluded that there was need for a periodical restricted to scientific matters, especially accounts of experiments conducted before the Society, and on 1 March 1664 the Society's Council ordered 'that the Philosophical Transactions, to be composed by Mr. Oldenburg [the Society's Secretary], be printed the first Monday of every month'. The first number of the *Philosophical Transactions: giving some Accompt of the present undertakings, Studies, and Labours of the Ingenious in many considerable parts of the World* appeared on Monday, 6 March 1665, and apart from a short break during 1676–83 publication has continued ever since. Although antedated by the at first rather similar but more broadly based *Le Journal des Sçavans* (begun in Paris the preceding January and from which extracts were published in the first number of the *Phil. Trans.*) the *Philosophical Transactions* may justifiably be claimed as the first scientific journal. For the first forty-six volumes (1665–1750) it was published in 497 numbered parts by the Secretary of the Royal Society, to whom any profits accrued. Then, on adoption as the Royal Society's official journal, a committee took over the editorial work and issues were no longer numbered consecutively.

A parallel development was taking place in France where informal meetings of scientists which began soon after the establishment, in 1635, of the Académie Française de la Littérature, resulted in the formation of the Académie des Sciences which held its first meeting in December 1666. Also under royal patronage – but the members received pensions so that they could devote all their time to research – and at first mainly concerned with mathematics and physics, the Academy's interests soon broadened to cover biological topics. From the first a record was kept of the proceedings of the Academy and from 1698 these were published as the *Mémoires de l'Académie Royale des Sciences*, the second earliest periodical from which mycological contributions (including those by de Tournefort, Marchant, and de Jussieu) have been noticed in the present review.

Towards the end of the eighteenth and during the early years of the nineteenth centuries, scientific periodicals multiplied.

Many were short lived and of limited circulation (and so are today only too frequently very difficult to locate for consultation) but increasing numbers contained articles of mycological interest. It was, however, the fifty years from about 1825 that saw the birth of the scientific periodicals in which the results of the major mycological investigations of the nineteenth century were reported. The Parisian *Annales des sciences naturelle*, begun in 1824, appeared in a number of concurrent and successive series and included classical contributions from, among others, Léveillé, the Tulasnes, van Tieghem, Raulin, and Noël Bernard. In Germany the *Botanische Zeitung* (subsequently edited by de Bary who was also a frequent contributor) started in 1843 and in the eighteen-fifties *Hedwigia* (1852) and Pringsheim's *Jahrbuch für wissenschaftliche Botanik* (1858). Other new periodicals included the North American *Botanical Gazette* (1875–), the German *Berichte der deutschen botanischen Gesellschaft* (1883–), and the English *Annals of Botany* (1887–), to mention but a few of the many botanical journals which have regularly or at one time or another provided major outlets for the results of mycological investigations.

In 1872 M. C. Cooke founded the journal *Grevillea*, 'a Quarterly Record of Cryptogamic Botany and its Literature'. This had a strong mycological bias and was published until 1894 (for the last few years under the editorship of George Massee) but the first journal to be exclusively devoted to mycology was the monthly *Revue Mycologique* started by C. Roumeguère of Bordeaux in 1879 and continued after his death in 1892 by his son-in-law, F. Ferry, until 1906. In 1885 W. A. Kellerman, J. B. Ellis, and B. M. Everhart in the United States began the *Journal of Mycology* which after a somewhat chequered career was replaced in 1909 by *Mycologia* published by the New York Botanic Garden, and the *Annales Mycologici* (1903) was started to meet the needs of German mycologists. It was, however, the formation of mycological societies in different countries that provided the major stimulus for the initiation of specifically mycological journals as can be seen from Table 7.

Some of these society journals such as those of the French, English, and American mycological societies have, though maintaining their regional flavours, become periodicals of international standing. One underlying reason for this pattern lies in a peculiarity of scientific literature. It has always been

Table 7. *Primary mycological journals published up to 1950*

1879–1906	*Revue Mycologique*	Toulouse
*1885–	*Bulletin Trimestriel de la Société Mycologique de France*	Paris
1885–1908	*Journal of Mycology* (replaced by *Mycologia*)	Manhattan
*1897–	*Transactions of the British Mycological Society*	London
1903–8	*Ohio Mycological Bulletin* (published by W. A. Kellerman)	Columbus, Ohio
1903–44	*Annales Mycologici* (continued as *Sydowia*)	Berlin
1908–24	*L'Amateur de Champignons*	Paris
*†1909–	*Mycologia*	New York
1912–15	*Mycologische Centralblatt*	Jena
*1914–36	*Bulletin de la Société Mycologique de Geneve*	Geneva
1915–22	*Pilz- und Krauterfreund* (continued as *Zeitschrift für Pilzkunde*)	Heilbronn
*1919–	*Mykologický Sborník*	Prague
*1922–	*Zeitschrift für Pilzkunde*	Heilbronn
*1923	*Schweizerische Zeitschrift für Pilzkunde*	Bern
1924–31	*Mykologia*	Prague
1925–	*Mycological Papers*	Kew
*1929–58	*Fungus*	Wageningen
*1931–	*Friesia*	Copenhagen
1936–	*Revue de Mycologie*	Paris
*1936–8	*Mitteilungen der Österreich Mykologisches Gesellschaft*	Vienna
1944–7	*Magyar Gombászati Lapok. Acta Mycologica Hungarica*	Budapest
*1947–	*Česka Mykologie*	Prague
1947–	*Sydowia*	Vienna
*1950–	*Karstenia*	Helsinki

* A society journal.
† Adopted as the journal of the American Mycological Society in 1931.

relatively easy to obtain financial support for scientific research, particularly from grant-giving bodies and indirectly from universities which, provided their staffs do a minimum of required teaching, have usually allowed considerable freedom in the matter of how the remainder of their employees' time is spent.

But always very little money has been made available for the necessary publication of the results of scientific research and this has encouraged scientists (including mycologists) to unite in co-operatives (the so-called 'publishing societies') to arrange for the publication of their own work. In general, research workers have benefited financially. They normally buy the journals of the societies of which they are members at prices lower than those of sales to libraries and the general public. This, combined with inflation, has usually allowed members of scientific societies on retirement to sell runs of the journals thus accumulated at a profit. Since the Second World War the realisation that the results of scientific research are a marketable commodity, particularly in currently fashionable areas such as microbiology where some societies have made very considerable incomes, has not escaped the notice of commercial publishers who have launched many new journals – some needed and of a high quality, others deficient on both counts.

For all science it has been estimated[4] that the total number of current scientific and technical serial publications in 1961 was of the order of 35000 and perhaps 15000 of these contained material of biological interest. Current mycological publication is widely scattered through some 3000–4000 journals and reports, it is believed.

Secondary publications

It has always been customary for writers to cite their sources and this practice has provided useful clues to the retrieval of earlier publications. With the increase in the size of the literature independent bibliographies were welcomed as useful tools and they became so numerous that by the twentieth century even a bibliography of bibliographies became necessary for the arts. The most famous early botanical bibliography was that of Albrecht von Haller (*Bibliotheca botanica*, 1771–2, Zürich) and, during the nineteenth century and since, G. A. Pritzel's *Thesaurus literaturae botanicae*, edn 2, 1871, with its 'supplement' by Daydon Jackson have proved invaluable compilations for practising botanists, librarians, second-hand booksellers, and historians. The first major mycological compilation was the 500-page survey by Paulet (1790) in the first volume of his *Traité des champignons* in which he summarised chronologically the literature on fungi from the earliest times to 1787 and attempted to identify the fungi mentioned and to update their nomenclature. It was a century later that G. Lindau and Paul Sydow published the five-volume *Thesaurus litteraturae mycologicae et lichenologicae* (1908–15), a compilation of all scientific books and papers dealing with fungi and lichens published up to 1910, which are listed by authors and also classified by subject. Ciferri (1957–60) extended the survey (with a strong taxonomic bias and the omission of much literature dealing with fungal disease in plants, animals, and man) from 1911 to 1930. This compilation is extraordinarily useful and the *Thesaurus* has been the work most frequently consulted during the writing of this history. It also marked the virtual end of private enterprise in this field. The amount of literature had become too great for an individual to scan in detail. In 1918 *Botanical Abstracts* was begun as an independent journal to survey the current world literature of the plant sciences. In 1928 it ceased publication on being absorbed by *Biological Abstracts* which is currently the only journal designed to survey the whole of biological literature excluding clinical medicine. *Biological Abstracts* now includes reference to approximately a quarter of a million items annually and since 1967 all abstracts referring to fungi and lichens have been re-issued as a separate publication – *Abstracts of Mycology.*

Normally the higher the quality of a paper reporting experimental results the shorter its effective life. The results are accepted or they suggest the next experiment and the paper is subsequently read only by historians. Descriptive work and taxonomic proposals have a very much longer life and taxonomists are thus more interested in the older literature and frequently use century-old publications at the bench. (This is reflected in the dictum that it is the experimental work that sells the current numbers of any journal, taxonomic papers the back numbers). Hence the tattered appearance of the *Thesaurus* in any working mycological herbarium.

Taxonomists have always been great compilers. The first compilation of generic and specific names of fungi, both accepted names and synonyms, was that by Wesceslao M. Streinz in his *Nomenclator fungorum* of 1862. Streinz used taxonomic judgement (mostly by accepting that of others) in his compilation as did Pier Andrea Saccardo of Padua (Fig. 85), where he was professor of botany, in the monumental *Sylloge fungorum hucusque cognitorum*. Begun in 1882 and continued by collaborators and successors as a series of 25 volumes, up to 1931, this covered fungal taxa proposed up to 1920. In 1972, a supplementary volume (vol. **26**) extended the cover by another thirty years. Saccardo gave short Latin descriptions of all the genera, species, and other taxa which he recognised and his compilation involved many new taxonomic arrangements made in the light of his wide first-hand experience of fungal taxonomy. The *Sylloge* will for long remain a basic work for mycological taxonomists.

In 1909 F. E. Clements brought out a volume, *The genera of fungi*, based on an English translation of the keys to the orders, families, and genera of fungi in Saccardo's *Sylloge* and in 1931, in collaboration with C. L. Shear, a second, enlarged and illustrated, edition appeared. The latter no doubt merited the savage review it received in *Mycologia* but twenty-three years later it had to be reprinted unchanged as it was one of the few books which every practising taxonomic mycologist kept within reach.

Meschinelli in 1896 compiled the fossil fungi and Zahlbruckner, no doubt inspired by Saccardo's *Sylloge*, between 1921 and 1940 brought out the ten volumes of his *Catalogus lichenum universalis*, a comprehensive compilation of lichen taxa and synonymy, which is a basic work for lichen taxonomy.

Modern taxonomists are more inclined to ask for facts on which they can base their own taxonomic judgements and two useful aids – the *Index of Fungi* and the *Bibliography of Systematic Mycology* – are now supplied by the Commonwealth Mycological Institute, Kew, on the original initiative of S. P. Wiltshire, director of the Institute from 1940 to 1956.

In the autumn of 1881 Charles Darwin decided to devote some of his money to the advancement of those sciences 'which had been the solace of what might have been a painful existence'. The *Index Kewensis*,[5] a compilation of the generic and specific names of flowering plants, was the somewhat unexpected result. This compilation by the Royal Botanic Gardens, Kew, began publication in 1893 and the two original volumes have (up to 1974) been supplemented by fifteen more, of decreasing size but each still showing a surprisingly wide scatter of the entries in time as long-overlooked names are recorded.[6] In 1940, Dr Wiltshire introduced the half-yearly *Index of Fungi* (with cumulative indexes each decade) to do for fungi what the *Index Kewensis* did for flowering plants and since the commencement of the fourth volume in 1971 lichen names have also been included. The *Index of Fungi* is a compilation of names (with their citations and typifications) proposed from 1940 onwards. Petrak had published lists of names proposed from 1920 (then the end of Saccardo's *Sylloge*) to 1939 and these have been republished by the Commonwealth Mycological Institute together with a long list of 'Petrak omissions'. The many names missed by Saccardo have yet to be compiled.

The *Bibliography of Systematic Mycology* has, beginning with the literature of 1943, provided yearly, or half-yearly, classified listings of books and papers on taxonomic mycology noted by the abstracting service of the Commonwealth Mycological Institute. It was again the indefatigable Franz Petrak who tried to fill the gap left by the *Thesaurus* by compiling the taxonomic fungal literature as lists of the 1922–35 publications which appeared in Just's *Jahrbuch* between 1930 and 1944.

Mycology has not the equivalent of the *Index Londonensis*, which supplements the *Index Kewensis* by listing illustrations of flowering plants, but this desideratum is in part met by Saccardo's *Index iconum fungorum* (*Sylloge*, 1910–11, **19–20**) and for larger fungi of Europe and North Africa by the *Dictionnaire iconographique* by M. C. de Laplanche, 1894.

Host indexes have always proved themselves popular and useful tools. Three of the most famous are the one in Saccardo's *Sylloge* (1898, **13**), Oudemans' *Enumeratio systematica fungorum*, 1919–24, and A. B. Seymour's, *Host index of fungi of North America*, 1929. Although confined to plants grown in Europe, Oudemans' five bulky volumes are of wider scope than might be imagined for if a tropical plant of economic importance had been grown in a greenhouse in Europe all the fungal parasites recorded for that host were compiled.

UNIVERSITIES

The student–teacher relationship has always played an important part in disseminating knowledge and stimulating research. Able teachers attract able students. Mycology owes much to these associations. For example, the American L. D. von Schweinitz went to the Theological Institution at Niesky, Prussia, in 1798 for his theological training as a pastor of the

Moravian church because of his own interests in natural history which attracted him to study under his co-religionist Professor J. B. de Albertini with whom during his residence in Europe he collaborated in producing the *Conspectus fungorum in Lusitiae*, 1805. De Bary was attracted to Tübingen by his admiration for von Mohl and no mycologist has ever had a more influential series of students and research workers pass through his laboratory than did de Bary. The outstanding mycologists who worked under de Bary included Magnus, Fischer von Waldheim, Millardet, and Woronin at Freiburg, Briosi, Rees, and Rostafinski at Halle, and among the many at Strasbourg, Farlow, Alfred Fischer, Eduard Fischer, Hartog, von Höhnel, Klebs, and von Tavel; also the microbiologists Winogradsky and Wakker. During the last quarter of the nineteenth century and well into the twentieth, the ambition of every aspiring young English-speaking botanist was to undertake postgraduate research in Germany and experience the 'new botany' as an escape from the dry systematic approach which then dominated British botany. Among the numerous well-known mycologists who studied in Germany may be mentioned Marshall Ward who studied under Sachs, R. A. Harper under Strasburger, Buller under Pfeffer at Leipzig and Robert Hartig at Munich, and Miss E. M. Wakefield under von Tubeuf at Munich. This fashion is long outmoded. Since 1945 North America has provided the vision of a New Atlantis.

Universities have traditionally established departments of zoology and botany. During the twentieth century, these biological departments have been supplemented by departments of bacteriology (usually with a medical bias), and more recently by departments of microbiology and virology. Particularly in North America plant pathology has achieved departmental status but mycology has rarely been given such recognition – in the United Kingdom never – and the emphasis on mycology in any one university has thus largely depended on the interests of the reigning professor of botany. This has in most universities prevented the establishment of a tradition for mycological teaching. The two exceptions among English universities have been Cambridge and the Imperial College, London, at both of which the mycological instruction has had a strong plant pathological bias. The first course of lectures and practical work on plant diseases at Cambridge was that on 'Fungoid diseases of plants' given in the autumn of 1893 by W. G. P. Ellis of St Catharine's College. There is little doubt that mycological work at Cambridge would have expanded and deepened but for the death of Marshall Ward in 1906 at the age of 52 just after the completion of the new botanical laboratories which he had designed. Plant pathological teaching was continued by a research student of Marshall Ward's, F. T. Brooks, who in 1936 succeeded to the chair of botany at Cambridge. At the Imperial College the initial inspiration came from V. H. Blackman, a fungal cytologist turned physiologist, who was joined in 1912 by William Brown who became the founder of a school, which achieved a worldwide reputation, for research into the physiology of fungal parasitism (see p. 159).

(see p. 159).

SOCIETIES

The first mycological society, the Société Mycologique de France, was founded in October 1884, under the patronage of the Société d'Emulation de Département des Vosges, by a small group of local naturalists including two doctors, Quélet and Mougeot, and two pharmacists, Boudier and Patouillard. Dr Lucien Quélet became the first president, Dr J.-B. Mougeot general secretary, and among the 128 foundation members more than a hundred were simple amateurs, 'dont le zèle constituait à peu près tout le bagage mycologique'. Medical men and pharmacists comprised the largest group of the original membership which included a number of eminent foreign mycologists (mostly from universities, in marked contrast to the almost total absence of members from the Université de France) and also, in what was clearly a predominantly middle-class society, two working men, probably patients of Dr Quélet, whose 'zèle fut recompensé par félicitations unanimes'. The society flourished but amateurs continued to predominate. According to Mangin, in 1907 these amateurs could be subdivided into three groups: the 'mycophages' (whose interest in fungi was gastronomic), the 'mycophiles' (interested in larger fungi for their own sake), and the 'mycologues' (amateurs who in effect attained the status of professional mycologists). The Society's journal, *Bulletin Trimestriel de la Société Mycologique de France*, has appeared continuously since 1885.

The next mycological society to be formed was the British, in

1.—The "Church in Danger" at Dinmore.   2.—Gathering the "Vegetable Beef Steak" and other Tree Fungi.   3.—Bog Fungi at Lyonshall Wood.   4.—The Drive to Dowaton.
5.—The Drive to Moor Court.   6.—Raking for Truffles.   7.—M. J. de Seynes, " Professeur agrégé à la Faculté de Médicine de Paris."   8.—M. Maxime Cornu, of the " Jardin des Plantes," Paris.

THE GREAT FUNGUS MEETING AT HEREFORD.

Fig. 102. Cartoon by Worthington G. Smith from *The Pictorial World*, 10 November 1877. (Original 16.5×23.5 cm.)

1896. The roots of this society go back to 1868 when Dr H. G. Bull of Hereford invited members of the Woolhope Naturalists' Field Club to 'a Foray among the Funguses', and by so doing introduced a term subsequently much used to designate mycological field excursions. The foray was a great success and became an annual event (culminating with a dinner at the Green Dragon Hotel in Hereford at which 'Fungi' always figured on the menu). It was on several occasions the subject of a cartoon in the national press (see Fig. 102)[8] and continued after Dr Bull's death in 1885 until 1892. Interest then moved to Yorkshire where the Yorkshire Naturalists' Union had initiated fungus forays in 1891 and it was at the Yorkshire Naturalists' meeting at Huddersfield in 1895 (Fig. 103) that a decision was made to found a national mycological society. The next year the inaugural meeting of the British Mycological Society[9] for 'the study of mycology in all its branches' was held at Selby in

Fig. 103. Founders of the British Mycological Society at Huddersfield, Yorks, September 1895. (*left to right*) *standing*: G. Massee
(1850–1917), Rev. W. W. Fowler (1835–1912), J. Needham
(1849–1914); *sitting*: Charles Crossland (1844–1916), M. C. Cooke
(1825–1914), Carlton Rea (1861–1946).

Yorkshire and the first autumn foray at Worksop, Nottingham-
shire, 1897. George Massee (Fig. 103) was the first president. The
'mycologue' Carleton Rea (Fig. 103), a barrister and, later, the
author of *British Basidiomycetae*, 1922, was elected secretary and
editor of the Society's *Transactions*; and from 1898 he was also
treasurer. No-one ever did more for the Society than Carleton
Rea during his thirty-four years of continuous service in one
office or another. Like the French society the British Mycological
Society was at first dominated by amateurs but as time passed the
Society became more and more professional, a change encouraged by the lack of a British plant pathological society so that
many plant pathologists joined the Society which developed
strong interests in aspects of plant disease.

The Mycological Society of America,[10] always the most professional, was not started until 1931 – twenty years after the
founding of the American Phytopathological Society – as a
development from the Botanical Society of America (founded
1893) which had had a section of mycology from 1919. William

H. Weston was the first president and the Society adopted *Mycologia* as its official organ.

Publishing mycological societies were also organised in other European countries – Czechoslovakia (1921), Germany[11] and Switzerland (1922), Holland (1929), Denmark (1931), Finland (1948) – and more recently in other parts of the world. Publishing societies are supplemented, particularly in Europe, by many smaller regional groups or societies, mainly confined to amateurs. These are often affiliated to their national societies. For example, more than fifty such groups are affiliated to the Société Mycologique de France. The two best known amateur societies in North America are the Boston Mycological Club and Le Cercle des Mycologiques Amateurs de Quebec, both now affiliated to the Mycological Society of America.

At the international level mycologists during the present century have met together at successive meetings of the International Botanical Congress (which has frequently had a special section for mycology) (Fig. 104) and it was not until 1971 that the

Fig. 104. Some mycologists at the International Congress for Plant Science, Ithaca, N.Y., 1926. (*left to right*): E. J. Butler (1874–1943); H. H. Whetzel (1877–1944); J. C. Arthur (1850–1942) and H. Klebahn (1859–1942), noted for their studies on the rust fungi; see 1934, 1904; A. F. Blakeslee (1874–1954).

first independent International Mycological Congress was held (in the U.K. at Exeter, Devon). This resulted in the formation of an International Mycological Association which is recognised by the International Union of the Biological Sciences (I.U.B.S.).

HERBARIA

There is no elementary biological material. Every specimen is unique. The completeness of descriptions of organisms depends on the current state of knowledge and the prevailing techniques. It is therefore frequently only possible to confirm with certainty the identity of two collections of one organism by a comparison of specimens. Thus, much biological information has been, and still is, stored as collections of dead or living organisms.

The herbarium as an adjunct to botanical studies is said to have been invented in Italy by Luca Ghini,[12] professor of botany at Bologna, who sent dried plants gummed to paper to Mattioli in 1551. Herbaria[13] were first assembled as books and only later was each specimen affixed to a separate loose sheet when the resulting flexibility gave collections great potential versatility for the devising and illustration of diverse taxonomic arrangements.

Many fungi when dried make excellent herbarium material, as do lichens, and although the colour and shape of fruit-bodies of larger fungi may be lost, the morphology of spores and other minute structures is frequently found on microscopic examination not to differ significantly from that of living material.

Most countries now have national herbaria and many of these, for example those at Kew, New York, Paris, Stockholm, and Leiden, to mention but a few, include notable collections of fungi and lichens from many parts of the world and much valuable historical material. These herbaria are, and always have been, dominated by vascular plants which because of their size and number occupy the greater part of the storage space and although mycologists are employed by these institutions they are always in a minority. It is only during recent years that independent mycological herbaria, such as those at the Commonwealth Mycological Institute at Kew (begun in 1922) and the National Fungus Collections of the United States Department of Agriculture at Beltsville, Washington, D.C. (1955),[14] have been established.

Mycologists have always been keen collectors and the personal herbaria of a number of the founders of taxonomic mycology are still available: for example, Persoon's at Leiden, E. M. Fries's at Uppsala, M. J. Berkeley's at Kew. Individual mycologists have also, since the early nineteenth century, issued sets of exsiccati of the fungi of their region or of groups in which they have had a special interest. These, to cite a few typical examples, included the regional issues of Berkeley's *British Fungi*, 1836–43, (comprising 350 dried specimens to illustrate Berkeley's account of fungi in Smith's *English Flora*, 1836, **5**, part 2), *Fungi europaei exsiccati*, 1859–1905 (4500 numbers) by Rabenhorst and others, and J. Ellis's *North American Fungi*, 1878–94 (3200); and the taxonomic series by Fries (*Scleromyceti sueciae*, 1819–34 (450) and Paul Sydow (*Uredineen*, 1888–1906 (2050), *Ustilagineen*, 1894–1904 (350)).

CULTURE COLLECTIONS

Culture collections, the microbiological equivalent of zoological and botanical gardens, were a relatively late development. The techniques for pure culture were a necessary prerequisite. The first 'public' collection of bacteria and fungi was that of Frantisek Král who after spending the first thirty-nine years of his life manufacturing chemical and physical instruments became professor of bacteriology, mycology, and microbiological techniques at the German Technical University of Prague where he started his collection of cultures which he made commercially available. After Král's death in 1911, the collection was acquired by E. Přibram of the Sero-physiological Institute at Vienna but with the growth of national collections business dwindled. Between the wars Přibram migrated to the United States taking part of the collection with him. The residue was destroyed during the Second World War. A number of the Král–Přibram cultures are, however, still to be found in some of the larger collections of today.

The most famous collection of fungus cultures is that of the Dutch Centraalbureau voor Schimmelcultures. The decision to assemble this collection was made at a meeting of the Association Internationale des Botanistes at Leiden in 1903 and the work begun three years later at Utrecht under the direction of Dr F. A. F. C. Went. The first catalogue appeared in the *Botanisches Centralblatt* as an unnumbered page in 1906 (Fig. 105). In the

Centralstelle für Pilzkulturen.

Die Kulturen folgender Pilze werden entweder im Tausch gegen andere Kulturen, welche die Centralstelle noch nicht besitzt, abgegeben, oder gegen Vorberbezahlung pro Kultur von fl. 1.50 (holl. Währung) für Mitglieder; und fl. 3 für Nichtmitglieder der Association. Die Anfragen sind zu richten an Herrn Prof. Dr. F. A. F. C. W e n t, Universität U.t r e c h t (Holland).

Acrothecium lunatum Wàkker.
Alternaria tenuis Nees.
Aspergillus ~~auricomus~~ Guéguen.
,, candidus Link.
,, clavatus Desmazières.
,, flavus Link.
,, ~~fumigatus~~ Fresenius.
,, glaucus Link.
,, niger v. Tieghem.
,, Oryzae (Ahlberg) Cohn.
,, ostianus Wehmer.
,, varians Wehmer.
,, Wentii Wehmer.
Botryosporium pyramidale Cost.
Botrytis Bassiana Balsamo.
,, cinerea Pers.
,, parasitica Cavara.
Chaetomella horrida Oudem.
Chaetomium indicum Corda.
,, Kunzeanum Zopf.
Chlamydomucor Oryzae Went et Prinsen Geerligs.
Cladosporium butyri Jensen.
,, herbarum Link.
Clasterosporium Lini Oudem.
Diplocladium minus Bon.
Dipodascus albidus de Lagerheim.
Endomyces Magnusii Ludwig.
Epicoccum purpurascens Ehrenb.
Fusarium Solani (Mach.) Sacc.
,, aquaeductum (Rabenh.) Radlkof.
Gymnoascus candidus Eidam.
,, Reessii Baranetzky.
,, setosus Eidam.
,, spec. von Miss Dale.
Hypocrea Sacchari Went.
Monascus Barkeri Dangeard.
,, purpureus Went.
Monilia candida Bon.
,, humicola Oudem.
,, javanica Went et Prinsen Geerligs.
,, sitophila (Mont.) Sacc.

Monilia variabilis Lindner.
Mortierella isabellina Oudem.
Mucor alternans van Tieghem.
,, circinelloides van Tieghem.
,, corymbifer Cohn.
,, javanicus Wehmer.
,, (Amylomyces Calm.) Rouxii Wehmer
,, spinosus v. Tieghem.
Mycogone puccinioides (Preuss.) Sacc.
Myxococcus ruber Baur.
Penicillium brevicaule Sacc.
,, olivaceum Wehmer.
,, purpurogenum Flerofi.
,, roseum Link.
Phycomyces nitens Kunze.
Pyrenochaeta humicola Oudem.
Pyronema confluens Tul.
Rhizopus arrhizus Fischer.
,, nigricans Ehrenb.
,. Oryzae Went et Prinsen Geerligs.
Saccharomyces rosaceus Frankl.
,, Vordermannii Went et Prinsen Geerligs.
Sordaria humicola Oudem.
Sporodinia grandis Link.
Sporotrichum bombycinum (Corda) Rabenh.
,, griseolum Oudem.
,, roseolum Oudem et Beijer.
Stachybotrys alternans Bon.
,, lobulata Berkeley.
Stemphylium botryosum Wallr.
,, macrosporoideum (B. et Br.) Sacc.
Thamnidium elegans Link.
Thielaviopsis ethaceticus Went.
Trichocladium asperum Harz.
Verticillium glaucum Bon.
,, rufum (Schwabe) Rabenh
Wurzelpilz (Symbiont) von Cattleya Beijerinck.

Fig. 105. First catalogue of the Centraalbureau voor Schimmelcultures, Baarn. Published in the *Botanisches Centralblatt*, 1906. The crossed-out words were deleted by hand before publication.

following year the collection was transferred to the care of Dr Johanna Westerdijk, director of the Wille Commelin Scholten Phytopathological Laboratory in Amsterdam and the name of the collection was changed to Centraalbureau voor Schimmelcultures. The growth of the collection into the internationally famous organisation it now is was due to the enthusiasm and drive of Dr, later Professor, Westerdijk (Fig. 106), an impressive and talented woman of great versatility and much energy, who, aided by a loyal staff, saw the collection safely through the hazards of two major wars. The collection was moved to Baarn in 1920 and now, including the yeasts (which since 1922 have been dealt with as a special division at the Technische Hogeschool at Delft) comprises more than 10000 strains.[15]

Fig. 106. Johanna Westerdijk (1833–1961).

Many other national collections were started.[16] In the United States the Museum of Natural History had from 1911 maintained a Bureau for the Distribution of Bacterial Cultures (under C.-E. A. Winslow) and after a period from 1922 to 1925 at the Army Medical Museum in Washington, D.C., the collection became the American Type Culture Collection under an arrangement by which Charles Thom and Margaret B. Church of the U.S.D.A. maintained the fungus cultures at Washington, the bacteria being held in Chicago. The idea was taken up in Britain where, in 1920, the Medical Research Council sponsored a National Collection of Type Cultures of bacteria at the Lister Institute and the next year the scope of the collection was extended to fungi in co-operation with the British Mycological Society which set up a standing advisory committee. In 1947 the N.C.T.C. again restricted its holdings to bacteria and the main collection of fungal cultures became the responsibility of the Commonwealth Mycological Institute, Kew, whose collection was supplemented by special collections of yeasts, wood-rotting fungi, and fungi of medical and veterinary importance at other centres.

Other well-known collections include those in France at the Institute Pasteur, Paris, and the Museum d'Histoire Naturelle and those in Japan[17] (including the important collection held by the Institute for Fermentation at Osaka) which since 1951 have been co-ordinated by the Japanese Federation of Culture Collections of Microorganisms. There are now so many official collections of cultures that in 1968 the first meeting of the International Conference of Culture Collections was held in Tokyo.

The traditional method of preserving cultures of fungi is on slopes of solid media either at laboratory temperature or in a refrigerator at about 4 °C. A simple variant of this method (first introduced by Ungerman in 1918) is to cover the culture with sterile high quality mineral oil (such as medicinal paraffin) when subcultures need to be made only at intervals of two to three years or longer (Buell & Weston, 1947). From about 1909 bacterial cultures were preserved by vacuum desiccation from the frozen state and later this method was found to be applicable to yeasts. Freeze-drying (or lyophilisation) was widely adopted for mycelial fungus cultures after the Second World War (see Raper & Alexander, 1945). More recently still storage at the temperature of liquid nitrogen has been finding increasing favour (see Onions (1971) for details). In this connection it is interesting to recall that as early as 1909 Buller freeze-dried sporophores of *Schizophyllum commune* and also found that such fruit-bodies survived subjection to the temperature of liquid nitrogen. More than fifty years later some of Buller's preparations were found to be still living.[18]

NOMENCLATURE

Since earliest times biologists have applied different names to the same organism and the same name to different organisms and the herbalists frequently compiled synonymies of the vernacular and latinised names of the organisms they described. By the nineteenth century the position had become acute, so many names had been and were being proposed that the regulation of biological nomenclature became a necessity. The first move was towards the regulation of plant names when in 1867 an International Botanical Congress in Paris adopted a set of rules, drawn up by Alphonse de Candolle, entitled *Lois de la nomenclature botanique*, which subsequently became the basis of the *International Rules* [since 1952, *Code*] *of Botanical Nomenclature* first approved by the Vienna International Botanical Congress in 1905 and modified by later Congresses. Similar but independent nomenclatural codes were established by zoologists in 1901 and bacteriologists in 1958.

Fungal nomenclature, fungi being traditionally classified as plants, is subject to the Botanical Code which like the Zoological and Bacteriological Codes is based on a few fundamental principles. The Codes are not penal and in the words of the third (1935) edition of the Botanical Code 'The rules of nomenclature should be simple and founded on considerations sufficiently clear and forcible for everyone to comprehend and be disposed to accept'. They thus tend to reflect current practice and to be slowly modified to meet new demands. The details of the Botanical Code are, however, far from simple and their interpretation can lead to heated argument for few subjects arouse stronger emotions than nomenclature – what is commonsense to one is pedantry to another.

The main provisions of the Botanical Code, which are designed to allow every taxonomist the freedom to designate any taxon by an internationally acceptable name which reflects his own taxonomic opinion, decree that specific names take the form of Latin binomials (as introduced by Linnaeus) composed of a generic name and a specific epithet and lay down the conditions that must be fulfilled when publishing new names and making name changes necessitated by change in rank or taxonomic position. When making a choice of names considered to be synonymous, priority of publication is the general guide, that is to say the earliest name in line with the current version of the Code takes precedence. The Code also deals with the ranks of taxa and the names denoting them.

In order to apply the principle of priority a starting point for nomenclature had to be set and for vascular plants this date was 1 May 1753, the day of publication of the first edition of the *Species plantarum* by Linnaeus. This date also marks the beginning of the nomenclature of lichens, myxomycetes, and fossil fungi (and also at present of bacteria) but because of the unsatisfactory treatment of fungi by Linnaeus in the second (1912) edition of the Code, as adopted by the 1910 Brussels International Botanical Congress, two later starting points for the nomenclature of fungi were intercalated: 31 December 1801 (Persoon, *Synopsis methodica fungorum*) for Uredinales, Ustilaginales, and Gasteromycetes and 1 January 1821 (Fries, *Systema mycologicum*, **1**) for other fungi. This decision was probably unwise but as during recent years taxonomic mycologists have paid increasing respect to the Code any radical change would now cause too much confusion for general acceptance.

Also in 1910, a special provision was included in the Code to cover the names of fungi having a pleomorphic life cycle. This Rule (Article 59 in the current (1972) edition of the Code) now reads:

In Ascomycetes and Basidiomycetes (inclusive of Ustilaginales) with two or more states in the life cycle (except those which are lichen–fungi), the correct name of all states which are states of any one species is the earliest legitimate name typified by the perfect state…However, the provisions of this Article shall not be construed as preventing the use of names of imperfect states in works referring to such states; in the case of imperfect states, a name refers only to the state represented by its type.

As has so often been recalled, nomenclature is subservient to taxonomy. Only after taxonomic decisions have been made are nomenclatural procedures invoked. However, codes of nomenclature by laying down taxonomic ranks or categories thereby greatly influence taxonomic practice. It is botanical practice to distinguish taxa on morphological grounds and in the 1952 edition of the Code (the 'Stockholm Code') the special non-morphological category of *forma specialis* was introduced for the convenience of mycologists. Thus the nomenclature of taxa of this rank are now excluded from the Code as the Note to Article 4 indicates:

In classifying parasites, especially fungi, authors who do not give specific, subspecific or varietal value to taxa characterized from a physiological standpoint but scarcely or not at all from a morphological standpoint may distinguish within the species special forms (*formae speciales*) characterized by their adaption to different hosts, but the nomenclature of *formae speciales* shall not be governed by the provisions of the Code.

Neither is the nomenclature of physiological races and biotypes.

In conclusion, attention may be drawn to one historical development which affected the theory underlying both botanical and mycological practice. The names regulated by the Code were originally assumed to be those of natural groupings, e.g. to the species of which Darwin speculated on the origin, but the development by taxonomists of the 'type method' during the early years of the present century was reflected in the 1935 edition of the Code (the 'Cambridge Rules') which regularised the procedure by which every taxon is based on a type – or as Principle II of the current (1972) Code has it 'The application of names of taxonomic groups is determined by nomenclatural types'; a nomenclatural type – a genus for a family, a species for a genus, and a specimen or in the absence of a specimen a description (with or without an illustration) for a species or taxon of subspecific rank – being 'that constituent element to which the name of the taxon is permanently attached whether as a correct name or as a synonym' (Article 7, 1972 Code). Although authors are now recommended to indicate the place where a type specimen is permanently conserved, many type specimens were not preserved (according to Corda who as a young man acted as his artist, Krombholz even ate the specimens on which a number of his species of larger fungi were based), or have been lost. Many early names have yet to be typified but since 1958 publication of the name of a new taxon of the rank of family or below is only valid if a type has been designated (Article 37). Names controlled by the Code are thus names of taxonomic groupings which may be considered to be only samples of the natural groupings from which they are derived and to which the same names are applied.[19]

Epilogue

It is abundantly clear from the history of mycology that fungi show diverse and intimate relationships to human affairs. If some fungi due to their poisonous, allergenic or pathogenic properties induce disease in man, his domesticated animals, and staple crops, while others cause deterioration of his possessions, fungi also make a major contribution to his survival. Some fungi can be eaten, others play an important part in the preparation of many types of food and drink and provide a wide range of stimulants, drugs, and useful organic products. They do much to ensure the maintenance of soil fertility and by symbiotic associations improve the growth of many plants. Indeed, by virtue of their ability to degrade organic residues fungi play a major role in the preservation of an environment which allows life to continue.

During the 250 years since the experimental study of fungi began, much has been learned regarding the essentials of fungal structure, nutrition, and life histories, knowledge which has frequently emphasised the distinction of fungi from both green plants and animals. Much however, still remains to be done. Only a small fraction of the total fungal population has been taxonomised and with the increasing pressures of modern civilisation on the natural environment, such as the exploitation of tropical rain forest, the expansion of agriculture and urban-isation, it is certain that many of the fungi, as with insects, of today will become extinct before they are described. But fungi are both versatile and adaptable always ready to colonise new substrata, whether it be aircraft fuel or optical equipment, and of a plasticity which enables them quickly to develop new forms able to parasitise the latest novelty evolved by the breeders of crop plants. Their study will of necessity continue as an increasingly important branch of biological science, both pure and applied.

Notes on the text

1. INTRODUCTION

1. The pronunciation of 'fungi' is various. The English tradition is that expressed by M. C. Cooke in *A plain and easy account of British fungi*, 1862, where he wrote (p. 4) 'in the singular the *g* should be hard as in *gum*, whilst in the plural *fungi* has the *g* soft, as *Fun-ji*'.

2. No serious attempt has been made to check or extend the references to fungi in the Greek and Roman classics compiled by Lenz (1859), the Rev. William Houghton (Houghton, 1885), E. Roze (in Richon & Roze, 1885–9), and Professor A. H. R. Buller (Buller, 1915*a, b*) who was aided in this field by W. B. Grove, schoolmaster, mycologist, and classical scholar. I am grateful to Charles Garton, professor of classics at the University of Buffalo, Ohio, for some perceptive comments on the notice given to fungi by classical writers.

3. The tabulations by Houghton (1885:48) and Buller (1915*a*:49–50) give additional details.

4. Rolfe & Rolfe (1925:294–300) list many of the variant spellings of 'mushroom' and 'toadstool' as does *The Oxford English Dictionary*.

5. Rolfe & Rolfe (1925: chap. 3) cite examples of references to fungi in fiction and literature.

6. For additional details of mushrooms in art see Wasson & Wasson (1957) **2**: 351–63.

7. Exactly when and by whom is uncertain, see R. S. Clay & T. H. Court, *The history of the microscope*, 1932 (London) for details. The instrument was first named *microscopium* in 1625 by John Faber, a member of the Italian Accademia dei Lincei, *fide* J. Needham & G. D. Lu, *Proc. Roy. microscop. Soc.* **2**: 128 (1967).

8. By J. C. Jacobsen, proprietor of the Carlsberg brewery, for research on chemical and physiological aspects of brewing, with Rasmus Pedersen as the first director. E. C. Hansen later became director of the Physiological Laboratory. Since 1878 the results of the research work in this laboratory have been published in the *Comptes rendus des travaux du Laboratoire Carlsberg*.

9. See P. H. A. Sneath in G. H. M. Lawrence (ed.), *Adanson* **2**:471–98 (1964).

2. THE ORIGIN AND STATUS OF FUNGI

1. Houghton (1885): 44.
2. *Ibid*. 43.
3. *Ibid*. 45.
4. Buller (1915*b*): 2.
5. Houghton (1885): 23–4.
6. *Ibid*. 25.
7. Buller (1915*b*): 2–3.
8. *Ibid*. 3.
9. G. Sarton, *Six wings: men of Science in the Renaissance*, 1957: 87. London.

10. Buller (1915*b*): 3.

11. First noted by Clusius (1601): 287 (*Fungus minimus*).

12. 'There is another kind of *Mushroom*, which rises from the Earth, which first of all appears shut, but in a days time is open, and then it represents the forme of a *Mushroom*, with a round dish or little house, like the cup out of which *Oake Acorns* are taken; in this cup is found Five round grains of seed, of the size of Raddish-seed, but pellucid; as *Christall*; after these grains fall out of the cup on to the superficies, or into chinks of the earth; they are cherished by the heat of the sun, till they begin to live, afterwards they get feet; and in three years time they attain to their full growth, all these things I tryed twice Two years together'. J. Goedart, *Metamorphosis et historia naturalis insectorum*, 1662; English transl. by M. L[ister] (1682): 137–8 (sect. 10, Of Spiders). York.

13. F. C. Medicus, Über den Ursprung und die Bildungsart der Schwämme, *Vorles. Churpfälz. Phys. Oek. Ges.* **3**: 331–86 (1788); French transl. by de Reynier, *J. de Physique* **34**: 241–7 (1789).
p. 246.
'De tout ce que j'ai rapporté je puis conclure,
1°. Tout ce que est dans un état pourriture, ne sauroit plus servir à la production des champignons.
2°. Tous les végétaux donnent naissance à des champignons, lorsqu'ils sont au premier degré de la décomposition.'
p. 247.
9°. Ansi lorsque des parties végétales mortes se trouvent au premier degré de décomposition, des qu'elle est accompagnée d'un certain degré de chaleur & d'humidité, il s'en échappe une matière durée d'une force élastique & attractive qui se forme en champignons. Voilà ce que je nomme *cristallisation végétale.*'

14. *Fide* Ramsbottom (1941*a*): 294; for the Linnaeus/Münchausen correspondence see *ibid.* pp. 364–5.

15. For the Linnaeus/Ellis correspondence from which this and the following extracts are taken, see Ramsbottom (1941*a*): 297–300.

16. O. F. Müller, *Berlinische Sammlungen* **1**: 41–52 (1708); see Ramsbottom (1941*a*): 304, for a quotation.

17. For an English translation of Torrubia's paper, see G. Edwards, *Gleanings of natural history* **3**: 262, pl. 335 (1763). Torrubia's original illustration is reproduced by Ramsbottom (1941*a*): pl. 7, fig. *e*.

18. Tulasne (1861–5; Grove transl.): 48.

19. *Ibid.* 61.

20. Berkeley (1857): 265.

21. *Ibid.* 242.

22. See Bulloch (1938): 188–92, for illustrations and a more detailed account.

23. A. L. de Jussieu, *Genera plantarum secundum ordines naturales disposita, juxta methodum in horto regio Parisiensi exaratum anno 1774*, 1774: 2; English transl. by W. B. Grove in Tulasne (1861–5) **1**: 3.

24. Bulliard (1791–1812) **1**: 65; English transl. by W. B. Grove in Tulasne (1861–5) **1**: 4.

25. In an interesting discussion on the kingdoms of organisms, G. F. Leedale (*Taxon* **23**: 261–70 (1974)) proposes a modification of Whittaker's system by which protista are treated not as a kingdom but as a state, or evolutionary stage, in the plant, fungus, and animal kingdoms.

3. FORM AND STRUCTURE

1. This fresco is illustrated in colour by Wasson & Wasson (1957) **1**: pl. lxxvi and by J.-M. Croisille, *Les natures mortes campaniennes*, 1968: pl. G (Brussels).

2. In passing it may be noticed that according to art historians the Tree of Good and Evil in the medieval fresco in the chapel of the Abbaye de Plaincourault, France, which mycologists have interpreted as a representation of a fly agaric (*Amanita muscaria*), has nothing to do with fungi; see Wasson (1968): 178–80.

3. See Wasson & Wasson (1957) **1**: fig. 10, for a reproduction of this woodcut.

4. See A. de Bary, Die Schrift des Hadrianus Junius über den *Phallus* und der *Phallus Hadriani*, *Bot. Z.* **22**: 114–16 (1864).

5. See Ramsbottom, *Proc. Linn. Soc. Lond. 1931–32*: 76–9 (1932), for a historical account of the fungus-stone; at one time believed to be the petrified urine (*lapis lyncurius*) of the lynx (Paulet (1790–3) **1**: 23).

6. For an account of Vaillant, including the *Botanicon Parisiense* (which was published posthumously under the editorship of H. Boerhaave who supplied a preface), see J. Rousseau in P. Smit & R. J. Ch. V. ter Laage (eds.), *Essays in biohistory*, 1970: 195–228.

7. The basic source for the life of Micheli, his publications, and his manuscripts is *Notizie della vita e delle opere di Pier' Antonio Micheli*, 1858 (Florence) by G. Targioni–Tozzetti (published by A. Targioni-Tozzetti).

8. Buller (The function and fate of the cystidia of *Coprinus atramentarius*, etc., *Ann. Bot. Lond.* **24**: 613–30 (1910)) confirmed the structural function of cystidia but showed that they do not fall to the ground but undergo digestion. See also Buller (1909–50) **3**: 285–92.

9. There are sets of these plates in the libraries of the British Museum (Nat. Hist.) and the Linnean Society of London, and also at Florence.

10. For further details of this volume see Ramsbottom, *Bull. Jard. bot. Bruxelles* **27**: 773–7 (1957).

11. For accounts of mycological illustration and illustrators, see Emma A. Rea, *Trans. Br. mycol. Soc.* **5**: 211–28 (1916); L. C. C. Krieger, *Mycologia* **14**: 311–31 (1922).

12. For further details and an account of Bulliard, his work and publications, see E.-J. Gilbert, *Bull. Soc. mycol. Fr.* **68**: 5–131 (1952).

13. See Chapter 1, note 7 above.

14. John Aubrey, *Brief lives and other selected writings* (ed. Antony Powell), 1949: 128. London.

15. *Fide* E. J. H. Corner, *Nature* **218**: 798 (1968).

16. One distinction between spores which is of historical interest is that made by Berkeley (1857: 269), as a basis for his two primary divisions of the fungi (which were also accepted by M. C. Cooke, 1871), into those produced in sacs, which he called 'sporidia' (Sporidiiferi – ascomycetes, mucors, and other phycomycetes) and naked 'spores' (Sporiferi – hyphomycetes, basidiomycetes).

17. Transl. by G. W. Keitt, *Phytopath. Classics* **6**: 50–1 (1939).

18. *Summa vegetabilium Scandanaviae*, sect. 2, 1849.

19. See F. Zernike, How I discovered phase contrast, *Science* **121**: 345–9 (1955).

20. First commercially available in 1965 as the 'Stereoscan' developed by Cambridge Scientific Instruments Ltd.

21. The most comprehensive modern review of fungal morphogenesis is in Ainsworth & Sussman (1965–73) **2**: chaps. 2–7. This may be supplemented by A. G. Morton, Morphogenesis in fungi, *Sci. Progress* **55**: 597–611 (1967); S. Bartnicki-Garcia, Cell wall chemistry, morphogenesis and taxonomy of fungi, *Ann. Rev. Microbiol.* **22**: 87–108 (1968); G. Turian, *Différenciation fongique*, 1969 (Paris).

22. See Rolfe & Rolfe (1925): 7–11; Ramsbottom (1953): 112–26.

23. 'So from dark clouds the playful lightning springs,
    Rives the dark oak, or prints the Fairy-rings.'
Erasmus Darwin, *The botanic garden*, Part 1, 1791: ll. 369–70; also Additional notes xiii.

24. W. Withering, *An arrangement of British plants*, 1792, edn 2, **3**: 336–7.

25. De Bary (1866): 147–55; (1887): 343–52. Sussman & Halvorson (1966) may be supplemented by Sussman in Ainsworth & Sussman (1965–73) **2**: chap. 23, while Margaret C. Ferguson (1902) gives a good summary of early work on the germination of basidiomycete spores.

4. CULTURE AND NUTRITION

1. Reissued in facsimile 1957, New York, Basic Books Inc.

2. G. de La Brosse, *De la nature, vertu et utilité des plantes, divise en cinq livres*, 1628. Paris (Livre II, Champignons: 168–77).

3. De Bary used symbiosis as a general term to cover parasitism, mutualism, and lichenisation and distinguished 'antagonistic' and 'mutualistic symbiosis' (de Bary, 1879).

4. G. W. Martin, *Bot. Gaz.* **93**: 427 (1932).

5. See the reviews by Knowles (1821), Dickson (1838), and Ramsbottom (1941*b*) for further details.

6. Printed for the first time by Ramsbottom (1941*b*): 251–5.

7. See *Gdnrs Chron.* 1860: 599, 646.

8. Boulton (1884): 9.

9. Boulton (1884): 11.

10. See Boulton (1884) for further details.

11. For the early history of lichenology, A. von Krempelhuber, *Geschichte und Litteratur der Lichenologie* 1867, **1**; 1869–72, **2**, may be consulted.

12. J. M. Crombie, *Popular Science Review* July 1874 (p. 18 of reprint).

13. *J. Queckett microscop. Cl.* **5**: 120 (1879). The 'Nineteenth Century' was a leading monthly literary periodical.

14. Mycobiont and phycobiont were introduced by G. D. Scott, *Nature* **179**: 486–7 (1957).

15. Rylands' figures are reproduced by Rayner (1927): figs. 1, 2.

16. The large literature on orchid mycorrhiza is reviewed by Rayner (1927): esp. chap. 5. See also Ramsbottom, *Trans. Br. mycol. Soc.* **8**: 28–61 (1922).

17. H. F. Link, *Icones selectae anatomico-botanicae* **2**: 10, tab. vii, figs. 10, 11 (1840).

18. Tyndall (1881): 214.

19. From the 'Parisian engineering firm Wiesnegg under the name Chamberland's autoclave, about 1884', *fide* Bulloch (1938): 234.

20. *Mitth. kaiserl. Gesundheitesante* **1**: 301–21 (1881), *fide* Bulloch (1938): 331.

21. *Ann. Chem. Pharm.* **89**: 232–43 (1854), *fide* Bulloch (1938): 88–90, 301.

22. C. Vittadini, Della natura del calcino o mal del segno, *Mem. I.R. Ist. Lombardo Sci., Lett., Art.* **3**: 447–512 (1852).

23. See particularly Brefeld (1872–1912), **4**.

24. *Fide* Bulloch (1938): 222–3.

25. See *Gesammdte Werke von R. Koch* **1**: 274–84 (1912).

26. For more details see Hawker (1950): 68–72.

27. For supplementary data on the study of growth factors during the period covered by this review see Hawker (1950): 72–94; *Roy. Coll. Sci. J.* **14**: 65–78 (1944): and B. C. J. G. Knight, *Vitamins and hormones*, **3**: 105–288b (1945).

5. SEXUALITY, CYTOLOGY, AND GENETICS

1. Quoted by Phoebus (1842): 217.

2. English transl. by Grove (Tulasne (1861–5) **1**: 181).

3. In a memoir submitted to the French Academy of Sciences for the Grand Prix des Sciences Naturelle and never published but reported on by Brongniart (Vuillemin (1912): 27; Ramsbottom (1941a): 349).

4. *Grevillea* **4**: 53–63 (1875).

5. *Bull. bot. Soc. Fr.* **22**: 99–101 (1875).

6. *Bot. Z.* **33**: 649–52, 665–7 (1875).

7. See centenary celebration of Robert Brown's discovery of the nucleus of the vegetable cell, *Proc. Linn. Soc. Lond. 1931–2*: 17–54 (1932).

8. B. Kisch, *Trans. Am. phil. Soc.* N.S. **44**: 258–61 (1954).

9. As 'maiosis'; 'We propose to apply the terms Maiosis or Maiotic phase to cover the whole series of nuclear changes included in the two divisions that were designated as Heterotype and Homotype by Flemming', Farmer & Moore, *Quart. J. microsp. Sci.* **48**: 489 (1905).

10. *Jahrb. wiss. Bot.* **42**: 62 (1905). Haploid and diploid do not terminate in the suffix -oid (derived from the Greek *eidos*, form) as does 'diploid' the crystallographic term. As Strasburger (*Progressus Rei Bot.* **1**: 137 (1907)) noted, he intended the final 'id' to connect the two terms with the 'Idioplasma' of C. Naegeli and the 'Id' and 'Idant' of A. Weismann. Haploid and diploid are thus really three-syllable words, as they are in French and German.

11. Dangeard's discovery was commemorated by the designation (introduced by Juel, 1898) 'fusion dangeardienne' for the fusion of the conjugate nuclei and the organ in which such karyogamy occurred became known as a 'dangardien' (the 'dangardium' of Moreau, 1949). See J. P. van der Walt & Elzbieta Johannsen, *Antonie van Leeuwenhoek* **40**: 185–92 (1974) for a discussion.

12. This fusion was interpreted as nutritive. Dangeard and Harper's conclusion that there was also nuclear fusion in the ascogonium was the beginning of a heresy, which persisted for more than fifty years, that a second reduction division (designated 'brachymeiosis' by Dame Helen Gwynne-Vaughan) occurred in the ascus; a view finally disproved by H. E. Hirsch (*Mycologia* **42**: 301–5 (1950)), Irene M. Wilson (*Ann. Bot. Lond.* N.S. **16**: 321–9 (1952)), and others.

13. For revised maps of the same eight linkage groups, see G. L. Dorn, *Genetics* **56**: 630–1 (1967).

14. C. M. Leach in C. Booth (ed.), *Methods in microbiology* **4**: 609–64 (1971), gives details and additional information on studies of light on fungi.

6. PATHOGENICITY

1. According to Ovid (*Fasti* **4**:905 *et seq.*) and Pliny (*Nat. Hist.* **18**:69), *fide* Buller (1915*a*): 30; *Encycl. Brit.*, 1910–11, edn 11, **22**: 416. The Robigalia was the forerunner of the early Christian *litania major* (instituted by Pope Liberius (352–66)) which took place on the same date (St Mark's Day) and involved a procession which followed the same route to the Milvian bridge but then returned to St Peter's where mass was celebrated.

2. J. S. Elsholz, *Neuangelegter Gartenbau*, Aufl. 3, 1684 (Berlin), *fide* Orlob (1964): 197–8.

3. *Fide* Orlob (1964): 221–2.

4. By Henry Baker, *Natural history of the polype insect*, 1744: chap. xi, pl. xxii, figs. 9, 10.

5. See Plowright (1889): 302–4, for a reprint of the whole Act.

6. As described by Cecil Woodham-Smith in *The great hunger, Ireland 1845–9*, 1962. London.

7. Including the first edition (in one volume) of Paul Sorauer's *Handbuch der Pflanzenkrankheiten*, 1874, now (Aufl. 7) a multi-author multi-volume reference text.

8. See L. Chiarappa, Phytopathological organizations of the world, *Ann. Rev. Phytopath.* **8**: 419–39 (1970), for details of the current position.

9. On a recommendation by the International Botanical Congress, Amsterdam, 1935.

10. Raulin found growth of *Aspergillus niger* to be inhibited by silver nitrate at 1:1.6 parts per million by weight, by mercuric chloride at 1:512000, and platinum dichloride at 1:8000. Copper sulphate allowed some growth at 1:160.

11. Published as a supplement to the *Rev. Path. vég. Ent. agric. France* **22**: 72 pp. (1935), which includes a biographical account and portrait of Millardet and a photograph of the Bordeaux monument.

12. Further details on the introduction of these and other fungicides are given in the successive editions of H. Martin's handbook (Martin, 1928) and in the historical review by McCallan (1967).

13. See A. Müller, *Die innere Therapie der Pflanzen*, 1926 (Monogr. der angewandten Entomologie 8). Berlin.

14. *Operi di Agostino Bassi*, 1925, 673 pp. Pavia (Società Medico-Chirugica di Pavia).

15. For details of Remak's life and work, see B. Kisch, *Trans. Am. phil. Soc.* N.S. **44**: 227–96 (1954).

16. For details of Gruby's life and work, see L. Le Leu, *Le Dr. Gruby. Notes et Souvenirs*, 1908 (Paris); B. Kisch, *Trans. Am. phil. Soc.* N.S. **44**: 193–226 (1954).

17. The text of Gruby's four classical papers on ringworm is reprinted by Sabouraud (1910):8, 21, 25, 28 and in English translation by S. J. Zakon & T. Benedek, *Bull. Hist. Med.* **16**: 155–68 (1944).

18. W. Dampier, *A new voyage round the world*, 1697. London (edn 7, 1729; reprint 1937: 228–9 (London)).

19. See L. Goldman, *Archs. Derm.* **98**:660–1 (1968); **101**:688 (1970).

20. J. Hogg, *Skin diseases, an inquiry into their parasitic origin, and connection with eye infections; also the fungoid germ theory of cholera*, 1873. London.

21. *Fide* H. G. Adamson, *Brit. J. Derm.* **22**:46–9 (1910).

22. J. Margarot & P. Devéze, *Ann. Derm. Syphil.* **6**: 581–608 (1929).

23. A. Whitfield, *Proc. R. Soc. Med.* **5**(1), Dermat. Sect.: 36–43 (1912).

24. See P. K. C. Austwick in Raper & Fennell (1965): chap. 7.

25. See *Sporotrichosis infection on mines of the Witwatersrand. A symposium,* 1947. Johannesburg (Transvaal Chamber of Mines).

26. See Schwarz & Baum (1965) for additional historical details.

27. See M. J. Fiese, *Coccidioidomycosis,* 1958. Springfield, Ill. (History pp. 10–22).

28. International symposium on opportunistic fungus infections, *Laboratory Investigation* **11** (11, part 2): 1017–241 (1962).

29. By Engelbert Kaempfer in his doctoral thesis, *fide* J. S. Bowers & R. W. Carruba, *J. Hist. Med. allied Sci.* **25**: 270–310 (1970).

30. By Elizabeth Hazen & Rachael Brown, *Proc. Soc. exp. Biol. N.Y.* **76**: 93–7 (1950).

31. By W. Gold *et al., Antibiotics Annual 1955–56*: 579–91 (1956).

32. For a review of fungal viruses, see P. A. Lemke & C. H. Nash, *Bact. Rev.* **38**: 29–56 (1974).

7. *POISONOUS, HALLUCINOGENIC, AND ALLERGENIC FUNGI*

1. Houghton (1885): 40.
2. *Ibid.* 25.
3. *Ibid.* 28.
4. *Ibid.* 29.
5. *Ibid.* 24.
6. *Ibid.* 26.
7. *Ibid.* 39.
8. *Ibid.* 28.
9. *Ibid.* 31–2.
10. *Ibid.* 26.
11. *Phil. Trans.* **43**(472): 96–101 (1746).
12. *Ibid.* **43**(473): 51–7 (1746).
13. For references to these recent studies, see Turner (1971); and for additional details on poisonous fungi, Dujarric & Heim (1938), who cite more than 600 references, may be consulted.
14. *Fide* Bové (1970): 162.
15. See E. L. Blackman, *Religious dances,* 1952. London.
16. *Fide* Garrison (1929): 243.
17. For a general account of facial eczema see, Microbiological aspects of facial eczema. *N.Z. D.S.I.R. Inform. Ser.* **37**, 62 pp. (1964).
18. By Ruth Alcroft & R. B. A. Carnaghan, *Vet. Rec.* **74**: 863 (1962).
19. Also, J. Needham, *Science and civilization in China,* vol. V. *Chemistry and chemical technology.* Part 2. *Spagyrical discovery and invention: Mysteries of gold and immortality,* 1974. Cambridge.
20. Wasson in Heim *et al.* (1967); B. Lowy, *Mycologia* **63**: 983–93 (1971). More recently, R. W. Kaplan (*Man* N.S. **10**: 72–9 (1975)) has drawn attention to mushroom motifs on Bronze Age (1100–700 B.C.) razors and in rock carving in Scandanavia suggestive of a mushroom cult.
21. See Turner (1971): 316.
22. A term coined by P. H. Gregory, *Nature* **170**: 475 (1952).
23. See also Buller (1909–50) **6**: 1–224.

24. In a letter to Linnaeus dated September 1740 von Haller wrote: 'I have detected a very curious elastic motion in the common sessile *Peziza*, of a dirty white hue. The whole plant contracts spontaneously, and discharges a powder upwards, with a sort of hissing noise. This doubtless is the seed' (Ramsbottom (1941*a*): 292).

25. *Linnaea* **17**: 474 (1843).

26. *Jahrb. wiss. Bot.* **3**: 93 (1863); *Handb. der experimental Physiologie der Pflanzen*, 1865: 93.

27. J. Bolton, *An history of British ferns*, 1790: appendix xxi.

28. Devised by R. L. Maddox, *Monthly microscop. J.* **3**: 286–90 (1870).

<h2 style="text-align:center">8. USES OF FUNGI</h2>

1. See R. W. Marsh, *Bull. Brit. mycol. Soc.* **7**: 34 (1973).

2. *Fide* Joseph Needham *in litt.*

3. The Chinese name *hsiang hsin* is mentioned in *Jih Yung Pen Tshao* (Pharmaceutical history for daily use) written by Wu Jui in 1329; *fide* Joseph Needham *in litt.*

4. H. E. Winlock, The Egyptian Expedition 1918–20. II. Excavations at Thebes, 1919–20, *Bull. Metropolitan Museum of Art* **12** (1920); *fide* W. J. Nickerson & A. H. Rose, *Encyclopedia of chemical technology*, 1956: 195–6.

5. *Fide* J. Sarton, *Introduction to the history of science* **1**: 339 (1927) (Carnegie Inst. Wash. Publ. 376).

6. 'Ubi notandum, nihil fermentare quod non sit dulce', *fide* Harden (1911–32, edn 2): 1.

7. In *Familiar letters on chemistry. Second series*, 1841: 200–1, 205. London.

8. (Wiegmann's) *Archiv. f. Naturgeschichte* **4**: 100 (1838).

9. Pasteur (1876): 271.

10. T. Schwann, *Mikroscopische Untersuchungen*, 1839: 234 (footnote).

11. See R. E. Kohler, The reception of Eduard Buchner's discovery of cell-free fermentation, *J. Hist. Biol.* **5**: 327–53 (1972).

12. 'Chhen Tshang-Chhi discussed ergot in his *Pên Tshao Shi I* (A supplement for the Pharmaceutical Natural Histories) published in +725. Ergot of rye was not discussed in Chinese works until the end of the +16 century and never became widely used there.', Joseph Needham *in litt.*

13. R. Hare, *The birth of penicillin and the disarming of microbes*, 1970: chaps. 3–4. London; also D. B. Colquhoun, *World Medicine* **10**: 41–3, 1975.

14. See, for example, the review by K. B. Raper, A decade of antibiotics in America, *Mycologia* **44**: 1–59 (1952).

15. See Tyndall (1881): 109; J. Friday, *Br. J. Hist. Sci.* **7**: 61–71 (1974) (Huxley and antibiosis).

16. First announced by A. Schatz, E. Bugie & S. A. Waksman, *Proc. Soc. Exp. Biol. & Med.* **55**: 66–69 (1944).

17. A pattern confirmed by the later survey of antibiotics from fungi by D. Broadbent *PANS* **14**(2): 120–41 (1968).

18. See E. P. Abraham & G. G. F. Newton, The cephalosporins, *Adv. Chemotherapy* **2**: 23–90 (1965).

9. DISTRIBUTION OF FUNGI

1. Elias Fries' autobiography, 'Historia studii mei mycologici', *Friesia* **5**: 135–60 (1955).

2. From English translation of footnote in *Abh. könig. Böhm. Ges. Wissensch.* **7**(5): 89–94 (1851–2), supplied by S. J. Hughes. The population of Houston now exceeds one million.

3. For much of his life Buller kept copies of letters he wrote and most of those he received. This correspondence together with his books and his ashes are deposited in Buller Memorial Library of the Research Branch, Canada Department of Agriculture, Winnipeg. His unpublished scientific MSS. are at the Royal Botanic Gardens, Kew.

4. See the review by E. J. Butler, *Trans. Br. mycol. Soc.* **14**: 1–18 (1929).

5. According to Dr Joseph Needham *in litt.*, a major historical study needs to be done by a mycologist with a good knowledge of classical Chinese on the Chinese pharmaceutical natural histories ranging from the second century B.C. to the eighteenth century A.D.

6. See the review by L. E. Hawker, *Biol. Rev.* **30**: 127–58 (1955).

7. See, for example, the review by J. Webster, *Trans. Br. mycol. Soc.* **54**: 161–80 (1970).

8. See *Trans. Br. mycol. Soc.* **35**: 188–9 (1952); *Bull. Br. mycol. Soc.* **6**: 17 (1972), for details.

9. See D. Botting, *Humboldt and the Cosmos*, 1973 (London), for a general account of Humbolt.

10. *Trans. Br. mycol. Soc.* **5**: 147–55 (1915); 257–63 (1916); *New Phytol.* **14**: 33, 183 (1915); **15**: 35 (1916).

11. *Beitr. Obst. v. Naturges. schedlicher Insecten* **4**, *fide* Ramsbottom (1953): 299.

12. So designated by L. R. Batra, *Trans. Kansas Acad. Sci.* **66**: 226 (1963).

13. First by J. G. Koenig in 1779, *fide* Ramsbottom (1953): 214.

14. See *Ainsworth & Bisby's Dictionary of the fungi*, 1970, edn 6: 233–40, for lists.

15. Apparently first by W. Nylander, *Bull. Soc. bot. Fr.* **13**: 364–72 (1886).

16. For reviews see O. L. Gilbert in Ahmadjian & Hale (1973), chap. 13; B. W. Ferry, M. S. Baddeley & D. L. Hawksworth (eds), *Air pollution and lichens*, 1973 (London).

10. CLASSIFICATION

1. S. T. Cowan, *A dictionary of microbial taxonomic usage*, 1968: 105. London.

2. J. Heslop-Harrison, *New concepts in flowering plant taxonomy*, 1953: 1. London.

3. V. H. Heywood, *Plant taxonomy*, 1967: 3. London.

4. Coined by A. Meyer in 1926; see Lam, *Taxon* **6**: 213–15 (1957), for history.

5. See W. Blunt (with the assistance of W. T. Stearn), *The compleat naturalist. A life of Linnaeus*, 1971 (London), for an excellent general account of Linnaeus. F. A. Stafleu, *Linnaeus and the Linnaeans*, 1971 (Utrecht), provides much background detail.

6. For further details of the treatment of fungi by Linnaeus see Ramsbottom (1941a): esp. pp. 290–300.

7. Translated into English by M. Åsberg & W. T. Stearn as 'Linnaeus's

Öland and Gotland journey'. *Biol. J. Linn. Soc. Lond.* **5**(1–2): 1–107, 109–220 (1973).

8. J. L. M. Franken, Uit die Lewe van'n Beroemde Afrikaner, Christiaan Hendrik Persoon, *Ann. Univ. Stellenbosch,* **15**B(4), 102 pp. (1937); A note on a visit to Christiaan Hendrik Persoon's grave, *J. South Afr. Bot.* **4**: 127 (1938); C. E. Hugo, The restoration of the grave of Christiaan Hendrik Persoon – 'Prince of mycologists'; *J. bot. Soc. S. Afr.*, part 51, 3 pp. (1965), give details of Persoon's life and work and references to most of the other papers consulted.

9. In A.-L. de Jussieu, *Genera plantarum,* 1789, see Stafleu (Chapter 10, note 5): 321–32, for background and comment.

10. In the Rijksherbarium, Leiden, Netherlands.

11. For an English translation, see Elias Fries' autobiography, 'Historiola studii mei mycologici', *Friesia* **5**: 135–60 (1955).

12. For details see, N. F. Buchwald, Elias Fries' aetlinge inden for mykologien, *Friesia* **9**: 348–54 (1970).

13. G. Eriksson, *Elias Fries och den romantiska biologien,* 487 pp., 1962 (Uppsala). (Swedish with English summary).

14. A view given wide circulation by Sachs in his *Text-book of botany* (English transl. 1875) and in modern times by Clements & Shear (1931) who gave algal names (Protococcales, Spirogyrales, etc.) to the six orders into which they divided the Phycomycetes.

15. Who also observed 'free cell formation' of ascospores. In Berkeley (1857): 25 he wrote – 'in those fungi in which, as in *Sphaeria* and *Peziza,* the reproductive bodies are generated by the endochrome of the fructifying cells, the Cryptogamist has the power of watching the development of spores from the very moment when the endochrome commences to be organised, and he can with confidence assert that they are not the creatures of previously-existing cells, but the produce of the endochrome itself.'

16. G. C. Ainsworth, *Taxon* **3**: 77–9 (1954).

11. ORGANISATION FOR MYCOLOGY

1. See W. T. Stearn, *Botanical Latin,* 1966 (edn 2, 1973). London.

2. G. C. Ainsworth, The pattern of mycological information. *Mycologia* **55**: 65–72 (1963).

3. M. Rees, *Ber. dtsch. bot. Ges.* **6**: viii–xxvi (1888).

4. By J. R. Porter, *Bact. Rev.* **28**: 228 (1964).

5. R. D. Meikle, The history of the *Index Kewensis, Biol. J. Linn. Soc. Lond.* **3**: 295–9 (1971).

6. For an analysis see F. A. Stafleu, *Taxon* **23**: 625–6 (1974).

7. See Guétrot (1934).

8. For reproductions of two other W. G. Smith cartoons, see *Trans. Br. mycol. Soc.* **46**: 161, pl. 12 (1963); **58** (suppl.): 14, pl. 1 (1972).

9. Historical details of the B.M.S. are given by Ramsbottom, *Trans. Br. mycol. Soc.* **30**: 1–12, 22–39 (1948).

10. See Fitzpatrick (1937) and Dearness, *Mycologia* **30**: 111–19 (1938).

11. See H. Haas, 50 Jahre Deutsche Gesellschaft für Pilzkunde, *Z. Pilzkde* **39**: 9–14 (1973).

12. *Fide* Arber (1938): 139.

13. The term 'herbarium' in the modern sense was first used by de Tournefort in his *Élémens*, 1694, *fide* Arber (1938):142.

14. See J. A. Stevenson, *Taxon* **4**: 181–5 (1955); Mary E. Lentz & P. L. Lentz, *BioScience* **18**: 194–200 (1968).

15. Reviews of the first 50 years of the C.B.S. are given by K. B. Raper, *Mycologia* **49**:884–92 (1957); and Agathe L. van Beverwijk, *Antonie v. Leeuwenhoek* **25**: 1–20 (1959).

16. W. A. Clarke & W. Q. Loegering, Function and maintenance of a type-culture collection, *Ann. Rev. Microbiol.* **5**:319–42 (1967), list important national collections.

17. See T. Hasegawa, Japanese culture collections of micro-organisms in the field of industry – their histories and actual state, *Ann. Rep. Inst. Ferment., Osaka* **3**: 139–43 (1967).

18. G. C. Ainsworth, *Nature* **195**: 1120–1 (1962); and the experiment is still in progress.

19. See E. W. Mason, Presidential address. Specimens, species and names, *Trans. Br. mycol. Soc.* **24**: 115–25 (1940), for further discussion.

Chronology and bibliography

Not all the easily identified publications cited in the text have been included in this chronology and bibliography. On the other hand a number of reviews and important secondary sources not specifically mentioned in the text have been compiled.

Useful general historical sources are:

General mycology. Paulet (1790–3, **1**), E. Roze in Richon & Roze (1885–9), Lindau & Sydow (1908–15), Rolfe & Rolfe (1925), Lütjeharms (1936), Ramsbottom (1941a, 1953), Morandi & Baldacci (1954), Raab (1965–75).

Regional mycology. Austria (Lowag, 1969); Brazil (Fidalgo, 1968); Czechoslovakia (Špaček, 1969); Denmark (Lind, 1913); France (Guétrot, 1934; Virville, 1954); Germany (Bresinsky, 1973; Bavaria); Great Britain (Ramsbottom, 1948; 1963, Scotland); Hungary (Ubrizsy, 1968); India (Das Gupta, 1958); Italy (Lazzari, 1973); Japan (Heim & Pellier, 1963; Tubaki, 1970); Switzerland (Blumer & Müller, 1971); U.S.A. (Fitzpatrick, 1937).

Plant pathology. Much has been written on the history of phytopathology, including fungal diseases of plants. The best introduction is still Whetzel (1918). Other general surveys include Sorauer (1914); Braun (1933), Wehnelt (1943), Keitt (1959), while Large (1940) provides much information in a popular form. The two reviews by Orlob (1964 (concepts of aetiology), 1971 (Middle Ages)) are valuable as are the chronologies by Mayer (1959) and Parris (1968). *Regional histories* include: Australia (Fish, 1970); Brazil (Puttemans, 1940); Canada (Conners, 1972); Denmark (Buchwald, 1967); Great Britain (Ainsworth, 1969); Hungary (Kiraly, 1972); India (Raychaudhuri *et al.*, 1972); Japan (Akai, 1974); U.S.A. (Stevenson, 1951, 1954, 1959; McCallan, 1959).

Medical mycology. Less has been written on the history of medical and veterinary mycology. Sabouraud (1910) reviews the history of dermatophyte infections; de Beurmann & Gougerot (1912), sporotrichosis. Other bibliographical starting points are Schwarz & Baum (1965; systemic mycoses) and Ainsworth (1951, Great Britain). A number of regional surveys of mycoses were published in *Mycopathologia et Mycologia Applicata* during the decade following 1958.

Industrial mycology. Lafar (1898–1910) and G. Smith (1938) provide many clues.

Lichenology. A. L. Smith (1921), Grummann (1974).

General botany: Jessen (1864), Sachs (1890), Arber (1938), Reed (1942). *Genetics:* Sturtevant (1965). *Medicine:* Garrison (1929). *Bacteriology:* Bulloch (1938), Grainger (1958).

An asterisk (*) preceding an entry indicates a secondary publication.

1400
1491   *Ortus sanitatis.* Cap. cciii, cccclxxxiii. Mainz.

1500
1526   *The grete herball which geveth parfyt knowledge and understandying of all maner of herbes and there gracyous vertues.* London.
1552   Bock, J. (H. Tragus). *De stirpium maxime earum quae in Germania nostra nascuntur, usitatis nomenclaturis.* Strasbourg.
1560   Mattioli, P. A. *Commentarii, in libros sex Pedacii Dioscoridis Anazarbei, de medica materia.* Venice.
1562   Jonghe, A. (H. Junius). *Phalli, ex fungorum genere in Hollandiae sabuletis passim crescentis descriptio et vivum expressa pictura.* Delft. (Also 1564, 1601.)
1564   Ciccarelli, A. *Opusculum de tuberibus, cum opusculo de Clitumno flumine.* Pavia. (P. J. Amoreux, *Opuscule sur les truffes, traduction libre du latin d'Aphonse Ciccarellus, auteur du xvᵉ siècle; avec des annotations sur le texte et un préamble historique,* 1813. Montpellier.)
1581   l'Obel, M. de (Lobelius). *Kruydtboeck.* Pp. 305–12. Antwerp. (For Latin transl. see Clusius, 1601:ccxcii.)
1582   Lonitzer, A. (Lonicerus). *Kreuterbuch.* P. 285 (ergot). Frankfurt-on-Main. (Also 1587.)
1583   Cesalpino, A. *De plantis libri xvi.* Pp. 28, 613–21. Florence.
       Dodoens, R. (Dodonaeus). *Stirpium historia pemptades sex. Sive libri xxx.* Pp. 474–8. Antwerp.
1588   Porta, G. della. *Phytognomonica.* Lib. vi, p. 240. Naples.

c. 1590–1600   Compound microscope invented.

1600
1601   Clusius, C. (J.-C. de l'Éscluse). *Rariorum plantarum historia.* Pp. 263–95 (Fungorum in Pannoniis observatorum). Antwerp.
1623   Bauhin, G. (C. Bauhinus). *Pinax theatri botanici.* (*Editio altera,* 1671.) Basle.
1633   Gerarde, J. *The herball or generall historie of plantes...very much enlarged and amended by Thomas Johnson.* Pp. 1365–6, 1578–84. London.
1640   Parkinson, J. *Theatrum botanicum.* Pp. 1316–24. London.
1651   Bauhin, J. & Cherler, J. H. *Historia plantarum universalis.* **3**:821–36. Yverdon, Switzerland.
1658   Bauhin, G. *Theatri botanici. Liber primus.* P. 434 (ergot). Basle.
1660   Royal Society of London founded.
1665   Hooke, R. *Micrographia.* London.
1673   van Leeuwenhoek, A. A specimen of some observations made by a microscope. *Phil. Trans.* **8**(94):6037.
1675   van Sterbeeck, F. *Theatrum fungorum.* Antwerp.

1679 Malpighi, M. *Anatome plantarum pars altera*. Pp. 64–7, tab. xxviii. London.
1686–1704 Ray, J. *Historia plantarum*. 3 vols. **1**:84–111 (1686); **3**: 16–26 (1704) (Fungi). London.
1689 Magnol, P. *Prodromus historiae generalis plantarum, in quo familiae plantarum per tabulas disponuntur*. Montpellier.
1690 Ray, J. *Synopsis methodica stirpium Britannicarum*. London. (Edn 3, 1724 by Dillenius; Ray Soc. reprint, 1974.)
1694 Sexuality in higher plants experimentally demonstrated by R. J. Camerarius.

Tournefort, J. P. de. *Élémens de botanique*. 3 vols (8vo). **1**: 438–42; **3**, tabs. 327–33 (Fungi). Paris.
1697 Boccone, P. *Museo di piante rare*. Venice.
1699 Morison, R. *Plantarum historia universalis Oxoniensis*. **3**: 631–5 (Musco-fungus), 635–43 (Fungus). Oxford.

1700

1700 Tournefort, J. P. de. *Institutiones rei herbariae*. 3 vols (4to). **1**: 556–76; **3**, tabs. 327–33 (Fungi). Paris.
1703 Loeselius, J. *Flora Prussica sive plantae in regno Prussicae sponte nascentes*. Kaliningrad, USSR.
1705 Tournefort, J. P. de. Observations sur les maladies des plantes. *Mém. Acad. Sci. Paris* 1705: 332–45.
1707 Linnaeus born.

Tournefort, J. P. de. Observations sur la naissance et la culture de champignons. *Mém. Acad. Sci. Paris* 1707: 58–66.
1711 Marchant *fils*, J. Observations touchant la nature des plantes et quelques uses de leur parties cachées, ou inconnues. *Mém. Acad. Sci. Paris* 1711: 100–9.
1714 Marsigli, L. F. *Dissertatio de generatio fungorum* with
Lancisi, J. M. *Dissertatio epistolaris de orto vegetatione, ac textura fungorum*. Rome.
1719 Dillenius, J. J. *Catalogus plantarum sponte circa Gissam nascentium. Cum appendice*. Frankfurt-on-Main.
1721 Buxbaum, J. C. *Enumeratio plantarum in agro Halensi*. Halle.
1723 Eysfarth, C. S. *Dissertatio de morbis plantarum*. Leipzig.
1726 Réaumur, R. A. [de]. Remarques sur la plante appelée à la Chine Hia Tsao Tom Tschom ou plante ver. *Mém. Acad. Sci. Paris* 1726: 302–5.
1727 Hales, S. *Vegetable staticks*. London. (Reprint Oldbourne, London, 1961.)
Marchant *fils*, J. Observation touchant une végétation particulière qui naît sur l'écorce du chêne battuë, et mise en poudre, vulgairement appelée du tan. *Mém. Acad. Sci. Paris* 1727: 335–9.
Vaillant, S. *Botanicon Parisiense*. Leyden and Amsterdam.
1728 Jussieu, A. de. De la necessité d'établir la methode nouvelle des plantes, une class particulière pour les fungus. *Mém. Acad. Sci. Paris* 1728: 377–82.
1728–40 Buxbaum, J. C. *Plantarum minus cognitarum centuriae*. Leningrad.
1729 Micheli, P. A. *Nova plantarum genera juxta Tournefortii methodum disposita*. Florence.

1733 Tull, J. *The horse-hoing husbandry.* Chap. x (Of smuttiness), xi (Of blight) (Edn 3, *Horse-hœing husbandry,* 1751). London.

1740 Gleditsch, J. G. *Consideratio epicriseos Siegesbeckianae.* Berlin.

1741 Dillenius, J. J. *Historia muscorum.* Oxford.

1742 Haller, A. von. *Enumeratio methodica stirpium Helvetiae indigenarum.* Göttingen.

1743 Mazzuoli, F. M. Dissertazione sopra l'origine dei funghi. *Memorie sopra la fisica e istoria di diversi Valentuomini.* **1**: 159–74.

1744 Seyffert, C. *De fungis.* Jena. (The Brit. Mus. (Nat. Hist.) copy is bound with 132 watercolours of fungi by Seyffert dated 1744.)

1745 Martyn, J. An account of a new species of fungus. *Phil. Trans.* **43**(475): 263–4.

1748 Arderon, W. The substance of a letter from Mr William Arderon F.R.S. to Mr Henry Baker F.R.S. *Phil. Trans.* **45**(487): 1321–3.

1749 Gleditsch, J. G. Expérience concernant le génération des champignons. (Traduit du Latin.) *Hist. Acad. Berlin* 1749: 26–32.

1750

1751 Hill J. *A history of plants.* Pp. 26–72, pl. 4 (Fungi). London.
Linnaeus, C. *Philosophia botanica.* Stockholm.

1753 Gleditsch, J. G. *Methodus fungorum exhibens genera, species et varietates.* Berlin.
Linnaeus, C. *Species plantarum.* 2 vols. Stockholm. (Facsimile reprint, 1957–9, Ray Soc., London.)

1754 Monti, G. De mucore. *Comment. Inst. Bononiens* **3**: 145–59.

1755 Battarra, A. J. A. *Fungorum agri Ariminensis historia.* Faenza.
Tillet, M. *Dissertation sur la cause qui corrompt et noircit les grains de bled dans les épis; et sur les moyens de prevenir ces accidens.* Bordeaux.
Suite des expériences et reflexions relatives à la dissertation sur la cause qui corrompt et noircit les grains de bled dans les épis. Paris. (English transl. of both by H. B. Humphrey, *Phytopathological Classics* **5** (1955).)

1759 Schaeffer, J. C. *Vorläufige Beobachtungen der Schwämme von Regensberg.* Regensberg.

1760 Scopoli, G. A. *Flora Carniolica.* Vienna. (Edn 2, 2 vols., 1772.)

1762 Schmidel, C. C. (Curante J. C. Keller) *Icones plantarum et analyses partium.* Pp. 42–4, tab. x.

1762–7 Schaeffer, J. C. *Fungorum qui in Bavaria et Palatinata circa Ratisbonam nascuntur icones.* 4 vols. Regensberg.

1763 Adanson, M. *Familles des plantes.* 2 vols. Paris.

1764 Watson, W. An account of the insect called the vegetable fly. *Phil. Trans.* **53**: 271–4.

1767 Fontana, F. *Osservazioni sopra la ruggine del grano.* Lucca. (English transl. by P. P. Pirone, *Phytopathological Classics* **2** (1932).)
Targioni-Tozzetti, G. *Alimurgia o sia modo di render meno gravi le carestie, proposto per sollievo de' poveri.* Florence. (Reprinted as *Reale Accademia d'Italia. Studi e Documenti* **12** (1943); English transl. of chap. 5 by L. R. Tehon, *Phytopathological Classics* **9** (1952).)

1768 Haller, A. von. *Historia stirpium indigenarum Helvetiae,* **3**: 110–80. Bern.

1772 Scopoli, J. A. Plantae subterraneae descriptae et delineatae, in

J. A. Scopoli, *Dissertationes ad scientiam naturalem pertinentes.* Pars I: 84–120. Prague.

1773 Zallinger, J. B. *De morbis plantarum.* Innsbruck.

1774 Fabricius, J. C. Forsøg til en Afhandling om Planternes Sygdomme. *Det. Kongelige Norske Vidensk. Selsk. Skrift.* **5**:431–92. (English transl. by M. K. Ravn, *Phytopathological Classics* **1** (1926).)

1776 Dryander, J. *Fungos regno vegetabile vindicans.* London. (Thesis, presided over by E. G. Lidbeck.)

Spallanzani, L. *Opuscoli di fisica animale e vegetabile.* Modena. (French transl. by J. Senebier, *Opuscules de physique, animale et vegetale,* 2 vols. 1777. Geneva.)

1783 Necker, N. J. de. *Traité sur la mycitologie.* Manheim.

Palisot de Beauvois, A. M. F. J. Champignons, Fungi. In Lamarck, *Encyclopédie Methodique, Botanique:* 691–4.

Tessier, H.-A. *Traité des maladies des grains.* Paris.

1783–9 Batsch, A. J. G. C. *Elenchus fungorum.* 3 vols. Halle.

1784 Villemet, R. Essai sur l'histoire naturelle du champignon vulgare. *Nouv. Mém. Acad. Dijon* 1783, 2e sem.: 195–211.

1788 Hedwig, J. *Stirpes cryptogamicae,* **2** (fasc. 1): 1–34, tabs. i–x. Leipzig. (= *Descriptio et adumbratio muscorum frondosum,* **2**: 1–34, tabs. i–x (1789).)

Picco, V. *Melethemata inauguralia.* Turin.

1788–91 Bolton, J. *An history of fungusses growing about Halifax.* 4 vols. Huddersfield. (German transl. by K. L. Willdenow, *Geschichte der merkwurdigsten Pilze,* 1795–1820. Berlin.)

1790–1 Tode, H. J. *Fungi Mecklenburgenses selecti.* 2 vols. Lüneburg.

1790–3 Paulet, J. J. *Traité des champignons.* 2 vols. Paris.

1790–9 Holmskiold, T. *Beata ruris otia fungis Danicis impensa.* 2 vols. Copenhagen.

1791–1812 Bulliard, P. *Histoire des champignons de la France.* 4 vols. Paris.

1793 Humboldt, F. A. von. *Flora Fribergensis specimen plantas cryptogamicas praesertim subterraneas exhibens.* Berlin. (Fungi pp. 71–132.)

1793 Persoon, C. H. Was sind eigentlich die Schwämme? (*Voight's*) *Magaz. Neuste Physik. Naturgesch.* **8**: 76–85.

1794 Persoon, C. H. Neuer Versuch einer systematischen Entheilung der Schwämme. (*Römer's*) *Neues Magaz. Bot.* **1**:63–128.

1796–9 Persoon, C. H. *Observationes mycologicae.* 2 parts. Leipzig and Lucerne.

1797 Blottner, C. *De origine ac modo generationis fungorum.* Halle.

Persoon, C. H. *Tentamen dispositionis methodicae fungorum in Classes, Ordines, Genera et Familias.* Leipzig.

1797–1815 Sowerby, J. *Coloured figures of English fungi or mushrooms.* 3 vols. & suppl. London.

1800

1801 Persoon, C. H. *Synopsis methodica fungorum.* Göttingen.

1802 Forsyth, W. *A treatise on the culture and management of fruit-trees.* London.

1803 Acharius, E. *Methodus qua omnes detectos Lichenes.* Stockholm.

1804–6 Trattinick, L. *Fungi Austriaci ad specimina viva cera expressi, descriptiones ac historiam completam addidit.* Vienna.

1805 Albertini, J. B. de & Schweinitz, L. D. de. *Conspectus fungorum in Lusitiae superioris agro Niskiensi crescentium.* Leipzig.

Banks, J. *A short account of the cause of the disease in corn. called by farmers the blight, the mildew, and the rust.* London.

1806 Windt, L. G. *Der Beberitzenstraunch, ein Feind des Wintergetreides. Aus Erfahrungen, Versuchen und Zeugnissen.* Bückeburg ü. Hannover. (MS. English transl. in Banksian library at Brit. Mus. (Nat. Hist.).)

1807 Candolle, A. P. de. Sur les champignons parasites. *Ann. Mus. Hist. nat. Paris* **9**: 56.

Prévost, B. *Mémoire sur la cause immédiate de la carie ou charbon des blés, et de plusieurs autres maladies des plantes, et sur les préservatifs de la carie.* Paris. (English transl. by G. W. Keitt, *Phytopathological Classics* **6** (1939).)

1808 Richard, L. C. M. *Démonstrations botaniques: ou analyse du fruit, considéré en général: publiées par H. A. Duval.* Paris.

1809–16 Link, H. F. Observationes in ordines plantarum naturales. Dissertatio I. *Mag. Ges. naturf. Freunde, Berlin* **3**: 3–42 (1809); Dissertatio II, *ibid.* **7**: 25–45 (1816).

1810 Acharius, E. *Lichenographia universalis.* Göttingen.

Willdenow, K. L. *Anleitung zum Selbststudium der Botanik, ein Handbuch zu öffentlichen Vorlesungen. Zweite Auflage.* Berlin. (English transl. as *Principles of botany, and vegetable physiology*, 1811. Edinburgh.)

1814 Acharius, E. *Synopsis methodica lichenum.* Lund.

1816 Nees von Esenbeck, C. G. *Das System der Pilze und Schwämme.* Würzburg. (Both copies seen dated 1817.)

1819 Richard, A. *Nouveaux élémens de botanique et de physiologie végétale.* Paris.

1820 Ehrenberg, C. G. (*a*) De mycetogenesi ad Acad. C.L.C.N.C. Praesidem epistola. *Nova Acta Physico-Medica Acad. Caes. Leop. Carol.*, **10**: 157–222.

Ehrenberg, C. G. (*b*) Syzygites, eine neue Schimmelpilzgattung. *Verh. GesNaturf. Freunde, Berlin* **1**: 98–109.

1821 Knowles, J. *An inquiry into the means which have been taken to preserve the British navy, from the earliest period to the present time, particularly from that species of decay, now denominated dry-rot.* London.

1821–32 Fries, E. M. *Systema mycologicum, sistens fungorum ordines, genera, et species.* 3 vols. Lund: Griefswald.

1822–8 Persoon, C. H. *Mycologia Europaea.* 3 vols. Erlangen.

1823–8 Greville, R. K. *Scottish cryptogamic flora.* 6 vols. Edinburgh.

1825–7 Wallroth, F. W. *Naturgeschichte der Flechten.* 2 vols. Frankfurt-on-Main.

1826 Desmazières, J. B. H. J. Recherches microscopiques sur le genre *Mycoderma. Rec. trav. Soc. Sci. Agric. Arts. Lille* **3**: 297–323 (1825); also *Ann. Sci. nat. Paris* **10**: 42–67 (1827).

1827 Schilling, J. Ueber die Bildung einer Schimmelart aus Saamen. *Kastner Archiv Naturl.* **10**: 429–42.

1830 Trattinick, L. *Fungi Austriaci delectu singulari iconibus xl observationibusque.* Vienna.

1831 Discovery of the nucleus in plant cells by Robert Brown.

Vittadini, C. *Monographia tuberacearum.* Milan.

1831–46 Krombholz, J. V. *Naturgetreue Abbildungen und Beschreibungen der essbären, schädlichen und verdächtigen Schwämme.* Prague.

1833 Hartig, T. *Abhandlung über die Verwandlung der polycotyledonischen Pflanzenzelle in Pilz- und Schwamm-Gebilde und die daraus hervorgehende sogennante Fäulniss des Holzes.* Berlin.

Secretan, L. *Mycographie Suisse.* Geneva.

Unger, F. *Die Exantheme der Pflanzen und einige mit diesen verwandte Krankheiten der Gewächse, pathogenetisch und nosographisch dargestellt.* Vienna.

1834 Corda, A. C. J. Ueber Micheli's Antheren der Fleischpilze. *Allg. Bot. Zeit.* **1**: 113–18.

Schweinitz, L. D. de Synopsis fungorum America Boreali media degentium. *Trans. Am. Phil. Soc.* **4**: 141–316.

1834–7 Biological nature of fermentation established independently by Cagniard-Latour (1837), Schwann (1837), and Kützing (1837).

1834–8 Viviani, D. *I funghi d'Italia.* Genoa.

1835 Birkbeck, G. *A lecture on the preservation of timber by Kyan's patent for preventing dry rot.* London.

Vittadini, C. *Descrizione di funghi mangerecci più communi dell'Italia.* Milan.

1835–6 Bassi, A. *Del mal del segno.* Lodi. (Facsimile, Novara, 1956; English transl. by P. J. Yarrow, *Phytopathological Classics* **10** (1958).)

1836 Ascherson, F. M. Ueber die Fructificationsorgane der höheren Pilze. (*Wiegmann's*) *Archiv f.Naturgesch.* **2**: 372.

Berkeley, M. J. 'Fungi' in *The English Flora of Sir James Edward Smith*, vol. **5** (= vol. **2** of *Dr Hooker's British Flora*), Part II. London.

Faraday, M. *On the practical prevention of dry rot in timber.* London.

Weinmann, J. A. *Hymeno- et Gastero-Mycetes hucusque in Imperio Rossico observatos recensuit.* Leningrad.

1836–8 Basidial structure elucidated independently by Ascherson (1836), Léveillé (1837), Corda (1837), Berkeley (1838), Klotzsch (1838), and Phoebus (1842).

1837 Cagniard-Latour, C. Mémoire sur la fermentation vineuse. *C. R. Acad. Sci. Paris* **4**: 905–6. (Also *Ann. Chimie Phys.* **68**: 206–22 (1838).)

Kützing, F. Microscopische Untersuchungen über die Hefe und Essigmutter nebst mehreren andern dazu gehörigen vegetabilischen. *J. prakt. Chemie* **11**: 385–409.

Léveillé, J. H. Recherches sur l'hymenium des champignons. *Ann. Sci. nat. Paris*, sér. 2, **8**: 321–38.

Schwann, T. Vorläufige Mittheilung betreffend Versuche über die Weingährung und Fäulniss. *Ann. Physik Chemie* **41**: 184.

1837–8 Meyen, F. J. F. Beiträge zur Pflanzenphysiologie. I. Ueber die Entwickelung des Getreidebrandes in der Mays-Pflanze. (*Wiegmann's*) *Archiv. f.Naturgesch.* **3**: 419–21 (1837); **4**: 162–3 (1838).

1837–54 Corda, A. C. J. *Icones fungorum hucusque cognitorum.* 6 vols. Prague.

1838 Audouin, V. Recherches anatomiques et physiologiques sur la maladie contagieuse qui attaque les vers à soie, et qu'on désigne sous le nom de muscardine. *Ann. Sci. nat. Paris*, sér. 2, Zool. **8**: 229–45.

Berkeley, M. J. (*a*) On the fructification of the pileate and clavate tribes of hymenomycetous fungi. *Ann. nat. Hist.* **1**: 81–101.

Berkeley, M. J. (*b*) On the existence of a second membrane in the asci of fungi. *Mag. Zool. Bot.* **2**: 222–5.

Dickson, R. *A lecture on the dry rot, and on the most effectual means of preventing it.* London.

Klotzsch, F. *Agaricus delequescens* Bull. In A. Dietrich, *Flora regni Borussici*, **6**, no. 375 (1838).

Quevenne, T. A. Sur la levure et la fermentation vineuse. *J. Pharm.* **24**: 265, 329.

Turpin, P. J. F. Mémoire sur la cause et les effets de la fermentation alcoolique et aceteuse. *C. R. Acad. Sci. Paris* **7**: 369–402. (Also *Mém. Acad. Sci. Paris* **17**: 93–180, pl. 9 (1840).)

1838–9 'Cell Theory' introduced by M. J. Schleiden and T. Schwann.

1839 Wiegmann, A. F. *Die Krankheiten und krankhaften Missbildungen der Gewächse.* Brunswick.

1841 Gruby, D. (*a*) Mémoire sur une végétation qui constitu la vrai teigne. *C. R. Acad. Sci. Paris* **13**: 72–5.

Gruby, D. (*b*) Sur les mycodermes qui constituent la teigne faveuse. *C. R. Acad. Sci. Paris* **13**: 309–12.

Lees, E. On the parasitic growth of *Monotropa hypopitys. Phytologist* **1**: 97–101.

Meyen, F. J. F. *Pflanzen-Pathologie.* Berlin.

Montagne, J. F. C. *Equisse organographique et physiologique sur la class des champignons.* Paris. (English transl. by M. J. Berkeley, *Ann. Mag. nat. Hist.* **9**: 1, 107, 230, 283 (1842).)

1842 Bennett, J. H. On the parasitic fungi found growing in living animals. *Trans. roy. Soc. Edinburgh* **15**: 277–94.

Corda, A. C. J. *Anleitung zum Studium der Mycologie, nebst kritischer Beschreibung aller bekannten Gattungen, und einer kurzen Geschichte der Systematik.* Prague.

Gruby, D. (*a*) Sur une espèce de mentagre contagieuse résultant du développement d'un nouveau cryptogame dans la racine des poils de la barbe de l'homme. *C. R. Acad. Sci. Paris* **15**: 512–13.

Gruby, D. (*b*) Recherches anatomiques sur une plant cryptogame qui constitue le vrai muguet des infants. *C. R. Acad. Sci. Paris* **14**: 634–6.

Knight, T. A. Upon the causes of the diseases and deformities of leaves of the peach-tree. *Trans. hort. Soc. London*, ser. 2, **2**: 27–9. (Paper read 15 July 1834.)

Naegeli, C. W. von. Sur les champignons vivant dans l'intérieur des celles végétales. *Ann. Sci. nat. Paris*, sér. 2, **19**: 86–91.

Phoebus, P. Über den Keimkörner-Apparat der Agaricinen und Helvellaceen. *Nova Acta Physico-Medica Acad. Caes. Leop.-Carol., Erlangen* **19**: 169–248.

Remak, R. Gelungene Impfung des Favus. *Med. Ztg.* **11**: 137.

Rylands, T. G. On the nature of the byssoid substance found investing the roots of *Monotropa hypopitys. Phytologist* **1**: 341–8.

Vittadini, C. *Monographia Lycoperdineorum.* Turin.

1843 Gruby, D. Recherches sur la nature, la siége et le développement du porrigo decalvans ou phyto-alopécie. *C. R. Acad. Sci. Paris* **17**: 301–3.

Léveillé, J. H. Mémoire sur le genre Sclerotium. *Ann. Sci. nat. Paris,* sér. 2, **20**: 218–48.

1844 Gruby, D. Recherches sur les cryptogames qui constituent la maladie contagieuse du cuir chevelu décrite sous le nom de teigne tondante (Mahon), herpes tonsurans (Cazenave). *C. R. Acad. Sci. Paris* **18**: 583–5.

Hartig, T. Ambrosia des *Bostrychus dispar. Allg. Forst-Jagdzeit.* **13**: 73–4.

1845 Remak, R. *Diagnostiche und pathogenetische Untersuchungen, in der Klinik des Herrn Geh. Raths Dr. Schoenlein.* Berlin. (*Achorion,* p. 193.)

Schmitz, J. Beiträge zur Anatomie und Physiologie der Schwämme. *Linnaea* **17**: 434.

1845–50 Trog, J. G. *Die essbaren, verdächtigen und giftigen Schwämme der Schweiz.* Bern.

1845–60 Venturi, A. *I myceti del agro Bresciano, descritti ed illustrati con figure tratte dal vero.* Brescia.

1846 Berkeley, M. J. Observations, botanical and physiological, on the potato murrain. *J. hort. Soc. London* **1**: 9–34. (Reprinted in *Phytopathological Classics* **8** (1948).)

Léveillé, J. H. *Considérations mycologiques, suivies d'une nouvelle classification des champignons.* Paris.

1847 Badham, C. D. *A treatise on the esculent fungusses of England.* London.

Reissek, S. Über Endophyten der Pflanzenzelle. *Naturwiss. Abhandl. von W. Haidinger* **1**: 31–46.

Tulasne, L.-R. & C. Mémoire sur les Ustilaginées comparées aux Urédinées. *Ann. Sci. nat. Paris,* sér. 3, **7**: 12–127. (See also 1854.)

1848 W. Hofmeister illustrates chromosomes.

1850

1851 Bonorden, H. F. *Handbuch der allgemeinen Mykologie.* Stuttgart.

Tulasne, L. R. & C. *Fungi hypogaei.* Paris. (Asher reprint, 1970.)

1852 de Bary, A. Beitrag zur Kenntnis der *Achlya prolifera. Bot. Z.* **10**: 473–9, 489–96, 505–11.

Schacht, H. *Physiologische Botanik. Die Pflanzenzelle, der innere Bau und das Leben der Gewächse.* Berlin.

Tulasne, L. R. Mémoire pour servir à l'histoire organographique et physiologique des lichens. *Ann. Sci. nat. Paris,* sér. 3, **17**: 5–128, · 153–249, pl. 16.

1853 de Bary, A. *Untersuchungen über die Brandpilze und die durch sie verursachten Krankheiten der Pflanzen.* Berlin. (English transl. by R. M. S. Heffner *et al., Phytopathological Classics* **11** (1969).)

Robin, C. *Histoire naturelle des végétaux parasites qui croissent sur l'homme et sur les animaux vivants.* 1 vol.+atlas. Paris.

Tulasne, L. R. Mémoire sur l'ergot des Glumacées. *Ann. Sci. nat. Paris,* sér. 3, **20**: 5–56.

1854 de Bary, A. Ueber die Entwicklung und den Zusammenhang von *Aspergillus glaucus* und *Eurotium. Bot. Z.* **12**: 425–34, 441–51, 465–71.

Tulasne, L. R. Second mémoire sur les Urédinées et les Ustilaginées. *Ann. Sci. nat. Paris,* sér. 4, **2**: 77–196. (See also 1847.)

1854–7 B[erkeley], M. J. Vegetable pathology. A series of 173 articles in the *Gardeners' Chronicle and Agricultural Gazette* between 7 Jan. 1854 and 3

Oct. 1857. (A selection reprinted in *Phytopathological Classics* **8** (1948).)

1856 Hoffmann, H. Die Pollinarien und Spermatien von Agaricus. *Bot. Z.* **14**: 137–48, 153–63.

1857 Berkeley, M. J. *Introduction to cryptogamic botany*. London.

1857–60 Pringsheim, N. Ueber die Befruchtung und Vermehrung der Saprolegnieen. *Monatsb. K. Akad. wiss. Berlin 1857*: 315–30; Beiträge zur Morphologie und Systematik der Algen. II Die Saprolegnieen. *Jahrb. wiss. Bot.* **2**: 205–36 (1860). (These and later papers are reprinted in *Gesammelte Abhandlungen von N. Pringsheim*, **2** (1895). Jena.)

1858 Kühn, J. *Die Krankheiten der Kulturgewächse, ihre Ursachen und ihre Verhütung*. Berlin.

Traube, M. *Theorie der Fermentwirkungen*. Berlin.

1859 *The Origin of Species* by Charles Darwin published 24 November.

*Lenz, H. O. *Botanik der alten Griechen und Römer*. Gotha. (Fungi, pp. 753–66.)

1860 Berkeley, M. J. *Outlines of British fungology*. London

Pasteur, L. Mémoire sur la fermentation alcoolique. *Ann. Chimie Phys.*, sér. 3, **58**: 323–426.

1861 Anderson, T. M'Call. *On the parasitic affections of the skin*. London.

de Bary, A. *Die gegenwärtig herrschende Kartoffelkrankheit, ihre Ursache und ihre Vehütung*. Leipzig.

1861–5 Tulasne, L. R. & C. *Selecta fungorum carpologia*. 3 vols. Paris. (English transl. by W. B. Grove, edited by A. H. R. Buller, 1931. Oxford.)

1862 Streinz, W. M. *Nomenclator fungorum*. Vienna.

1863 de Bary, A. (*a*) *Über die Fruchtentwicklung der Ascomyceten. Eine pflanzenphysiologische Untersuchung*. Leipzig.

de Bary, A. (*b*) Recherches sur le développement de quelques champignons parasitaires. *Ann. Sci. nat. Paris*, sér. 4, **20**: 5–148.

Fox, W. Tilbury. *Skin diseases of parasitic origin*. London.

1864 *Jessen, K. F. W. *Botanik der Gegenwart und Vorzeit in culturhistorischer Entwicklung*. Leipzig. (Chronica Botanica reprint, 1948.)

1864–81 de Bary, A. & Woronin, M. Beiträge zur Morphologie und Physiologie der Pilze. (ex *Abhandl. Senckenberg Naturf. Ges.*, **5**, **7**, **12**) **I** (de Bary only), 1864; **2**, 1866; **3**, 1870; **4**, 1881.

1865 Ørsted, A. S. Jagtagelser anstillede i löbet af vinteren, 1863–64, som have ledet til opdagelsen af de hidtil ukjendte befruchtnigsorganer hos bladswampene. *Oversigt. K. danske Vidensk. Selsk. Forlandl. Kjobenhavn 1865*: 11–23. (English transl. *Quart. J. microscop. Sci.* N.S. **8**: 18–26 (1868).)

1865–6 de Bary, A. Neue Untersuchungen über Uredineen, insbesondere die Entwicklung der *Puccinia graminis*. *Monatsbr. Kon. Akad. Wiss. Berlin 1865*: 15–49; 1866: 205–15.

1866 Gregor Mendel published his work on plant hybridisation.

de Bary, A. *Morphologie und Physiologie der Pilze, Flechten, und Myxomyceten*. Leipzig. (Edn 2, 1884; English transl. see de Bary, 1887.)

Karsten, H. Zur Befruchtung der Pilze. *Bot. Unters. Lab. landwirtsch. Lehranstalt Berlin* **I**: 160–9. (English transl. *Ann. Mag. nat. Hist.*, ser. 3, **19**: 72–80 (1867).)

1866–7   Willkomm, M. *Die mikroscopische Feinde des Waldes*. Dresden.
1867   Kickx, J. (published by J. J. Kickx *fils*) *Flore cryptogamique des Flandres*. 2 vols. Gand and Paris.

Lois de la Nomenclature Botanique adoptées par le Congres International de Botanique tenu à Paris en Août 1867. Paris. (Edn 11 as *International Code of Botanical Nomenclature*, 1972. Utrecht.)

Schwendener, S. Ueber den Bau des Flechtenthallus. *Verh. Schweiz Naturf. Ges. Aarau* **51**: 88–9.

Tulasne, L. R. & C. Note sur les phénomènes de copulation que présentent quelques champignons. *Ann. Sci. nat. Paris*, sér. 5, **6**: 211–20.

van Tieghem, P. Recherches pour servir à l'histoire physiologique des Mucédinées. Fermentation galliques. *Ann. Sci. nat. Paris*, sér. 5, **8**: 210–44.

1867–9   de Bary, A. Zur Kenntnis insectentötender Pilze. *Bot. Z.* **25**: 1–13 (1867); **27**: 585–93, 601–6 (1869).
1867–1915   Peck, C. *Reports of the New York State Museum*, 1867–1908.
1868   Seynes, J. de. Sur le Mycoderma vini. *C. R. Acad. Sci. Paris* **67**: 105–9.
1869   Raulin, J. Études chimiques sur la végétation. *Ann. Sci. nat. Paris*, sér. 5, **11**: 93–299. (As thesis 1870; reprinted 1905, Paris.)

Schmiedeberg, O. & Koppe. *Das Muskarin*. Leipzig.

1869–75   Fuckel, L. *Symbolae mycologicae. Beiträge zur Kenntnis der rheinischen Pilze*. 4 vols. Wiesbaden.
1870   Rees, M. *Botanische Untersuchungen über die Alkolgäringspilze*. Leipzig.
1871   Cooke, M. C. *Handbook of British fungi*. 2 vols. London.

Fox, W. Tilbury. (Clinical Society of London.) *Lancet* **1**: 412.

*Pritzel, G. A. *Thesaurus literaturae botanicae*. Leipzig. (Reprint, 1950, Milan.) See also B. D. Jackson, *Guide to the literature of botany*, 1881. London.

1872   Engel. Les ferments alcooliques. Thèse. Paris. (*Fide* Lafar, 1898–1910).
1872–6   Quélet, L. *Les champignons du Jura et les Voges*. 3 vols. Montbéliard and Paris.
1872–1912   Brefeld, O. Untersuchungen aus dem Gesammtgebiete der Mykologie. Heft **1** (1872); **2** (1874); **3** (1877); **4** (1881); **5** (1883); **6** (1884); **7** (1888); **8** (1889); **9**, **10** (1891); **11**, **12** (1895); **13** (1905); **14** (1908); **15** (1912). Leipzig; Munster; Berlin.
1873   Blackley, C. H. *Experimental researches on the causes and nature of* Catarrhus aestivus (*hay-fever or hay-asthma*). London. (Facsimile 1959. London.)

Cunningham, D. D. *Microscopic examination of air*. Calcutta. (Ex *Ninth Annual Rep. of the Sanitary Commissioner with the Government of India*, 1872.)

Fitz, A. Ueber alkolische Gährung durch *Mucor mucedo*. *Ber. dtsch. chem. Ges.* **6**: 48–58.

Reinke, J. Zur Kenntniss des Rhizoms von *Corallorhiza* u. *Epipogon*. *Flora* **31**: 145, 161, 177, 209.

van Tieghem, P. & Le Monnier, G. Recherches sur les Mucorinées. *Ann. Sci. nat. Paris*, sér. 6, **17**: 261–399.

1873–7   Kalchbrenner, K. *Icones selectae hymenomycetum Hungariae*. Pest.
1874   Carter, H. V. *On mycetoma or the fungus disease of India*. London.

Fries, E. M. *Hymenomycetes Europaei*. Uppsala.

Fuckel, L. Endophytische Pilze, in *Zweite Deutsche Nordpolfahrt*, **2**: 90–6 (Leipzig); Fungi, in M. T. von Heuglin, *Reisen nach dem Nordpolarmeer in den Jahren 1870–1871*, **3**: 317–23. Brunswick.

Hartig, R. *Wichtige Krankheiten der Waldbäume*. Berlin.

Roberts, W. Studies on biogenesis. *Phil. Trans.* **169**: 457–77.

Sorauer, P. C. M. *Handbuch der Pflanzenkrankheiten*. 1 vol. Berlin. (Currently (Aufl. 7, 1962–) a multi-author multi-volume work.)

1874–5 Rostafinski, J. T. *Śluzowce (Mycetozoa) monografia*. Paris.

1875

1875 Tulasne, L. R. Étude de la fécondation dans la classe des champignons. *C. R. Acad. Sci. Paris* **80**: 1466.

van Tieghem, P. (*a*) Sur la fécondation des Basidiomycètes. *C. R. Acad. Sci. Paris* **80**: 373–7.

van Tieghem, P. (*b*) Sur le développement du fruit des *Coprinus*, et la prétendue sexualité des Basidiomycètes. *C. R. Acad. Sci. Paris* **81**: 877–80.

1876 Boudier, E. Du parasitisme probable de quelques espèces du genre *Elaphomyces* et de la recherche de ces Tuberacées. *Bull. Soc. bot. Fr.* **23**: 115.

Brefeld, O. Die Entwicklungsgeschichte der Basidiomyceten. *Bot. Z.* **24**: 49–62.

Pasteur, L. *Études sur la bière*. Paris. (English transl. by F. Faulkner & D. C. Robb as *Studies on fermentation*, 1879. London.)

Rees, M. Ueber den Befruchtungsvorgang bei den Basidiomyceten. *Jahrb. wiss. Bot.* **10**: 179–98.

1876–7 Hansen, E. C. De danske Gjodnings-svampe (Fungi fimicoli Danici). *Vidensk. Meded. Nath. For. Copenhagen* 1876: 207–354.

1877 Cooke, M. C. *The myxomycetes of Great Britain*. London.

Frank, A. B. Über die biologischen Verhältnisse des Thallus einiger Krustflechten. (Cohn's) *Beitr. Biol. Pflanz.* **2**: 123–200.

Stahl, E. *Beiträge zur Entwickelungsgeschichte der Flechten*. Leipzig.

1878 Hartig, R. *Die Zersetzungserscheinungen des Holzes, der Nadelbäume und der Eich.* Berlin.

1878–90 Gillet, C. C. *Les champignons (Fungi, Hyménomycètes) qui croissent en France*. Paris.

1879 de Bary, A. *Die Erscheinung der Symbiose*. Strasbourg. (De la symbiose, *Brébissonia* **2**: 17–19, 38–42, 99–103 (1880).)

Schroeter, J. Entwickelung einiger Rostpilze. (Cohn's) *Beitr. Biol. Pflanz.* **3**: 69–70.

Stevenson, J. *Mycologia Scotica or the fungi of Scotland*. Edinburgh.

1879–1906 *Revue mycologique*, the first mycological journal, founded by Roumeguère.

1880 Robinson, W. *Mushroom culture: its extension and improvement*. London. [n.d.]

Schmitz, F. Ueber die Zellkerne der Thallophyten. *Verh. naturhist. Vereins preuss. Rheinl. Westfal.* **37**: 195.

1881 Thin, G. On *Trichophyton tonsurans* (the fungus of ringworm). *Proc. roy. Soc.* **33**: 234–46.

Tyndall, J. *Essays on the floating-matter of the air in relation to putrefaction and infection.* London.

1881–1900 Bresadola, G. *Fungi Tridentini novi, vel nondam delineati, descripti et iconibus illustrati.* Trento.

1882 Kamienski, F. Les organs végétatifs du *Monotropa hypopitys. Mém. Soc. nat. Sci. nat. Math. Cherburg* **24**: 5–40.

1882–1972 Saccardo, P. A. *Sylloge fungorum hucusque cognitorum.* 26 vols. Pavia.

1883 Hansen, E. C. Recherches sur la physiologie et le morphologie des ferments alcooliques. II. Les ascospores, chez le genre *Saccharomyces. C. R. Lab. Carlsberg* **2**: 13–47.

Miquel, P. Des organismes vivants de l'atmosphère. Thesis. Paris.

1883–9 Patouillard, N. *Tabulae analyticae fungorum.* Paris.

1884 *Société Mycologique de France* founded.

Boulton, S. B. *On the antiseptic treatment of timber.* London (Inst. Civil Engineers).

*Gautier, L.-M. *Les champignons considérés dans leur rapports avec la médecine, l'hygiène publique et privée, l'agriculture et l'industrie.* Paris. (History and bibliography, pp. 175–88.)

Strasburger, E. *Das botanische Praktikum.* Jena.

1884–1960 *Rabenhorst's Kryptogamenflora von Deutschsland, Oesterreich und der Schweiz.* 11 vols. Leipzig.

1885 Boudier, E. Nouvelle classification naturelle des Discomycètes charnus, connus généralement sur le nom de Pézizés. *Bull. Soc. mycol. Fr.* **2**: 135.

Frank, A. B. Ueber die auf Wurzelsymbiose beruhende Ernährung gewissen Bäume durch unterirdische Pilze. *Ber. deutsch. bot. Ges.* **3**: 128–45.

Hartig, R. *Der ächte Hausschwamm* (Merulius lacrimans). Berlin.

*Houghton, W. Notices of fungi in Greek and Latin authors. *Ann. Mag. nat. Hist.,* ser. 5, **15**: 22–49, 153–4.

Millardet, P. M. A. Traitment du mildou et du rot. *J. Agric. pract.* **2**: 513–16. (For English transl., of this and other papers by Millardet on Bordeaux mixture, by F. J. Schneiderhan see *Phytopathological Classics* **3** (1933).)

1885–9 Richon, C. & Roze, E. *Atlas des champignons comestibles et vénéneux de la France et des pays circonvoisins.* Paris. (Chap. 1, historical review by Roze.)

1885–1908 Schroeter, J. Pilze. In F. Cohn, *Kryptôgamen-Flôra von Schlesien,* **3**. Breslau.

1886 Adametz, L. *Untersuchungen über die niederen Pilze der Ackerkrume.* Inaug. Dissert. Leipzig. (*Fide* Chesters, 1949.)

de Bary, A. Ueber einige Sclerotinien und Sclerotienkrankheiten. *Bot. Z.* **44**: 377, 392, 408, 432, 448, 464.

Bonnier, G. Recherches expérimentales sur la synthèse des lichens dans un milieu privé de germes. *C. R. Acad. Sci. Paris* **103**: 942–4.

1887 de Bary, A. *Comparative morphology and biology of the Fungi, Mycetozoa and Bacteria.* Oxford. (= English transl. of de Bary (1866). Edn 2, 1884, by H. E. F. Garnsey & I. B. Balfour.)

Frank, A. B. Ueber neue Mykorriza–formen. *Ber. deutsch. bot. Ges.* **5**: 395–409.

Patouillard, N. *Les Hyménomycètes d'Europe. Anatomie générale et classification des champignons supérieurs.* Paris.

Petri, R. J. Eine kleine Modification des Koch'schen Plattenverfahrens. *Zbl. Bakt.* **1**: 279–80.

Thin, G. *Pathology and treatment of ringworm.* London.

1888   Ward, H. M. A lily disease. *Ann. Bot. London* **2**: 319–82.

1889   Fayod, V. Prodrome d'une histoire naturelle des Agaricinés. *Ann. Sci. nat. Paris*, sér. 7, **9**: 181–411.

1889   Plowright, C. B. *A monograph of the British Uredineae and Ustilagineae.* London.

1890   *Sachs, J. von. *History of botany (1530–1860).* English transl. by H. E. F. Garnsey & I. B. Balfour. Oxford. (J. R. Green, *A history of botany, 1860–1900*, 1909. Oxford.)

Zopf, W. *Die Pilze in morphologischer, physiologischer und systematischer Beziehung.* Breslau.

1891   Hansen, E. C. Recherches sur la physiologie et la morphologie des ferments alcooliques. VIII Sur la germination des spores chez les Saccharomyces. *C. R. Carlsberg Lab.* **3**: 44–66.

Kobert, R. Matières toxiques dans les champignons. *Petersb. med. Wochenschr.* 1891: 51–2.

Wehmer, C. Entstehung und physiologische Bedeutung der Oxalsäure im Stoffwechsel einiger Pilze. *Bot. Z.* **49**: 233–638 (in 22 parts).

1892   Chatin, A. *La truffe.* Paris.

Cooke, M. C. *Handbook of Australian fungi.* London.

Rosen, F. Beiträge zur Kenntnis der Pflanzenzellen. (Cohn's) *Beitr. Biol. Pflanz.* **6**: 237–64.

Tavel, F. von. *Vergleichende Morphologie der Pilze.* Jena.

Wager, H. On the nuclei of the Hymenomycetes. *Ann. Bot. London* **6**: 146–8.

1892–5   Massee, G. *British fungus flora.* 4 vols. London.

1893   Wager, H. On nuclear division in hymenomycetes. *Ann. Bot. London* **7**: 489.

Wehmer, C. *Beiträge zur Kenntnis einheimischer Pilze.* Hanover and Leipzig.

1894   Bommer, C. Sclérotes et cordons mycéliens. *Mém. couron. Mém. sav. étrang. Acad. roy. Sci. Let. Beau-Arts Belg.* **54**, 116 pp.

Costantin, J. & Matruchot, L. Culture d'un champignon lignicole. *C. R. Acad. Sci. Paris* **119**: 752–3.

Dangeard, P. A. La reproduction sexuelle des Ascomycètes. *C. R. Acad. Sci. Paris* **118**: 1065–6.

Eriksson, J. Ueber die Specialisirung des Parasitismus bei den Getreiderostpilzen. *Ber. deutsch. bot. Ges.* **12**: 292–331.

Haeckel, E. H. P. A. *Systematische Phylogenie der Protisten und Pflanzen* **1**: 90–1. Berlin.

Laplanche, M. C. de. *Dictionnaire iconographique des champignons supérieurs (Hyménomycètes) qui croissent en Europe, Algérie & Tunisie.* Paris.

Lister, A. *A monograph of the Mycetozoa.* London. (Edn 2, 1911; edn 3, 1925, revised by G. Lister.)

1895   Cooke, M. C. *Introduction to the study of fungi.* London.

Dangeard, P. A. Mémoire sur la reproduction sexuelle des Basidiomycètes. *Le Botaniste*, sér. 4, 1895: 119–81.

Harper, R. A. Die Entwicklung des Peritheciums bei *Sphaerotheca
castagnei. Ber. deutsch. bot. Ges.* **13**: 475–81.

Maddox, F. *Experiments at Enfield, Dept Agric. Tasmania.*

Poirault, G. & Račiborski, M. Sur les noyaux des Urédinées. *C. R. Acad.
Sci. Paris* **121**: 308–10.

Schiønning, H. Nouvelle et singulière formation d'ascus dans une
levure. *C. R. Carlsberg Lab.* **4**: 30–5.

1896   British Mycological Society founded.

Klebs, G. *Die Bedingungen der Fortpflanzung bei einigen Algen und Pilzen.*
Jena. (Fungi pp. 446–532.)

Marchand, L. *Énumération methodique et raisonnée des familles et genres de la
class des Mycophytes* (champignons & lichens). Paris.

Meschinelli, L. *Fungorum fossilium omnium hucusque cognitorum iconographia xxxi tabulis exornata.* Vicenza.

Sappin-Trouffy, P. Recherches histologiques sur la famille des
Uredinées. *Le Botaniste*, sér. 5, 1896: 59–68.

1896–1931   Thaxter, R. Contributions towards a monograph of the Laboul-
beniaceae I–V. *Mem. Am. Acad. Arts Sci.* **12**: 187–249 (1896);
13: 217–649 (1908); **14**: 309–426 (1924); **15**: 427–580 (1926); **16**: 1–435
(1931). (Cramer reprint 1971.)

1897   Buchner, E. Alkoholische Gährung ohne Hefzellen. *Ber. deutsch. chem.
Ges.* **30**: 117–24, 1110–13.

Lucet, A. *De l'Aspergillus fumigatus chez animaux domestiques et dans les
oeufs en incubation. Étude clinique et expérimentale.* Paris.

Rénon, L. *Étude sur l'aspergillose chez les animaux et chez l'homme.* Paris.

1897–1904   Penzig, O. & Saccardo, P. A. Diagnoses fungorum novorum in
insula Java collectorum. *Malpigia* **11**: 387–409, 491–530 (1897);
15: 201–60 (1902). *Icones fungorum Javanicorum*, 1904. Leiden.

1898   Juel, H. O. Die Kerntheilungen in den Basidien und die Phylogenie der
Basidiomyceten. *Jahrb. wiss. Bot.* **32**: 361–88.

Morris, M. *Ringworm in the light of recent research.* London.

Takamine, J. Diastische Substanzen aus Pilzculturen. *J. Soc. chem.
Industr.* **17**: 118–20.

1898–1900   Klebs, G. Zur Physiologie der Fortpflanzung einiger Pilze. *Jahrb.
wiss. Bot.* **32**: 1–70 (1898); **33**: 513–93 (1899); **35**: 80–203 (1900).

1898–1910   Lafar, F. *Technical mycology* (transl. by C. T. C. Salter of *Handbuch
für technische Mykologie*, 1896–1906; edn 2, 1910–14). 3 vols. London.

1899   MacBride, T. H. *The North American slime-moulds.* New York. (Edn 2,
1922.)

1900

1900   Rediscovery of Mendel's Laws independently by De Vries, Correns, and
von Tschermak.

Harper, R. A. Sexual reproduction in *Pyronema confluens* and the
morphology of the ascocarp. *Ann. Bot. Lond.* **14**: 321–400.

Hektoen, L. & Perkins, C. F. Refractory subcutaneous abscesses caused by *Sporothrix schenckii*, a new pathogenic fungus. *J. exp. Med.* **5**: 77–89.

Hoffmeister, C. Zum Nachweise des Zellkerns bei Saccharomyces. *Sitzber. deutsch. Naturw-med. Ver.* (*Lotos*) N.F. **20**: 250–62.

Istvánffi, G. *Études et commentaires sur le Code de l'Éscluse.* Budapest.

Magnus, W. Studien an der endotrophen Mycorrhiza von *Neottia nidus avis* L. *Jahrb. wiss. Bot.* **35**: 205–72.

Maire, R. Sur la cytologie des Hyménomycètes. *C. R. Acad. Sci. Paris* **131**: 121–4; des Gasteromycètes, **131**: 1246–8.

Ophüls, W. & Moffitt, H. C. A new pathogenic mould (formerly described as a protozoon: *Coccidioides immitis pyogenes*): preliminary report. *Philadelphia med. J.* **5**: 1471–2.

1901   Wildiers, E. Nouvelle substance indispensable au développement de la levure. *La Cellule* **18**: 313–32.

1901–2   Massee, G. & Salmon, E. S. Researches on coprophilous fungi. *Ann. Bot. London* **15**: 313–57 (1901); **16**: 57–93 (1902).

1902   Elenkin, A. A. Zur Frage der Theorie des 'Endosaprophytismus' bei Flechten. *Bull. Jard. bot. St Petersburg* **2**: 65–84.

Ferguson, Margaret C. A preliminary study of the germination of the spores of *Agaricus campestris* and other basidiomycetous fungi. *Bull. U.S.D.A. Bur. Pl. Industr.* **16**, 40 pp.

Maire, R. Recherches cytologiques et taxonomiques sur les Basidiomycètes. *Bull. Soc. mycol. Fr.* **18**, appendice: 1–209.

1903   Publication of *Annales Mycologici*, Berlin, began.

1904   Biffen, R. H. Experiments with wheat and barley hybrids illustrating Mendel's Laws of heredity. *J. roy. Agric. Soc.* **65**: 337–45.

Blackman, V. H. On the fertilization, alternation of generations, and general cytology of the Uredineae. *Ann. Bot. London* **18**: 323–73.

Blakeslee, A. F. Sexual reproduction in the Mucorineae. *Proc. Am. Acad. Arts Sci.* **40**: 205–319. (Results summarised *Science* **19**: 863–6 (1904).)

Gallaud, G. Études sur les mycorhizes endotrophes. Thèse, Paris. *Rev. gen. Bot.* **17**: 5, 66, 123, 223, 313, 423, 479 (1905).

Klebahn, H. *Die Wirtswechselnden Rostpilze.* Berlin.

Oudemans, C. A. J. A. *Catalogue raisonné des champignons des Pays-Bas.* Amsterdam.

Pallin, W. A. *A treatise on epizootic lymphangitis.* London.

1905   Butler, E. J. The bearing of Mendelism on the susceptibility of wheat to rust. *J. agric. Sci.* **1**: 361–3.

Duggar, B. M. Some principles of mushroom growing and mushroom spawn making. *Bull. U.S.D.A. Bur. Pl. Industr.* **85**, 60 pp.

Harper, R. A. Sexual reproduction and organization of the nucleus in certain mildews. *Carnegie Inst. Wash., Publ.* **37**, 104 pp.

Shirai, M. *A list of Japanese fungi hitherto known.* Tokyo.

1905–10   Boudier, É. *Icones mycologicae.* 4 vols. Paris.

1906   Centraalbureau voor Schimmelcultures founded, as Centralstelle für Pilzkulturen, at Utrecht.

Blakeslee, A. F. Zygospore germination in the Mucorineae. *Ann. mycol. Berlin* **4**: 1–28.

Brumpt, É. Les mycétomes. *Arch. Parasitol* **10**: 489–572. (Also as Thèse Fac. Méd. Paris, 1906, 94 pp.)

Guéguen, M. F. La moisissure des caves et des celliers; étude critique, morphologique et biologique sur le *Racodium cellare* Pers. *Bull. Soc. mycol. Fr.* **22**: 77–95, 146–63.

Hedgcock, G. G. Zonation in artificial cultures of *Cephalosporium* and other fungi. *Rep. Mo. bot. Gard.* **17**: 115–17.

McAlpine, D. *The rusts of Australia, their structure, nature and classification.* Melbourne.

Massee, G. *Text-book of fungi.* London.

Thom, C. Fungi in cheese ripening: Camembert and Roquefort. *Bull. U.S.D.A. Bur. Anim. Industr.* **82**, 39 pp.

1907   Ternetz, Charlotte. Über die Assimilation des atmosphärischen Stickstoffes durch Pilze. *Jahrb. wiss. Bot.* **44**: 353–408.

Vuillemin, P. *La bases actuelles de la systématique en mycologie.* Jena. (Ex *Lotsy Progressus Rei Botanicae* 2(1).)

Zellner, J. *Chemie der höheren Pilze. Eine Monographie.* Leipzig.

1908   Lutz, A. & Splendore, A. Über eine an Menschen und Ratten beobachtete Mycose. Beitrag zur Kenntnis der Sporotrichosen. *Zbl. Bakt.* (Abt. I) **46**: 21–97.

Mez, C. *Der Hausschwamm und die übrigen holzzerstörenden Pilze der menschlichen Wohnungen. Ihre Erkennung, Bedeutung und Bekämpfung.* Dresden.

1908–15   *Lindau, G. & Sydow, P. *Thesaurus litteraturae mycologicae et lichenologicae.* 5 vols. Leipzig. (Johnson reprint, n.d. New York.) (*Supplementum 1911–1930* by R. Ciferri, 4 vols., 1957–60. Pavia.)

1909   Bernard, N. L'évolution dans la symbiose. Les orchidées et leurs champignons commensaux. *Ann. Sci. nat. Paris*, sér. 9, **9**: 1–196.

Burgeff, H. *Die Wurzelpilze der Orchideen, ihre Kultur und ihr Leben in der Pflanze.* Jena.

Clements, F. E. *The genera of fungi.* Minneapolis.

Cotton, A. D. Notes on marine pyrenomycetes. *Trans. Br. mycol. Soc.* **3**: 92–9.

Wakefield, Elsie M. Über die Bedingungen der Fruchtkörperbildung, sowie das Auftreten fertiler und steriler Stämme bei Hymenomyzeten. *Naturw. Z. Forst.-u. Landw.* **7**: 521–51.

1909–50   Buller, A. H. R. *Researches on fungi.* **1** (1909); **2** (1922); **3** (1924); **4** (1931); **5** (1933); **6** (1934). London. **7** (1950). Toronto.

1910   Morgan announces his gene theory.

McAlpine, D. *The smuts of Australia, their structure, life history, treatment, and classification.* Melbourne.

Sabouraud, R. *Maladies du cuir chevelu. III. Les maladies cryptogamiques. Les Teignes.* Paris.

Thom, C. Cultural studies of species of *Penicillium. Bull. U.S.D.A. Bur. Anim. Industr.* **118**, 109 pp.

1911   Bayliss, Jessie M. Observations on *Marasmius oreades* and *Clitocybe gigantea*, as parasitic fungi causing 'fairy rings'. *J. econ. Biol.* **6**: 111–32.

Bertrand, G. & Javillier, M. Influence du manganèse sur le développement de l'*Aspergillus niger. C. R. Acad. Sci. Paris* **152**: 225–8.

1911–13   Fink, B. The nature and classification of lichens. I. Views and arguments of botanists concerning classification. *Mycologia* **3**: 231–69 (1911); II. The lichen and its algal host. *ibid.* **5**: 97–166 (1913).

1911–15   Minden, M. von. Chytridiineae. Anclystineae. Monoblepharidineae. Saprolegniineae. *KryptogamFl. Mark Brandenburg* 5 (2–3) (1911); (4) (1912); (5) (1915). Leipzig.

1911–32   Harden, A. *Alcoholic fermentation.* London. (Edn 2, 1914; edn 3, 1923; edn 4, 1932.)

1912   Beurmann, L. de & Gougerot, H. *Les sporotrichoses.* Paris.

Claussen, P. Zur Entwichlungegeshichte der Ascomyceten (*Pyronema confluens*). *Z. Bot.* 4: 1–64.

Guilliermond, A. *Les levures.* Paris.

Falk, R. Die Merulius Fäule des Bauholzes. *Hausschwammforsch.* 6: 1–405.

*Lister, Gulielma. The past students of the Mycetozoa and their work. *Trans. Br. mycol. Soc.* 4: 44–61.

Vuillemin, P. *Les champignons. Essai de classification.* Paris.

1913   Blakeslee, A. F. & Gortner, R. A. On the occurrence of a toxin in the juice expressed from the bread mould, *Rhizopus nigricans* (*Mucor stolonifer*). *Biochem. Bull.* 11: 1–2.

Lind, J. *Danish fungi as represented in the herbarium of E. Rostrup.* Copenhagen.

Stakman, E. C. A study of cereal rusts: physiological races. Thesis, Univ. Minnesota. (= *Bull. Minn. Exp. Stn* 138, 56 pp.)

1913-22   Lagarde, J. Champignons in biospeologica. *Arch. Zool. exp. gen.* 53: 277–307 (1913); 56: 279–314 (1917); 60: 593–625 (1922).

1914   *Sorauer, P. Historical survey. In P. Sorauer, *Manual of plant diseases*, pp. 41–70, 1914. (English transl. by Frances Dorrance of *Handbuch der Pflanzenkrankheiten* 1: 37–68 (1909).)

1915   Brown, W. Studies in the physiology of parasitism. I. The action of *Botrytis cinerea. Ann. Bot. Lond.* 29: 313–48.

*Buller, A. H. R. (*a*) The fungus lore of the Greeks and Romans. *Trans. Br. mycol. Soc.* 5: 21–66.

*Buller, A. H. R. (*b*) Micheli and the discovery of reproduction in fungi. *Trans. roy. Soc. Canada*, ser. 3, 9: 1–25.

Buller, A. H. R. (*c*) Die Erzeugung und Befreiung der Sporen bei *Coprinus sterquilinus. Jahrb. wiss. Bot.* 56: 299–329.

Kneip, H. Beiträge zur Kenntnis der Hymenomyceten. III. Über der konjugierten Teilungen und die phylogenetische Bedeutung der Schnallenbildungen. *Z. Bot.* 7: 369–98.

Tolaas, A. G. A bacterial disease of cultivated mushrooms. *Phytopathology* 5: 51–3.

1916   Brown, H. B. The life history and poisonous properties of *Claviceps paspali. J. agric. Res.* 7: 401–6.

Waksman, S. A. Do fungi live and produce mycelium in the soil? *Science* N.S. 44: 320–2.

1917   Bensaude, Mathilde. Sur la sexualité chez les champignons Basidiomycètes. *C. R. Acad. Sci. Paris* 165: 286–9.

Currie, J. N. The citric acid fermentation of *Aspergillus niger. J. biol. Chem.* 31: 15–37.

Schantz, H. L. & Piemeisel, F. J. Fungus fairy rings in Eastern Colorado and their effect on vegetation. *J. agric. Res.* 11: 191–245.

Stakman, E. C. & Piemeisel, F. J. Biologic forms of *Puccinia graminis* on cereals and grasses. *J. agric. Res.* 10: 429–95.

Waksman, S. A. Is there any fungus flora of the soil? *Soil Sci.* **3**: 565–89.

1918 Bensaude, Mathilde. *Recherches sur le cycle évolutif et la sexualité chez les Basidiomycètes.* Nemours.

Buder, J. Die inversion des Phototropismus bei *Phycomyces. Ber. deutsch. bot. Ges.* **36**: 104–5.

Ungermann, E. Eine einfache Method zur Gewinnung von Dauerkulturen empfindlicher Bakterienarten and zur Erhaltung der Virulenz tierpathogener Keime. *Arb.a.d. Reichgesundheitsamte* **51**: 180–99.

*Whetzel, H. H. *An outline of the history of phytopathology.* Philadelphia and London.

1919 Connstein, W. & Ludecke, K. Über Glycerin Gewinnung durch Gärung. *Ber. deutsch. chem. Ges.* **52**: 1385–91.

Kniep, H. Untersuchungen über den Antherenbrand (*Ustilago violacea* Pers.). Ein Beitrag zum sexualitätproblem. *Z. Bot.* **11**: 257–84.

Paine, S. G. Studies in bacteriosis II. A brown blotch disease of cultivated mushrooms. *Ann. appl. Biol.* **5**: 206–19.

Steinberg, R. A. A study of some factors in the chemical stimulation of growth of *Aspergillus niger. Am J. Bot.* **6**: 330–72.

1919–24 Oudemans, C. A. J. A. *Enumeratio systematica fungorum.* 5 vols. The Hague.

1920 Commonwealth Mycological Institute founded, as Imperial Bureau of Mycology, at Kew.

Dodge, B. O. The life history of *Ascobolus magnificus.* Origin of the ascocarp from two strains. *Mycologia* **12**: 115–34.

Kniep, H. Über morphologische und physiologische Geschlechtsdifferenzierung. (Untersuchungen an Basidiomyceten.) *Verh. phys.-med. Ges. Würzburg* **46**: 1–18.

1921 Mounce, Irene. Homothallism and the production of fruit-bodies in the genus *Coprinus. Trans. Br. mycol. Soc.* **7**: 198–217.

Smith, Annie Lorrain. *Lichens.* Cambridge.

1921–40 Zahlbruckner, A. *Catalogus lichenum universalis.* 10 vols. Leipzig. (Johnson reprint, 1951.)

1922 Conn, H. J. A microscopic method for demonstrating fungi and actinomycetes in soil. *Soil Sci.* **14**: 149–51.

Molliard, M. Sur une nouvelle fermentation acide produite par le *Sterigmatocystis nigra. C. R. Acad. Sci. Paris* **174**: 881–3.

Rea, C. *British Basidiomycetae.* Cambridge.

Stakman, E. C. & Levine, M. N. The determination of biologic forms of *Puccinia graminis* on *Triticum* spp. *Tech. Bull. Univ. Minn. agric. Exp. Stn* **8**, 10 pp.

1922–5 Knudsen, L. Non-symbiotic germination of orchid seed. *Bot. Gaz.* **73**: 1–25 (1922). Additional papers *ibid.* **77**: 212–19 (1924); **79**: 345–79 (1925).

1923 Coker, W. C. *The Saprolegniaceae, with notes on other water molds.* Chapel Hill, N.C.

Lehfeldt, W. Über die Entstehung des Paarkernmycel bei heterothallischen Basidiomyceten. *Hedwigia* **64**: 30–51.

Vandendries, R. Nouvelle recherches sur la sexualité des Basidiomycètes. *Bull. Soc. Bot. Belg.* **56**: 73–97.

1924 Brunswick, H. Untersuchungen über die Geschlechts und Kernverhältnisse bei der Hymenomyzetengattung *Coprinus. Bot. Abhandl.* **5**, 152 pp.

Burgeff, H. Untersuchungen über Sexualität und Parasitismus bei Mucorineen. I. *Bot. Abhandl.* **4**:5–135.

Cadham, F. T. Asthma due to grain rusts. *J. Am. med. Assn* **83**:27.

Funke, G. L. Über die Isolierung von Basidiosporen mit dem Mikromanipulator nach Janse und Péterfi. *Z. Bot.* **16**:619–23.

Haehn, H. & Kintoff, W. Beitrag über den chemischen Mechanismus der Fettbildung aus Zucker. *Chem. Zelle* **12**:115–56.

Hanna, W. F. The dry-needle method of making monosporous cultures of hymenomycetes and other fungi. *Ann. Bot. London* **38**:791–5.

van Leeuwen, W. S. Bronchial asthma in relation to climate. *Proc. roy. Soc. Med.* (Sect. Therap. & Pharmacol.) **17**:19–26.

*Woolman, H. M. & Humphrey, H. B. Summary of literature on bunt, or stinking smut, of wheat. *Dep. Bull. U.S.D.A.* **1210**, 44 pp.

1925

1925 Hanna, W. F. The problem of sex in *Coprinus lagopus. Ann. Bot. London* **39**:431–57.

*Rolfe, R. T. & F. W. *The romance of the fungus world.* London.

1926 Gäumann, E. A. *Vergleichende Morphologie der Pilze.* Jena. (English transl., *Comparative morphology of the fungi*, by C. W. Dodge, 1928. New York.)

Newton, Dorothy E. The bisexuality of individual strains of *Coprinus rostrupianus. Ann. Bot. London* **40**:105–28.

Thom, C. & Church, Margaret B. *The aspergilli.* Baltimore. (See also 1965.)

1927 Brierley, W. B., Jewson, S. T. & Brierley, M. The quantitative study of soil fungi. *Proc. Papers First internat. Congr. Soil Sci.* **3**:24–48.

Craigie, J. H. (*a*) Experiments on sex in rust fungi. *Nature* **120**:116–17.

Craigie, J. H. (*b*) Discovery of the function of pycnia of the rust fungi. *Nature* **120**:765–7.

Gwynne-Vaughan, H. C. I. & Barnes, B. *The structure and development of the fungi.* Cambridge.

Nannizzi, A. Ricerche sull'origine saprofitica dei funghi delle tigne. *Atti Acad. Fisiocr. Sienna* **10**:89–97.

Rayner, M. C. *Mycorrhiza.* London. (New Phytologist Reprint 15.)

Shear, C. L. & Dodge, B. O. Life histories and heterothallism of the red bread-mold fungi of the *Monilia sitophila* group. *J. agric. Res.* **34**:1019–42.

Theiler, A. Die Diplodiosis der Rinder und Schafe in Süd-Afrika. *Deutsch. tierarztl. Wschr.* **35**:395–9.

Waksman, S. A. *Principles of soil microbiology.* London. (Edn 2, 1931.)

1927–60 Bresadola, G. *Iconographia mycologica.* 28 vols. Milan.

1928 Dodge, B. O. Production of fertile hybrids in the ascomycete Neurospora. *J. agric. Res.* **36**:1–14.

Eastcott, E. V. Wildier's bios. The isolation and identification of bios I. *J. physiol. Chem.* **32**:1093–111.

Martin, H. *The scientific principles of plant protection.* London. (Edn 6, 1973.)

Moreau, F. *Les lichens.* Paris.

Rossi, G. M. Il terreno agario nella teoria e nelle realta. *L'italia Agric.* no. 4 (1928). (*Fide* Chesters, 1949.)

Stevens, F. L. Effects of ultra-violet radiation on various fungi. *Bot. Gaz.* **86**: 210–25.

1929 Fleming, A. On the antibacterial action of cultures of a *Penicillium*, with special reference to the isolation of *B. influenzae. Br. J. exp. Path.* **10**: 226–36.

*Garrison, F. H. *An introduction to the history of medicine.* Philadelphia and London.

Kinoshita, K. Formation of itaconic acid and mannitol by a new filamentous fungus. *J. chem. Soc. Japan* **50**: 583–93.

Sass, J. E. The cytological basis for homothallism and heterothallism in the Agaricaceae. *Am. J. Bot.* **16**: 663–701.

Seymour, A. B. *Host index of fungi of North America.* Cambridge, Mass.

Thom, C. *The penicillia.* Baltimore. (See also 1949.)

1930 Bernton, H. S. Asthma due to a mold – *Aspergillus fumigatus. J. Am. med. Assn* **95**: 189–91.

Buller, A. H. R. The biological significance of conjugate nuclei in *Coprinus lagopus* and other hymenomycetes. *Nature* **126**: 686–9. (Also Buller (1909–50) **4**: 187–293.)

Cholodny, N. Über eine neue Methode zur Untersuchungen der Boden-mikroflora. *Arch. Mikrobiol.* **1**: 620–52.

Hopkins, J. G., Benham, Rhoda W. & Kesten, Beatrice M. Asthma due to a fungus – Alternaria. *J. Am. med. Assn* **94**: 6.

Keissler, K. Die Flechtenparasiten. *Rabenh. Krypt.-Fl.* **8**.

1931 Mycological Society of America founded.

Barger, G. *Ergot and ergotism.* London and Edinburgh.

Brodie, H. J. The oidia of *Coprinus lagopus* and their relation to insects. *Ann. Bot. London* **45**: 315–44.

Butler, E. J. & Bisby, G. R. *The fungi of India.* Calcutta. (Imp. Counc. Agric. Res. Sci. Monogr. **1**.)

*Chapman, A. C. The yeast cell: what did Leeuwenhoeck see? *J. Inst. Brewing* **37**: 433–6.

Clements, F. E. & Shear, C. L. *The genera of fungi.* New York.

Kinnear, J. Wood's glass in the diagnosis of ringworm. *Br. med. J.* **1**: 791–3.

Matthews, Velma D. *Studies on the genus Pythium.* Chapel Hill, N.C.

1931–64 Raistrick, H. *et al.* Studies in the biochemistry of micro-organisms. *Phil. Trans.* **220B**: 1–367 (the first 18 parts; subsequent parts appeared in *Biochem. J.* **25** (1931)–93 (1964)).

1932 Brown, A. M. Diploidisation of haploid by diploid mycelium of *Puccinia helianthi* Schw. *Nature* **130**: 777.

Cobe, H. M. Asthma due to a mold. *Cladosporium fulvum J. Allergy* **3**: 389.

Corner, E. J. H. (*a*) The fruit-body of *Polystictus xanthopus*, Fr. *Ann. Bot. London* **46**: 71–111.

Corner, E. J. H. (*b*) A *Fomes* with two systems of hyphae. *Trans. Br. mycol. Soc.* **17**: 51–81.

Drayton, F. C. The sexual function of microconidia in certain discomycetes. *Mycologia* **24**: 345–8.

*Dobell, C. *Antony van Leeuwenhoek and his 'little animals'.* London. (Dover reprint, 1960.)

1933   *Braun, K. Überblick über die Geschichte der Pflanzenkrankheiten und Pflanzenschadlinge (bis 1880). In *Sorauer Handbuch der Pflanzenkrankheiten,* Aufl. 6, **1**(1): 1–79.

Craigie, J. H. Union of pycniospores and haploid hyphae in *Puccinia helianthi* Schw. *Nature* **131**: 25.

Drechsler, C. Morphological diversity among fungi capturing and destroying nematodes. *J. Wash. Acad. Sci.* **23**: 138–41. (See also *ibid.*: 200–2, 267–70, 355–7; and the many later papers from 1934, mostly in *Mycologia.*)

Tanaka, S. Studies in the black spot disease of the Japanese pears (*Pyrus serotina*). *Mem. Coll. Agric. Kyoto Imp. Univ.* **28**: 1–31.

1934   Arthur, J. C. *Manual of the rusts of the United States and Canada.* Lafayette.

DeMonbreun, W. A. The cultivation and cultural characteristics of Darling's *Histoplasma capsulatum. Am. J. trop. Med.* **14**: 93–125.

Drayton, F. L. The sexual mechanism of *Sclerotinia gladioli. Mycologia* **26**: 46–72.

Emmons, C. W. Dermatophytes. Natural grouping based on the form of the spores and accessary organs. *Arch. Derm. Syph. Chicago* **30**: 337–62.

*Guétrot, M. Le quarantenaire de la Société Mycologique de France (1884–1924). Paris (Soc. mycol. Fr.).

Lind, J. Studies on the geographical distribution of arctic circumpolar micromycetes. *Det. Kgl. Danske Vidensk. Selskab. Biol. Medd.* **11**(2): 1–152.

Schopfer, W. H. Les vitamines crystallisées B commes hormones de croissance chez un microorganisme (*Phycomyces*). *Arch. Mikrobiol.* **5**: 502–10.

1935   Bessey, E. A. *A text-book of mycology.* (Edn 2, *Morphology and taxonomy of fungi,* 1950.) Philadelphia.

Dodge, C. W. *Medical mycology.* St Louis, Mo. and London.

Herrick, H. T., Hellbach, R. & May, O. E. Apparatus for the application of submerged mold fermentations under pressure. *Industr. Engng Chem.* **27**: 681–3.

Jensen, H. L. Contributions to the microbiology of Australian soils. III. The Rossi-Cholodny method as a quantitative index of the growth of fungi in the soil. *Proc. Linn. Soc. N.S.W.* **60**: 145–54.

Winge, Ø. On haplophase and diplophase in some Saccharomycetes. *C. R. Lab. Carlsberg,* sér. Physiol., **21**: 77–112.

1935–41   Lange, J. *Flora agaricina Danica.* 5 vols. Copenhagen.

1936   *Brown, W. The physiology of host–parasite relations. *Bot. Rev.* **5**: 236–81. (See also *Ann. Rev. Phytopath.* **3**: 1–18 (1965).)

Christensen, J. J. & Kernkamp, H. C. H. Studies on the toxicity of blighted barley to swine. *Tech. Bull. Univ. Minn. agric. Exp. Stn* **11**, 28 pp.

Lindegren, C. C. A six-point map of the sex chromosome of *Neurospora crassa. J. Genetics* **32**: 243–56 (April). (Revised map *J. Heredity* **27**: 251–9 (July 1936).)

*Lütjeharms, W. J. Zur Geschichte der Mykologie das XVIII. Jahrhundert. *Meded. Nederl. Mycol. Vereen.* **23**: 1–262.

Steinberg, R. A. Relation of accessory growth substances to heavy

metals, including molybdenum, in the nutrition of *Aspergillus niger.*
J. agric. Res. **52**:439–48.

Ward, G. E., Lockwood, L. B., May, O. E. & Herrick, H. T. Studies in
the genus *Rhizopus.* I. The production of dextro-lactic acid. *J. Am.
chem. Soc.* **58**:1286.

Weindling, R. & Emerson, C. H. The isolation of a toxic substance from
the culture filtrate of *Trichoderma. Phytopathology* **26**:1068–70.

1937 *Fitzpatrick, H. M. Historical background of the Mycological Society of
America. *Mycologia* **29**:1–25.

Quintanilha, A. Contribution à l'étude génétique du phénomène de
Buller. *C. R. Acad. Sci. Paris* **205**:745–7.

Robbins, W. J. The assimilation by plants of various forms of nitrogen.
Am. J. Bot. **24**:243–50.

*Wehnelt, B. Mathieu Tillet. Tilletia. Die Geschichte einer Entdeckung.
Nachr. Schädlingsbekämpfung **12**(2):45–146.

1938 *Arber, Agnes. *Herbals. Their origin and evolution.* Edn 2. Cambridge.

Buller, A. H. R. Fusions between flexuous hyphae and pycniospores
in *Puccinia graminis. Nature* **141**:33.

*Bulloch, W. *The history of bacteriology.* London.

Couch, J. *The genus Septobasidium.* Chapel Hill, N.C.

Dickson, E. C. & Gifford, Myrnie A. *Coccidioides* infection (coccidioidomycosis), the primary type of infection. *Arch. int. Med.* **62**:853–71.

*Dujarric de la Rivière, R. & Heim, R. *Les champignons toxiques.* Paris.
(600 refs.)

Durham, O. C. An unusual shower of fungus spores. *J. Am. med. Assn*
111:24–5.

Mulder, E. G. Influence of copper on growth of microorganisms. *Ann.
Ferm.* **4**:513–33.

Smith, G. *An introduction to industrial mycology.* London. (Edn 6, 1969.)

Steinberg, R. A. The essentiality of gallium to growth and reproduction
of *Aspergillus niger. J. agric. Res.* **57**:569–74.

1939 Benham, Rhoda H. The cultural characters of *Pityrosporum ovale. J.
invest. Dermat.* **2**:187–202. See also *Proc. Soc. exp. Biol. Med.* **46**:176–8
(1941).

Oxford, A. E., Raistrick, H. & Simonart, P. Studies in the biochemistry
of micro-organisms. LX. Griseofulvin, $C_{17}H_{17}O_6Cl$, a metabolic product of *Penicillium griseo-fulvum* Dierckx. *Biochem. J.* **33**:240-8.

Thomas, E. A. Über die Biologie von Flechtenbildnern. *Kryptog-
Flora KryptogFlora Schweiz* **9**(1), 208 pp.

1940 *Index of Fungi,* Kew, began.

*Large, E. C. *The advance of the fungi.* London.

*Puttemans, A. History of phytopathology in Brazil. *J. Agric. Univ.
Puerto Rico* **24**:77–107.

Quintanilha, A. & Balle, Simonne. Étude génétique des phénomènes de
nanisme chez les Hyménomycètes. *Bull. Soc. Broteriana,* sér. 2,
14:17–46.

Su, U. Thet & Seth, L. N. Cultivation of the straw mushroom. *Indian
Farming* **1**:332–3.

1941 Beadle, G. W. & Tatum, E. L. Genetical control of biochemical reactions
in *Neurospora. Proc. nat. Acad. Sci. Wash.* **27**:499–506.

*Buller, A. H. R. The diploid cell and the diploidisation process in plants and animals with special reference to the higher fungi. *Bot. Rev.* **7**: 335–431.

*Ramsbottom, J. (*a*) The expanding knowledge of mycology since Linnaeus. *Proc. Linn. Soc. Lond.* **151**: 280–367.

*Ramsbottom, J. (*b*) Dry rot in ships. *Essex Naturalist* **25**: 231–67.

1942 Dobbs, C. G. On the primary dispersal and isolation of fungal spores. *New Phytol.* **41**: 63–9.

Ingold, C. T. Aquatic hyphomycetes of decaying alder leaves. *Trans. Br. mycol. Soc.* **25**: 339–417.

*Reed, H. S. *A short history of the plant sciences.* Waltham, Mass. (Chap. 18, Mycology; 19, Plant pathology.)

1943 Ainsworth, G. C. & Bisby, G. R. *A dictionary of the fungi.* Kew. (Edn 6, 1971.)

Lindegren, C. C. & Gertrude. Segregation, mutation and copulation in *Saccharomyces cerevisiae. Ann. Mo. bot. Gdn* **30**: 453–68.

Sparrow, F. K. *Aquatic phycomycetes exclusive of the Saprolegniaceae and Pythium.* Ann Arbor, Mich. (Edn 2, 1960.)

*Wehnelt, B. *Die Pflanzenpathologie der deutschen Romantik als Lehre vom kranken Leben und Bilden der Pflanzen.* Bonn.

1944 Barghoorn, E. S. & Linder, D. H. Marine fungi: their taxonomy and biology. *Farlowia* **1**: 395–467.

Conant, N. F., Martin, D. S., Smith, D. T., Baker, R. D. & Callaway, J. L. *Manual of clinical mycology.* Philadelphia; London. (Edn 3, 1971.)

Garrett, S. D. *Root disease fungi.* Waltham, Mass.

Hagelstein, R. *The Mycetozoa of North America.* New York.

1945 Beadle, G. W. & Tatum, E. L. *Neurospora.* II. Methods of producing and detecting mutations concerned with nutritional requirements. *Am. J. Bot.* **32**: 678–86.

*Horsfall, J. G. *Fungicides and their action.* Waltham, Mass.

Langeron, M. *Précis de mycologie.* Paris.

Raper, K. B. & Alexander, D. F. Preservation of molds by the lyophil process. *Mycologia* **37**: 499–525.

1946 Hyde, H. A. & Williams, D. A. A daily census of *Alternaria* spores caught from the atmosphere at Cardiff in 1942 and 1943. *Trans. Br. mycol. Soc.* **29**: 78–85.

1947 Buell, Caroline B. & Weston, W. H. Application of mineral oil conservation method to maintaining collections of fungal cultures. *Am. J. Bot.* **34**: 555–61.

Le Mense, E. H., Gorman, I., van Lanen, J. M. & Langlykke, A. F. The production of mold amylases in submerged culture. *J. Bact.* **54**: 149–59.

Meehan, F. & Murphy, H. C. Differential phytotoxicity of metabolic byproducts of *Helminthosporium victoriae. Science* **106**: 270–1.

1948 Feinberg, S. M. *Allergy in practice.* Chicago. (Allergy to fungi, pp. 216–84.)

Nobles, Mildred K. Studies in forest pathology VI. Identification of cultures of wood-rotting fungi. *Can. J. Res.* **C26**: 281–431. (See also *Can. J. Bot.* **43**: 1097–1139 (1965).)

*Ramsbottom, J. Presidential address [Mycology then and now]. *Trans. Br. mycol. Soc.* **30**: 22–39.

1949 Bullen, J. J. The yeast-like form of *Cryptococcus farciminosus* (Rivolta) (*Histoplasma farciminosum*). *J. Path. Bact.* **61**: 117–20.

*Chesters, C. G. C. Presidential address. Concerning fungi inhabiting the soil. *Trans. Br. mycol. Soc.* **32**: 197–216.

Emmons, C. W. Isolation of *Histoplasma capsulatum* from soil. *U.S. Pub. Hlth Rep.* **64**: 892–6.

Foster, J. W. *Chemical activities of fungi.* New York.

Raper, K. B. & Thom, C. *A manual of the penicillia.* Baltimore.

Whitehouse, H. L. K. (*a*) Heterothallism and sex in fungi. *Biol. Rev.* **24**: 411–47.

Whitehouse, H. L. K. (*b*) Multiple-allelomorph heterothallism in the fungi. *New Phytol.* **48**: 212–44.

1950

1950 Andersson, O. Larger fungi on sandy grass heaths and sand dunes in Scandanavia. *Bot. Notiser*, suppl. vol. **2**(2): 89 pp.

Corner, E. J. H. *A monograph of Clavaria and allied genera.* Oxford.

Gregory, P. H. & Nixon, H. L. Electron micrographs of spores of some British gasteromycetes. *Trans. Br. mycol. Soc.* **33**: 359–62.

Hawker, Lilian E. *Physiology of fungi.* London.

*Kelley, A. P. *Mycotrophy in plants.* Waltham, Mass.

Petch, T. & Bisby, G. R. *The fungi of Ceylon.* Colombo. (Peradeniya Mannual **6**.)

Warcup, J. H. The soil-plate method for isolation of fungi from soil. *Nature* **166**: 117–18.

1951 *Ainsworth, G. C. Presidential address. A century of medical and veterinary mycology in Britain. *Trans. Br. mycol. Soc.* **34**: 1–16.

*Brian, P. W. Antibiotics produced by fungi. *Bot. Rev.* **17**: 357–430. (See also D. Broadbent, *PANS*, Sect. B, **14**: 120–41 (1968).)

Lilley, V. G. & Barnett, H. L. *Physiology of fungi.* New York.

Luttrell, E. S. Taxonomy of the Pyrenomycetes. *Univ. Miss. Stud.* **24**(3): 1–120.

Manier, J.-F. Recherches sur les Trichomycètes. *Ann. Sci. nat. Bot.*, sér. 11, **11**: 53–162.

*Stevenson, J. A. A résumé of the activities of the mycological collections of the United States Department of Agriculture, with a phytopathological slant, 1885–1950. *Plant Dis. Reptr*, suppl. **200**: 21–9.

1952 Hirst, J. An automatic volumetric spore trap. *Ann. appl. Biol.* **39**: 257–65.

Lodder, Johanna & Kreger-van Rij, Nellie W. J. *The yeasts. A taxonomic study.* Amsterdam. (Edn 2, J. Lodder (ed.), 1970.)

Manton, I., Clarke, B., Greenwood, A. D. & Flint, E. A. Further observations on the structure of plant cilia, by a combination of visual and electron microscopy. *J. exp. Bot.* **3**: 204–15.

Mitchell, Mary B. & H. K. A case of 'maternal' inheritance in *Neurospora crassa. Proc. nat. Acad. Sci. Wash.* **38**: 442–9.

Peterson, D. H. & Murray, H. C. Microbial oxidation of steroids at carbon 11. *J. Am. chem. Soc.* **74**: 1871–2.

Pontecorvo, G. & Roper, J. A. Genetic analysis without sexual reproduction by means of polyploidy in *Aspergillus nidulans. J. gen. Microbiol.* **6**: vii–viii.

Raper, J. R. Chemical regulation of sexual processes in the Thallophytes. *Bot. Rev.* **18**: 447–545.
Vanbreuseghem, R. Technique biologique pour l'isolement des dermatophytes du sol. *Ann. Soc. belge Méd. trop.* **32**: 173–8.
Wickerham, L. J. & Burton, K. A. Occurrence of yeast mating types in nature. *J. Bact.* **63**: 449–51.

1953   Elucidation of the structure of DNA by J. D. Watson and F. H. C. Crick.

Dimond, A. E. & Waggoner, P. E. On the nature and role of vivotoxins in plant disease. *Phytopathology* **43**: 229–35.
Ephrussi, B. Nucleo-cytoplasmic relations in micro-organisms. London.
Hughes, S. J. Conidiophores, conidia, and classification. *Can. J. Bot.* **31**: 577–659.
*Ramsbottom, J. *Mushrooms and toadstools.* London. (History, pp. 12–24.)

1954   International Society for Human and Animal Mycology founded.

*Morandi, L. & Baldacci, E. *I funghi. Vita, storia, leggende.* Milan.
Sarkisov, A. K. [Mycotoxicoses (Fungus poisons).] Moscow. [Russian.]
*Stevenson, J. A. Plants, problems, and personalities: the genesis of the Bureau of Plant Industry. *Agric. Hist.* **28**: 155–62.
*Virville, A. D. de (ed.). *Histoire de la botanique en France.* Paris. (Mycologie: 219–34; Lichenologie: 235–42; Mycologie medicale: 307–12.)

1955   Martin, G. W. Are fungi plants? *Mycologia* **47**: 779–92.
1956   Copeland, H. F. *The classification of lower organisms.* Palo Alto, Calif.
Flor, H. H. The complementary gene systems in flax and flax rust. *Adv. Genetics* **8**: 29–54.
*Horsfall, J. G. *Principles of fungicidal action.* Waltham, Mass.
Levi, J. D. Mating reaction of yeast. *Nature* **177**: 753–4.
Pontecorvo, G. The parasexual cycle in fungi. *Ann. Rev. Microbiol.* **10**: 393–400.
1956–7 Bistis, G. N. Sexuality in *Ascobolus stercorarius.* I. II. *Am. J. Bot.* **43**: 389–94; **44**: 436–43.
1957   *Fischer, G. W. & Holton, C. S. *Biology and control of the smut fungi.* New York.
Gäumann, E. A. Fusaric acid as a wilt toxin. *Phytopathology* **47**: 342–57.
*Hawker, Lilian E. *The physiology of reproduction in fungi.* Cambridge.
Henry, B. S. & O'Hearn, Elizabeth M. The production of spherules in a synthetic medium by *Coccidioides immitis. Proc. Symp. Coccidioidomycosis 1957*: 183–8. (U.S. Public Hlth Service Publ. 575.)
*Wasson, Valentina P. & R. G. *Mushrooms, Russia and history.* 2 vols. New York.
1957–8 Cejp, K. *Houby.* 2 vols. Prague.
1958   *Das Gupta, S. N. *History of botanical researches in India, Burma and Ceylon. Part I. Mycology and plant pathology.* Bangalore.
*Grainger, T. H. *A guide to the history of bacteriology.* New York.
Heim, R. & Wasson, R. G. *Les champignons hallucinogènes du Mexique.* Paris. (Mus. nat. Hist. natur.)
Käfer, Etta. An 8-chromosome map of *Aspergillus nidulans. Adv. Genetics* **9**: 105–45.

Sparrow, F. K. Interrelationships and phylogeny of the aquatic phycomycetes. *Mycologia* **50**: 797–813.

1959   Bonner, J. T. *The cellular slime molds.* Princeton, N.J. (Edn 2, 1967.)

Cutter, V. M. Jr. Studies on the isolation and growth of plant rusts in host tissue culture and on synthetic media. I. Gymnosporangium. *Mycologia* **51**: 248–95. (See also *ibid.* **52**: 726–42 (1961).)

Harley, J. L. *The biology of mycorrhizas.* London. (Edn 2, 1970.)

*Keitt, G. W. History of plant pathology. In J. G. Horsfall & A. E. Dimond (eds.) *Plant pathology* **1**: 61–97 (1959). New York.

*McCallan, S. E. A. The American Phytopathological Society – the first fifty years. In C. S. Holton *et al.*, *Plant pathology, problems and progress, 1908–1959*: 24–31.

*Mayer, K. *4500 Jahre Pflanzenschutz.* Stuttgart. (A chronology.)

*Stevenson, J. A. The beginnings of plant pathology in North America. In C. S. Holton *et al.*, *Plant pathology, problems and progress, 1908–1959*: 14–23.

1960   Gandy, D. G. 'Watery stipe' of cultivated mushrooms. *Nature* **185**: 482–3.

Plempel, M. Die zygotropische Reaktion bei Mucorinieen. I. Mitteilung. *Planta* **55**: 254–8.

1961   Dawson, Christine O. & Gentles, J. C. The perfect states of *Keratinomyces ajelloi* Vanbreuseghem, *Trichophton terrestre* Dury & Frey and *Microsporum nanum* Fuentes. *Sabouraudia* **1**: 49–57.

Gregory, P. H. *Microbiology of the atmosphere.* London. (Edn 2, 1973.)

*Johnson, T. W. & Sparrow, F. K. *Fungi in oceans and estuaries.* Weinham.

Sargeant, K., Sheridan, A., O'Kelly, J. & Carnaghan, R. B. A. Toxicity associated with certain samples of groundnuts. *Nature* **192**: 1096–7.

Scheffer, R. P. & Pringle, R. B. A selective toxin produced by *Periconia circinata. Nature* **191**: 912–13.

*Singer, R. *Mushrooms and truffles. Botany, cultivation and utilization.* London.

Stockdale, Phyllis M. *Nannizzia incurvata*, gen.nov., sp.nov., a perfect state of *Microsporum gypseum. Sabouraudia* **1**: 41–8.

1962   Bartnicki-Garcia, S. & Nickerson, W. J. Induction of yeastlike development in Mucor by carbon dioxide. *J. Bact.* **84**: 829–40.

Moore, R. T. & McAlear, J. H. Fine structure of mycota. 7. Observations on septa of ascomycetes and basidiomycetes. *Am. J. Bot.* **49**: 86–94.

Stakman, E. C., Stewart, D. M. & Loegering, W. Q. Identification of physiological races of *Puccinia graminis* var. *tritici. U.S.D.A. Agric. Res. Serv. E616,* 53 pp.

1963   Cunningham, G. H. *The Thelephoraceae of Australia and New Zealand.* Wellington. (N.Z.D.S.I.R. Bull. 145.)

Fincham, J. R. S. & Day, P. R. *Fungal genetics.* Oxford.

*Heim, R. & Pellier, Jeanne. La mycologie au Japon. Revue historique. *Rev. Mycol. Paris* **28**: 68–81.

Nickerson, W. J. Symposium on biochemical bases of morphogenesis in fungi. IV. Molecular basis of form in yeasts. *Bact. Rev.* **27**: 305–24.

*Ramsbottom, J. History of Scottish mycology. *Trans. Br. mycol. Soc.* **46**: 161–78.

Reijnders, A. F. M. *Les problèmes du developpement des carpophores des*

Agaricales et de quelques groupes voisins. The Hague.
*Wheeler, H. & Luke, H. H. Microbial toxins in plant disease. *Ann. Rev. Microbiol.* **17**: 223–42.

1964 Olive, L. S. Spore discharge mechanism in basidiomycetes. *Science* **146**: 542–3.
*Orlob, G. B. The concepts of etiology in the history of plant pathology. *Pflanzenschutz-Nachr.* **17**(4): 185–268.
*Pringle, R. B. & Scheffer, R. P. Host specific plant toxins. *Ann. Rev. Phytopath.* **2**: 133–56.

1965 Cunningham, G. H. *Polyporaceae of New Zealand*. Wellington. (N.Z.D.S.I.R. Bull. 164.)
Esser, K. & Kuenen, R. *Genetik der Pilze*. Berlin. (English transl. by E. Steiner, *Genetics of fungi*, 1968.)
*Hawker, Lilian E. Fine structure of fungi as revealed by electron microscopy. *Biol. Rev.* **40**: 52–92.
*Hesseltine, C. W. A millenium of fungi, food and fermentation. *Mycologia* **57**: 149–97.
*Müller-Kögler, E. *Pilzkrankheiten bei Insekten*. Berlin.
Raper, K. B. & Fennell, Dorothy I. *The genus* Aspergillus. Baltimore.
*Schwarz, J. & Baum, G. L. Pioneers in the discovery of deep fungus diseases. *Mycopath. Mycol. applic.* **25**: 73–81.
*Sturtevant, A. H. *A history of genetics*. New York.

1965–73 *Ainsworth, G. C. & Sussman, A. S. (eds). *The fungi*. **1** (*The fungal cell*) (1965); **2** (*The fungal organism*) (1966); **3** (*The fungal population*) (1968); **4** (with F. K. Sparrow) (*Taxonomic review with keys*), **A** (*Ascomycetes, Fungi imperfecti*), **B** (*Other groups*) (1973). New York.

1965–75 *Raab, H. Aus der Geschichte der Mykologie. *Schweiz. Z. Pilzkde* **43**: 81–4 (1965); **44**: 149–54 (1966); **46**: 17–19, 105–10 (1968); **48**: 13–16 (1970); **49**: 153–8 (1971); **50**: 41–5 (1972); **51**: 53–9 (1973); **53**: 22–8 (1975).

1966 Burnett, J. H. & Evans, E. J. Genetical homogeneity and stability of mating-type factors in 'fairy rings' of *Marasmius oreades*. *Nature* **210**: 1368–9.
*Machlis, L. Sex hormones in fungi. In Ainsworth & Sussman **2**: 415–33 (1965–73).
*Sussman, A. S. & Halvorson, H. O. *Spores: their dormancy and germination*. New York.

1966–7 Williams, P. G., Scott, K. J. & Kuhl, J. L. Vegetative growth of *Puccinia graminis* f. sp. *tritici in vitro*. *Phytopathology* **56**: 1418–19 (1966); & Maclean, D. J. Sporulation and pathogenicity of *Puccinia graminis* f. sp. *tritici* grown on an artificial medium. *Ibid.* **57**: 326–7 (1967).

1967 Ahmadjian, V. *The lichen symbiosis*. Waltham, Mass.
Batra, L. R. Ambrosia fungi: a taxonomic revision and nutritional studies of some species. *Mycologia* **59**: 976–1017.
*Buchwald, N. F. Die Entwicklung der Pflanzenpathologie in Dänemark. *Acta Phytopathologica* **2**: 183–94.
*Hale, M. E. *The biology of lichens*. London.
Heim, R., Cailleux, R., Wasson, R. G. & Thévenard, P. *Nouvelle investigations sur les champignons hallucinogènes*. Paris. (Mus. nat. Hist. natur.)
*McCallan, S. E. A. History of fungicides. In D. C. Torgeson (editor), *Fungicides; an advanced treatise*, **1**: 1–37. New york.
*Wood, R. K. S. *Physiological plant pathology*. Oxford & Edinburgh.

1968 *Fidalgo, O. Introdução à historia da micologia brasileira. *Rickia* 3: 1–44. (See also *Ibid.* 5: 1–3 (1970).)

Martin, G. W. The origin and status of fungi (with a note on the fossil record). In Ainsworth & Sussman 3: 635–48 (1965–73).

*Parris, G. K. *A chronology of plant pathology.* Starkville, Miss.

Person, C. Genetical adjustment of fungi to their environment. In Ainsworth & Sussman 2: 395–415 (1965–73).

*Ubrizsy, G. A Magyarországi mykológiai kutatások a múltban es jelenleg. *Herba Hung.* 7: 11–16.

Wasson, R. G. *Soma: divine mushroom of immortality.* New York. (Reprint, 1971.)

1969 *Ainsworth, G. C. History of plant pathology in Great Britain. *Ann. Rev. Phytopath.* 7: 13–30.

Jellison, W. L. *Adiaspiromycosis* (= *Haplomycosis*). Missoula, Mont.

Kreisel, H. *Grundzüge eines natürlichen Systems der Pilze.* Lehre.

*Lowag, K. Ein Beitrag zur Geschichte der Mykologie in Österreich. *Sydowia* 22: 311–22.

*Špaček, J. Historische Skizze der Mycologie in Mähren und Schleisien bis 1945. *Spisy přir Univ. J. E. Purkyně* 37: 171–91.

Whittaker, R. H. New concepts of kingdoms of organisms. *Science* 163: 150–60.

1970 Barkley, F. A. *Outline classification of organisms.* Edn 3. Providence, Mass.

*Bové, F. J. *The story of ergot.* Basel & New York.

*Fish, S. The history of plant pathology in Australia. *Ann. Rev. Phytopath.* 8: 13–36.

*Tubaki, K. Historical survey of the studies on microfungi in Japan. *Proc. 1st. Internat. Congr. Culture Coll.*: 57–61.

1971 First International Mycological Congress, Exeter, Devon, England.

*Blumer, S. & Müller, E. Mykologie und Mykologen in der Schweiz. *Schweiz. Z. Pilzkde* 49: 97–108.

Ellis, M. B. *Dematiaceous hyphomycetes.* Kew.

*Onions, Agnes, H. S. Preservation of fungi. In C. Booth (ed.), *Methods in microbiology*, 4: 113–51. London and New York.

*Orlob, G. B. History of plant pathology in the Middle Ages. *Ann. Rev. Phytopath.* 9: 7–20.

Pegler, D. N. & Young, T. W. K. Basidiospore morphology in the Agaricales. Lehre. (Beih. *Nova Hedwigia* 35.)

*Turner, W. B. *Fungal metabolites.* London.

1972 *Conners, I. L. (ed.). *Plant pathology in Canada.* Winnipeg. (Can. Phytopath. Soc.)

Heim, R. Mushroom madness in the Kuma. *Hum. Biol. Oceania* 1: 170–8.

*Kiraly, Z. Main trends in the development of plant pathology in Hungary. *Ann. Rev. Phytopath.* 10: 9–20.

Lowy, B. Mushroom symbolism in the Maya codices. *Mycologia* 64: 816–21.

*Marsh, R. W. (ed.). *Systemic fungicides.* London.

*Raychaudhuri, S. P., Verma, J. P., Nariani, T. K. & Sen, B. The history of plant pathology in India. *Ann. Rev. Phytopath.* 10: 21–36.

1973 *Ahmadjian, V. & Hale, M. E. (eds.). *The Lichens.* New York.

*Bresinsky, A. 200 Jahre Mykologie in Bayern. *Z. Pilzkde* 39: 15–38.

*Lazzari, G. *Storia della micologia italiana. Contributo dei botanica italiana allo sviluppo delle scienze micologiche.* Trento.

Mahgoub, El S. & Murray, I. G. *Mycetoma.* London.

1974 *Akai, S. History of plant pathology in Japan. *Ann. Rev. Phytopath.* **12**:13–26.

*Grummann, V. *Biographisch-bibliographisches Handbuch der Lichenology.* Lehre.

Lowy, B. *Amanita muscaria* and the thunderbolt legend in Guatemala and Mexico. *Mycologia* **66**: 188–91.

Names index

† indicates author deceased but no date(s) ascertained.

Abbé, E. (1840–1908), 61
Abraham, E. P., 304
Acharius, E. (1757–1819), 94, 96, 312, 313
Adametz, L. (†), 232, 320
Adamson, H. G. (1865–1955), 172, 302
Adanson, M. (1729–1806), 10, 254, 311
Agardh, C. A. (1785–1859), 259, 261
Ahlburg, H. (?–1878), 214
Ahmadjian, V., 98, 305, 335–6
Ainsworth, G. C., 305–8, 331–2, 335–6
Akai, S., 308, 337
Albertini, J. B. de (1769–1831), 255, 259, 263, 283, 313
Alcroft, R., 303
Alexander, D. F., 292, 331
Allegro, J., 193
Amici, G. B. (1784–1860), 60, 114
Amoreux, P. J. (1741–1824), 309
Anderson, T. M'Call (1836–1908), 172, 317
Andersson, O., 233, 332
Arber, A. (1879–1960), 308, 330
Arderon, W. (1703–67), 175–6, 311
Aristotle (384–322 B.C.), 35
Arthur, J. C. (1850–1942), 287, 329
Åsberg, M. 305
Ascherson, F. M. (1798–1879), 71, 314
Aubrey, J. (1626–97), 56
Aubriet, C. (1665–1742), 48, 50
Audouin, V. (1797–1841), 166–7, 170, 314
Austen, R. (?–1676), 157
Austwick, P. K. C., 303

Baddeley, M. S., 305
Badham, C. D. (1806–57), 1, 183, 316
Bail, C. A. E. T. (1833–1922), 29–30
Baker, H. (1698–1744), 302
Baker, R. D., 331
Baldacci, E., 308, 331
Balfour, I. B. (1853–1922), 320, 321
Balle, S., 80, 330
Balsamo-Crivelli, G. G. (1800–74), 166
Banks, J. (1743–1820), 151, 152, 313
Barclay, A. (1852–91), 230
Barger, G. (1878–1939), 186, 188, 328
Barghoorn, E. S., 237, 331
Barkley, F. A., 33, 336
Barnes, B. F. (1888–1965), 275, 327
Barnett, H. L., 275, 332

Barnici-Garcia, S., 31, 300, 334
de Bary, H. A. (1831–88), 5, 25, 29, 30, 61–2, 65, 74, 79, 89, 96, 98, 107, 117, 120, 122, 139, 149, 151, 153–4, 156, 159, 191, 213, 234, 236, 264–5, 273–4, 275, 277, 283, 299, 300, 316, 317, 318, 319, 320
Bassi, A. (1773–1856), 11, 163–6, 314
Bateman, F. (1788–1821), 171
Batista, A. C. (1916–67), 228
Batra, L. R., 239, 305, 335
Batsch, A. J. G. C. (1761–1802), 226, 254, 255, 257, 312
Battarra, G. A. (1714–89), 56, 183, 225, 311
Batthyány, B. de (1538–90), 41, 42
Bauer, F. A. (1758–1840), 151
Bauhin, G. (C. Bauhinus) (1560–1624), 14, 35, 45, 186, 243, 246, 309
Bauhin, J. (1541–1613), 5, 40, 45–6, 49, 186, 187, 246–7, 270, 309
Baum, G. L., 303, 308, 335
Bayliss, J. M. (1869–1957), 78, 324
Bazin, A. P. E. (1807–78), 171
Beadle, G. W., 136, 330, 331
Becher, J. J. (1635–82), 210
Beijerinck, M. W. (1851–1931), 214
Benedek, T. (1892–1974), 302
Benham, R. H. (1894–1957), 111, 203, 328, 330
Bennett, J. H. (1812–75), 173, 175, 313
Bensaude, M. (1890–1969), 127, 128, 130, 325, 326
Berg, F. T. (1806–87), 170
Berkeley, M. J. (1803–89), 2, 28–9, 71, 73, 89, 114, 116, 143, 155, 160, 206, 227–8, 266, 289, 299, 306, 314, 315, 316, 317
Berlese, A. N. (1864–1903), 237
Bernard, N. (1874–1911), 103–4, 277, 324
Bernton, H. S., 203, 328
Bertrand, G. (1867–1962), 110, 324
Berzelius, J. J. (1779–1848), 212
Bessey, E. A. (1877–1957), 266, 275, 329
Bethell, J. (1804–67), 93
de Beurmann, L. (1851–†), 177, 181, 308, 325
van Beverwijk, A. L. (1907–63), 307
de Beythe, I. (?–1611), 41
Biffen, R. H. (1874–1949), 157, 323
Birkbeck, G. (1776–1841), 93, 314

Birkinshaw, J. H., 112
Bisby, G. R. (1889–1958), 230, 305, 328, 331, 332
Bistis, G. N., 139, 333
Bitancourt, A. A., 228
Blackley, C. H. (1820–1900), 202, 318
Blackman, E. L., 303
Blackman, V. H. (1872–1967), 124, 284, 323
Blakeslee, A. F. (1874–1954), 6, 119, 124–6, 287, 323, 325
Blanchard, R. (1857–1919), 171
Blaxall, F. R. (1866–1930), 172
Blottner, C. L. (1773–1802), 18, 312
Blumer, S., 308, 336
Blunt, W., 305
Boccone, P. (1633–1703), 49, 310
Bock, J. (H. Tragus) (1498–1554), 13, 145, 309
Bolton, J. (?–1799), 21, 54, 55, 56, 182, 200, 234, 272, 304, 312
Bommer, C. B. J. P. (1866–1938), 65, 321
Bonnefons, N. de (fl. 1650), 82
Bonner, J. T., 80, 334
Bonnet, C. (1720–93), 20
Bonnier, G. E. M. (1853–1922), 98, 320
Bonorden, H. F. (1801–44), 61, 316
Botting, D., 305
Boudier, J. L. É. B. (1828–1920), 56, 102, 226, 266, 284, 319, 320, 323
Boulton, S. B. (†), 300, 320
Bové, F. J., 336
Bowers, J. S., 303
Bradley, R. (1688–1732), 145
Braun, A. C. H. (1805–77), 117, 236, 274
Braun, K., 308, 329
Brefeld, O. (1839–1925), 62, 106, 120, 197, 198, 199, 212, 265, 275, 318, 319
Bresadola, G. (1847–1929), 226, 320, 327
Bresinsky, A., 308, 336
Brian, P. W., 221, 332
Brierley, M., 232, 327
Brierley, W. B. (1889–1963), 232, 327
Briosi, G. (1846–1919), 283
Broadbent, D., 304, 332
Brodie, H. J., 128, 131, 328
Brongniart, A. T. (1801–76), 28, 301
Brooks, F. T. (1882–1952), 143, 284
Brosse, G. de la (1586–1641), 300
Brown, A. M., 128, 132, 328
Brown, H. B. (1876–1962) 188, 325
Brown, Rachael, 303
Brown, Robert (1773–1858), 121, 301
Brown, W. (1888–1975), 159, 284, 325, 329
Brugnatelli, L. V. (1761–1818), 168
Brumpt, É. (1877–1951), 66, 323
Brunswik, H., 128, 129, 326–7
Buchner, H. (1850–1902), 7, 214, 322

Buchwald, N. F., 306, 308, 335
Buder, J. (1884–1966), 198, 326
Buell, C. B., 292, 331
Buffon, G.-L. L. Comte de (1707–88), 20, 242
Bugie, E., 304
Bull, H. G. (c. 1818–85), 285
Bullen, J. J., 31, 332
Buller, A. H. R. (1874–1944), 50, 84, 128, 130, 131, 132, 182, 197, 198, 229–30, 275, 283, 292, 324–5, 328, 330, 331
Bulliard, J. B. F. (dit Pierre) (1752–93), 21, 32, 56, 116, 182, 197–8, 226, 233, 272, 292, 297, 299, 305, 312
Bulloch, W. (1868–1941), 18, 211–12, 308, 330
Burgeff, H., 103, 139, 324, 327
Burnett, J. H., 78, 335
Burnett, W. (1779–1861), 93
Burton, K. A., 133, 333
Butler, E. J. (1874–1943), 157, 230, 287, 305, 323, 328
Butler, G. M., 80
Büttner, D. S. A. (1724–68), 24
Buxbaum, J. C. (1693–1730), 225, 310

Cadham, F. T., 203, 327
Cagniard-Latour, C. (1777–1859), 211, 314
Cailleux, R., 335
Callaway, J. L., 331
Calmette, A. (1863–1933), 186, 223
Camerarius, R. J. (1665–1721), 114
de Candolle, A. P. (1778–1841), 151, 154, 187, 241–2, 293, 313
Cantino, E. C., 80
Carnaghan, R. B. A., 303, 334
Carruba, R. W., 303
Carter, H. V. (1831–97), 66, 180, 318
Cartwright, K. St G. (1891–1964), 94
Castle, E. S., 79
Cazenave, A. (1795–1877), 171
Cejp, K., 266, 333
Celsus, A. C. (1st cent. A.D.), 170, 184
Cesalpino, A. (1519–1603), 13, 309
Chabrey (Chabré), D. (1607–67), 46
Chain, E. B., 220
Chapman, A. C., 58, 328
Chatin, G. A. (1813–1901), 207, 321
Cherler, J.-H. (c. 1570–c. 1610), 45, 309
Chesters, C. G. C., 232, 332
Chevalier, C.-L. (1804–47), 60
Chevalier, J. L. V. (1770–1847), 60
Chiarappa, L., 302
Cholodny, N., 232, 328
Christensen, J. J. (?–1964), 188, 329
Church, M. B. (1889–1949), 291, 327
Ciccarelli, A. (?–1580), 232, 270, 309
Ciferri, R. (1895–1964), 280, 324

Clarke, B., 332
Clarke, W. A., 307
Claussen, P. (1877–1959), 138, 325
Clay, R. S., 297
Clements, F. E. (1874–1945), 281, 324, 328
Clusius, C. (J.-C. l'Écluse, l'Éscluse) (1526–1609), 8, 40–4, 46, 49, 183, 225, 227, 243, 246, 270, 309
Cobe, H. M., 204, 328
Cohn, F. J. (1828–98), 105, 226
Coker, W. C. (1872–1953), 236, 326
Conant, N. F., 181, 331
Configliachi, L. (1787–1864), 168
Conn, H. J. (1886–1952), 232, 326
Conners, I. L., 308, 336
Connstein, W., 224, 326
Cooke, M. C. (1825–1914), 97, 227, 230, 234, 274, 277, 286, 297, 299, 318, 319, 321
Copeland, H. F., 33, 34, 333
Corda, A. C. J. (1809–49), 57, 61, 63, 71, 74, 116, 228, 273, 314, 315
Corner, E. J. H., 75, 299, 328, 332
Cornu, M. M. (1843–1901), 236, 285
Costantin, J. N. (1857–1936), 94, 209, 321
Cotton, A. D. (1879–1962), 237, 324
Couch, J., 239, 330
Court, T. H., 297
Cowan, S. T., 241, 305
Craigie, J. H., 128, 132, 327, 329
Crick, F. H. C., 136
Croisille, J.-M., 299
Crombie, J. M. (1831–1906), 97, 300
Crossland, C. (1844–1916), 286
Cunningham, D. D. (1843–1914), 200–1, 230, 318
Cunningham, G. H. (1892–1962), 75, 231, 334, 335
Currie, J. N., 222, 325
Cutter, V. M. (1917–62), 11, 112, 334

Dampier, W. (1652–1715), 170, 302
Dangeard, P.-A. C. (1862–1947), 123, 236, 321, 322
Darling, S. T. (1872–1925), 179
Darwin, C. (1809–82), 10, 191, 206, 228, 264
Darwin, E. (1731–1802), 77, 300
Das Gupta, S. N., 308, 333
David, A. (†), 48–9
Dawson, C. O., 174, 334
Day, P. R., 135, 275, 334
Dearness, J. (1852–1954), 306
Delbruck, M., 79
DeMonbreun, W. A., 179, 329
Desmazières, J. B. H. J. (1786–1862), 70, 156, 211, 313
Devéze, P., 174, 302
Dickson, E. C., 178, 330
Dickson, R. A. (†), 300, 315

Dillenius, J. J. (1684–1747), 17, 52, 95, 96, 160, 246–8, 252, 310, 311
Dimond, A. E. (1914–72), 191, 333
Dioscorides (fl. 50–100), 38, 82, 160, 183, 184, 216
Ditmar, L. P. F. (†), 70
Dobbs, C. G., 197, 331
Dobell, C. (1886–1949), 58, 329
Dodge, B. O. (1872–1960), 129, 132, 134–6, 326, 327
Dodge, C. W., 173, 327, 329
Dodoens, R. (1517–85), 40, 309
Donk, M. A. (1908–72), 269
Dorn, G. L., 301
Dorrance, F., 325
Drayton, F. L. (1892–1970), 132, 328, 329
Drechsler, C., 239, 240, 329
Dryander, J. (1748–1810), 24, 312
Duclaux, É. P. (1840–1904), 106–7, 209, 214
Duggar, B. M. (1872–1956), 209, 323
Dujarric de la Rivière, R. (1885–1969), 186, 330
Durham, O. C., 203, 330
Dusch, T. von (1824–90), 105

Eastcott, E. V., 111, 327
Edwards, G. (1694–1773), 298
Ehrenberg, C. G. (1795–1876), 5, 6, 18, 21, 117, 118, 211, 274, 313
Eidam, M. E. H. (1845–1901), 120
Elenkin, A. A. (1873–1942), 98, 323
Ellis, J. (c. 1710–76), 23, 24
Ellis, J. B. (1829–1905), 277, 289
Ellis, M. B., 64, 336
Ellis, W. G. P. (1863/4–1925), 284
Elsholtz, J. S. (1623–88), 141, 302
Emerson, C. H., 221, 330
Emerson, R., 80
Emmons, C. W., 173, 178, 329, 332
Engel (†), 213, 318
Engler, H. G. A. (1844–1930), 273
Ephrussi, B., 80, 333
Eriksson, G., 306
Eriksson, J. (1848–1931), 157, 321
Esser, K., 135, 335
Euripides (480–406 B.C.), 183
Evans, E. J., 78, 335
Evelyn, J. (1620–1706), 82
Everhart, B. M. (1818–1904), 277
Eysfarth, C. S. (†), 141, 310

Faber, J. (1570–†), 297
Fabricius, J. C. (1745–1808), 143, 147, 312
Falk, R. (1873–1955), 94, 325
Faraday, M. (1791–1867), 93, 314
Farlow, W. G. (1844–1919), 283
Farmer, J. B. (1865–1944), 121, 301

Faulkner, F. (†), 319
Fayod, V. (1860–1900), 75, 197, 321
Feinberg, S. M., 203–4, 331
Fennell, D. I., 335
Ferguson, M. C. (1863–1951), 209, 323
Ferry, B. W., 305
Ferry, F. (†), 277
Fidalgo, O., 228, 308, 336
Fiese, M. J., 303
Fincham, J. R. S., 135, 275, 334
Fink, B. (1861–1927), 98, 324
Fischer, A. (1858–1913), 214, 284
Fischer, E. (1861–1939), 283
Fischer, G. W., 145, 333
Fischer von Waldheim, A. (1839–1920), 283
Fish, S., 308, 336
Fitz, A. (1842–85), 214, 318
Fitzpatrick, H. M. (1886–1949), 306, 308, 330
Fleming, A. (1881–1955), 217–20, 328
Flemming, W. (1843–1915), 121
Flint, E. A., 332
Flor, H. H., 158–9, 333
Florey, H. W. (1898–1968), 220
Floyer, J. (1649–1734), 204
Fontana, F. (1731–1805), 150–1, 311
Ford, W. W. (1871–1914), 185
Forsyth, W. (1737–1804), 160, 312
Foster, J. W., 113, 275, 332
Fowler, W. W. (1835–1912), 286
Fox, T. Colcott (1848–1906), 172, 174
Fox, W. Tilbury (1836–79), 30, 171, 173, 317, 318
Fraenkel, E. M., 204
Frank, A. B. (1839–1900), 89, 98, 100–1, 319, 320
Franken, J. L. M., 306
Freeman, J., 219
Frenzel, J. S. T. (1740–1807), 18
Fresenius, J. B. G. W. (1808–66), 176
Friday, J., 303
Fries, E. M. (1794–1878), 21, 26, 60, 61, 65, 70, 74, 154, 187, 226, 227–8, 259–63, 271, 273, 289, 294, 305, 313, 318
Fries, N. T. E., 128, 259, 260
Fries, R. E. (1876–1966), 259, 260
Fries, T. M. (1832–1913), 97, 259, 260
Fuckel, K. W. G. L. (1821–76), 226, 231, 31–9
Funke, G. L., 130, 327
Furthmann, W. (†), 172

Gaertner, J. (1732–61), 115
Galen, (130–200), 50, 184
Gallaud, G., 104, 323
Gandy, D. G., 182, 334
Garnsey, H. E. F. (1826–1903), 320, 321
Garrett, S. D., 80, 232, 331

Garrison, F. H. (1870–1935), 308, 328
Gäumann, E. A. (1893–1963), 192, 266, 275, 327, 333
Gautier, L. M. (†), 320
Gayon, L. U. (1845–1929), 162
Gentles, J. C., 174, 334
Geoffroy, E. F. (1672–1731), 187
Gérard, F. (†), 184–5
Gerarde (Gerard), J. (1545–1612), 38, 40, 309
Ghini, L. (1500–56), 288
Gifford, M. A., 178, 330
Gilbert, E.-J. (1888–1954), 299
Gilbert, J. H. (1817–1901), 78
Gilbert, O. L., 305
Gilchrist, T. C. (1862–1927), 178
Gillet, C. C. (1806–96), 226, 319
Gleditsch, J. G. (1714–86), 18, 115, 200, 226, 252–4, 311
Goedart, J. (1620–68), 14, 298
Goidànich, G., 151
Gold, W., 303
Goldman, L., 302
Gorman, I., 331
Gorton, R. A., 126, 325
Gougerot, H. (1881–1955), 177, 181, 308, 325
Goulden, C. H., 128
Grainger, T. H., 308, 333
Greaney, F., 128
Green, J. R. (1848–1914), 214, 321
Greenwood, A. D., 332
Gregory, P. H., 76, 201–2, 303, 332, 334
Greville, R. K. (1794–1866), 71, 227, 313
Grove, W. B. (1848–1938), 297, 317
Gruby, D. (1810–98), 169–70, 171, 315, 316
Grummann, V. J. (1898–1967), 308, 337
Guéguen, F. P. J. (1872–1915), 236, 324
Guétrot, M. (1873–1941), 308, 329
Guilliermond, A. (1876–1945), 213, 325
Gwynne-Vaughan, H. C. I. (1879–1967), 275, 301, 327

Haas, H., 306
Haeckel, E. H. P. A. (1834–1919), 33, 321
Haehn, H., 224, 327
Hagburg, W. A. F., 128
Hagelstein, R. (1870–1945), 234, 331
Hale, M. E., 100, 305, 335, 336
Hales, S. (1677–1761), 17, 91, 144, 310
Haller, A. von (1708–77), 115, 197, 226, 251, 254, 263, 304, 311
Hallier, E. (1831–1904), 30
Halvorson, H. O., 79, 335
Hanna, W. F. (1892–1972), 128, 130, 327
Hansen, E. C. (1842–1909), 132, 213, 233, 297, 319, 320, 321
Hansen, K., 204

Harden, A. (1865–1940), 123, 214, 325
Hare, R., 217, 304
Harley, J. L., 334
Harper, R. A. (1862–1946), 123, 283, 301, 322–3
Hartig, H. J. A. R. (1839–1901), 93–4, 144, 319, 320
Hartig, T. (1805–80), 93, 100, 154, 239, 283, 314, 316
Hartog, M. M. (1851–1924), 283
Hasegawa, T., 307
Hawker, L. E., 138, 275, 301, 305, 332–3, 335
Hawksworth, D. L., 305
Hazen, E. (?–1975), 303
Hedgcock, G. G. (1863–1946), 138, 324
Hedwig, J. (1730–99), 62, 66, 68, 312
Heffner, R. M. S., 316
Heim, R.-J., 186, 193–5, 239, 308, 330, 333, 334, 335, 336
Hektoen, L. (1863–1951), 176, 322–3
Hellbach, R., 223, 329
Helmont, J. B. von (1577–1644), 18
Hennings, P. C. (1841–1908), 228
Henry, B. S., 31, 333
Herrick, H. T., 223, 329, 330
Heslop-Harrison, J. 241, 305
Hesseltine, C. W., 214, 335
Heywood, V. H., 241, 305
Hill, J. (1716–75), 20, 25, 311
Hirsch, E. F. (1886–1972), 178
Hirsch, H. E., 178, 301
Hirst, J. M., 202, 332
Hoffmann, H. C. H. (1819–91), 29, 74, 116, 120, 127, 317
Hoffmeister, W. (1824–77), 132, 198, 316, 323
Hogg, J. (1817–99), 171, 302
Höhnel, F. X. R. von (1852–1920), 283
Holmskiold (Holmskjold), T. (1731–93), 25, 226, 312
Holton, C. S., 145, 333
Hooke, R. (1635–1703), 5, 7, 15, 48, 56, 58, 59, 61, 64–5, 144, 228, 309
Hooker, J. D. (1817–1911), 71, 206, 228
Hooker, W. J. (1785–1865), 228, 314
Hopkins, J. G., 203, 328
Horace (65–8 B.C.), 183
Horsfall, J. G., 162, 331, 333
Houghton, W. (1829?–95), 183, 252, 297, 303, 320
Hughes, S. J., 64, 305, 333
Hugo, C. E., 306
Humboldt, F. H. A. von (1769–1859), 235, 312
Humphrey, H. B. (1873–1955), 145, 147, 311, 327
Huxley, T. H. (1825–95), 176, 221, 230
Hyde, H. A. (1892–1974), 204, 331

Ingold, C. T., 236–7, 331
Istvánffi, G. (1860–1930), 42, 49, 267, 323

Jackson, B. D. (1846–1927), 279, 318
Jacobsen, J. C. (1811–87), 297
Janse, J. M. (1860–†), 103
Javillier, M. (1875–1955), 110, 324
Jellison, W. L., 178, 336
Jenner, W. (1815–98), 171
Jensen, H. L., 232, 329
Jessen, K. F. W. (1821–89), 308, 317
Jewson, S. T. (?–c. 1930), 232, 327
Johannsen, E., 301
Johnson, T., 128
Johnson, T. W., 237, 334
Jonghe, A. de (H. Junius) (1511–75), 38, 41, 270, 309
Juel, H. O. (1863–1931), 123, 322
Junius, H., see Jonghe, A. de
Jussieu, A. de (1686–1758), 31–2, 310
Jussieu, A. L. de (1748–1836), 32, 257, 276, 298, 306
Juvenal (60–140), 12

Kaempfer, E. (1651–1716), 303
Käfer, E., 137, 333
Kalchbrenner, K. (1807–86), 226, 318
Kamienski, F. (1851–1912), 102, 320
Kaplan, R. W., 303
Karsten, G. K. W. H. (1817–1908), 120, 317
Karsten, P. A. (1834–1917), 226
Keissler, K. von (1872–1965), 182, 328
Keitt, G. W., 299, 308, 313, 334
Kellerman, W. A. (1850–1908), 277
Kelley, A. P., 332
Kernkamp, H. C. H., 188, 329
Kesten, B. M., 203, 328
Kickx, J. J. (1803–64), 226, 318
Kinnear, J. 175, 328
Kinoshita, K., 223, 328
Kintoff, W., 224, 327
Kiraly, Z., 308, 336
Kisch, B., 301–2
Kitayima, K., 207
Klebahn, H. (1859–1942), 287, 323
Klebs, G. A. (1857–1918), 78, 138, 283, 322
Klebs, T. A. E. (1834–1913), 106
Klotzsch, J. F. (1805–60), 70, 74, 116, 315
Kniep, H. (1881–1930), 127, 128, 129–30, 131, 325–6
Knight, B. C. J. G., 301
Knight, T. A. (1759–1838), 152, 157, 160, 315
Knowles, J. (1781–1841), 91, 300, 313
Knudsen, L. (1884–1958), 104, 326
Kobert, R. (†), 185, 321
Koch, R. (1843–1910), 61, 105–6, 172, 301
Koenig, J. G. (1728–85), 305

Kohler, R. E., 304
Koppe, F. (†), 85, 318
Kossel, A. (1853–1927), 136
Král, F. (1846–1911), 289
Kreger-van Rij, N. W. J., 213, 332
Kreisel, H., 266, 366
Krempelhuber, A. von (1813–82), 300
Krieger, L. C. C. (1873–1940), 299
Krombholz, J. V. von (1782–1843), 70, 226, 295, 314
Kuenen, R., 135, 335
Kuhl, J. L., 112, 335
Kühn, J. G. (1825–1910), 143, 317
Kühne, W. (†), 211
Kützing, F. T. (1807–93), 28–9, 211, 314
Kyan, J. H. (1774–1850), 93

Laboulbène, A. (1825–98), 237
Lafar, F. (1865–†), 308, 322
Lagarde, J.-J. (1866–1933), 236, 325
Lam, H. J., 305
Lancisi, G. M. (1654–1720), 17, 310
van Lanen, J. M., 331
Lange, J. E. (1864–1941), 56
Langeron, M. C. P. (1874–1950), 275, 331
Langlykke, A. F., 331
Laplanche, M. C. de (1843–1904), 282, 321
Large, E. C., 308, 330
La Touche, C. J., 219
Lawes, J. B. (1814–1900), 78
Lawrence, G. H. M., 297
Lazzari, G., 308, 337
Leach, C. M., 301
l'Écluse (l'Éscluse), see Clusius
Leedale, G. F., 298
Lees, E. (1800–87), 102, 315
van Leeuwen, S., 203, 204, 219, 327
van Leeuwenhoek, A. (1632–1723), 5, 58, 60, 61, 211, 270, 309
Lehfeldt, W., 130, 326
Le Leu, L., 302
Le Mense, E. H., 224, 331
Lemké, P. A., 303
Le Monnier, A.-A.-G. (1843–1931), 106, 318
Lentz, M. E., 307
Lentz, P. L., 307
Lenz, H. O. (1799–1870), 297, 317
Léveillé, J. H. (1796–1870), 61, 65, 71, 72, 116, 154, 187, 263–4, 277, 314, 316
Levi, J. D., 139, 333
Levine, M. N. (1886–1962), 158, 326
Lichenstein, G. R. (1745–1807), 32
Liebig, J. von (1803–73), 111, 210, 212
Lilly, V. G., 275, 332
Lind, J. V. A. (1874–1939), 226, 231, 308, 325, 329
Lindau, G. (1866–1923), 9, 280, 308, 324

Lindegren, C. C., 133, 135, 329, 331
Lindegren, G., 133, 331
Linder, D. H. (1899–1946), 237, 331
Lindley, J. (1799–1865), 156
Link, J. H. F. (1767–1851), 61, 70, 102, 300, 313
Linnaeus, C. (1707–78), 11, 23–4, 26, 31, 32, 45, 96, 116, 142, 250–2, 254, 259, 294, 305, 310, 311
Lister, A. (1830–1908), 234, 321
Lister, G. (1860–1949), 60, 234, 321, 325
Lister, J. (1827–1912), 106, 221
Lister, M. (1638–1712), 298
l'Obel, M. de (Lobelius) (1538–1616), 4, 38–40, 43, 46, 49, 243, 309
Lockwood, L. B., 330
Lodder, J., 213, 332
Loegering, W. Q., 158, 307, 334
Loeselius, J. (1607–55), 225, 310
Lonitzer, A. (Lonicerus) (1528–86), 186, 187, 217, 309
Lowag, K. (1913–1970), 308, 336
Lowy, B., 13, 194, 303, 336–7
Lu, G. D., 297
Lucet, A. (1852–1916), 176, 322
Ludecke, K., 224, 326
Luke, H. H., 192, 335
Lütjeharms, W. J., 17, 24, 25, 308, 329
Luttrell, H., 266, 332
Lutz, A. (1855–1940), 31, 324

McAlear, J. H., 76, 334
McAlpine, D. (1849–1932), 230–1, 324
MacBride, T. H. (1848–1934), 234, 322
McCallan, S. E. A., 308, 334, 335
Machacek, J. E. (1902–70), 128
Machlis, L., 139, 335
Maclean, D. J., 335
Maddox, F. (1856–1937), 149, 304, 322
Maddox, R. L. (1816–1902), 304
Magnol, P. (1638–1715), 244, 310
Magnus, P. W. (1844–1914), 102–3, 283, 323
Mahgoub, El S., 66, 337
Maire, R. C. J. E. (1878–1949), 123, 323
Malmsten, P. H. (1811–83), 170
Malpighi, M. (1628–94), 5, 15, 60, 309
Mangin, L. (1852–1937), 284
Manier, J.-F., 239, 332
Manson, P. (1844–1922), 171
Manton, I. 76, 332
Marchand, N. L. (1833–1911), 266, 322
Marchant, J. (c. 1650–1738), 16–17, 32, 60, 276, 310
Marchant, N. (?–1678), 65, 83
Margarot, J., 174, 302
Marsh, R. W., 163, 304, 366
Marsigili, L. F. (1658–1730), 17, 65, 310

Martin, D. S., 331
Martin, G. W. (1886–1971), 34, 89, 234, 300, 333, 336
Martin H., 302, 327
Martyn, J. (1699–1768), 234, 311
Mason, E. W. (1890–1975), 64, 307
Massee, G. E. (1850–1917), 227, 233, 266, 274, 277, 286, 321, 324
Matruchot, L. (1863–1921), 94, 209, 321
Matthews, V. D. (1904–58), 236, 328
Mattioli, P. A. (1500–77), 37, 38, 243, 309
Maty, M. (1718–76), 24
May, O. E., 223, 329, 330
Mayer, K., 308, 334
Mazzuoli, F. M. (?–1756), 18, 311
Medicus, F. C. (1736–1808), 18, 298
Meehan, F., 192, 331
Meikle, R. D., 306
Meschinelli, A. L. (1865–†), 281, 322
Meyen, F. J. F. (1804–40), 123, 154, 212, 314, 315
Meyer, A., 305
Mez, C. C. (1866-1944), 94, 324
Micheli, P. A. (1679–1737), 4, 5, 7, 10, 18, 49, 50–5, 60, 62, 66–70, 77, 84–9, 96, 114–16, 181, 196–7, 225, 233, 236, 248–50, 251, 252, 299, 310
Mieschner, J. F. (1844–95), 136
Millardet, P. M. A. (1838–1902), 161–2, 283, 320
Miller, P. (1691–1771), 83
Minden, M. D. von (1871–†), 236, 325
Miquel, P. (1850–1922), 200–1, 320
Mitchell, H. K. & M. B., 80, 332
Moffitt, H. C. (1867–1951), 178, 323
Mohl, H. von (1805–72), 121, 274, 283
Moll, F. (†), 93
Molliard, M. (1866–†), 223, 326
Montagne, J. P. F. C. (1784–1866), 61, 116, 155, 237, 315
Montagu, G. (1751–1815), 176
Monti, G. (1682–1760), 18, 311
Moore, J. E., 121, 301
Moore, R. T., 76, 334
Morandi, L., 308, 333
Moreau, F., 98, 328
Morison, R. (1620–83), 96, 310
Morren, C. J. E. (1833–86), 155
Morris, M. (1847–1924), 172, 322
Morton, A. G., 300
Mougeot, J.-B. (1776–1858), 284
Mounce, I., 128, 129, 326
Mulder, E. G., 110, 330
Müller, A., 302
Müller, E., 308, 336
Müller, J. H. H. (1855–1912), 97
Müller, Johannes P. (1801–58), 211, 274

Müller, O. F. (1730–84), 24, 68, 117, 298
Müller-Kogler, E., 237, 335
Münchausen, O. von (1716–74), 23, 24, 187
Murphy, H. C., 192, 331
Murray, H. C., 224, 332
Murray, I. G. (1928–71), 66, 337

Naegeli, K. (1817–91), 21, 117, 121, 274, 301, 315
Nannizzi, A. (1877–1961), 173, 327
Nariani, T. K., 336
Nash, C. H., 303
Necker, N. J. de (1729–93), 32, 65, 312
Neebe, C. H. (†), 172
Needham, James (1849–1914), 286
Needham, John T. (1713–81), 18
Needham, Joseph, 297, 303–5
Nees von Esenbeck, C. G. D. (1776–1858), 18, 21, 61, 66, 70, 117, 261, 313
Newton, D. E. (Mrs Swales), 128, 129, 327
Newton, G. G. F., 304
Newton, M. (1887–1971), 128
Nicander (2nd. cent. B.C.), 13, 184
Nickerson, W. J., 31, 79, 304, 334
Nicolai, O. (†), 100
Nixon, H. L., 76, 332
Nobles, M. K., 94, 331
Noire, H., 174
Nylander, W. (1822–99), 97, 305

O'Hearn, E. M., 31, 333
O'Kelly, J., 334
Oken, L. (1779–1851), 261
Oldenburg, H. (1615–77), 276
Olive, L. S., 197, 335
Onions, A. H. S., 292, 336
Oort, A. J. P., 128
Ophüls, W. (1871–1933), 178, 323
Orlob, G. B., 308, 335, 336
Ørsted, A. S. (1816–72), 120, 317
Oudemans, C. A. J. A. (1825–1906), 227, 282, 323, 326
Ovid (43 B.C.–A.D. 17), 13, 302
Owen, R. (1804–92), 168
Oxford, A. E., 221, 330

Paine, S. G. (1881–1937), 182, 326
Pallin, W. A. (1873–1956), 180, 323
Pallisot de Beauvois, A. M. F. J. (1752–1820), 312
Papin, D. (†), 105
Parkinson, J. (1567–1650), 40, 309
Parris, G. K., 308, 336
Pasteur, L. (1822–95), 5, 8, 18, 105–6, 111, 200, 212, 214, 224, 317, 319
Patouillard, N.-T. (1854–1926), 74, 226, 266, 284, 320–1

Paulet, J. J. (1740–1826), 184, 226, 279–80, 308, 312
de Payen, A. (1795–1871), 155
Peck, C. H. (1833–1917), 229, 318
Pedersen, R. (1840–1905), 297
Pegler, D. N., 77, 336
Pellier, J., 308, 334
Penzig, A. G. O. (1856–1929), 231, 322
Pepys, S. (1633–1703), 58, 90
Percebois, G., 242
Perkins, C. F., 176, 322–3
Person, C., 159, 336
Persoon, C. H. (1761–1836), 21, 32, 60, 61, 70, 71, 117, 197, 211, 255–8, 268, 271, 273, 289, 294, 306, 312, 313
Petch, T. (1870–1948), 230, 332
Peterson, D. H., 332
Peterson, R., 128
Petrak, F. (1886–1973), 282
Petri, R. J. (1852–1921), 106, 321
Peturson, B., 128
Peyritsch, J. J. (1835–89), 237
Pfeffer, W. F. P. (1845–1920), 102, 283
Phoebus, P. (1804–80), 73, 74, 315
Picco, V. (†), 32, 312
Pickering, R. (c. 1720–55), 185
Piemeisel, F. J. (1891–1925), 78, 157–8, 325
Pirone, P. P., 311
Plantin, C. (1514–89), 40
Plempel, M., 139, 334
Pliny, (23–79), 2, 12, 35, 38, 184, 216, 302
Plowright, C. B. (1848–1910), 302, 331
Plunkett, B. E., 80
Plutarch (46–120), 12
Poirault, G. (1858–1936), 123, 332
Pontecorvo, G., 136–7, 332–3
Popp, W., 128
Porta, G. della (1535/6–1615), 5, 13–14, 43, 62, 82, 309
Porter, J. R., 306
Posadas, A. (1870–1902), ≈78
Prantl, K. A. E. (1849–93), 273
Prévost, I.-B. (1755–1819), 11, 62, 117, 147–9, 151, 161, 236, 313
Přibram, E. (?–1940), 289
Pringle, R. B., 192, 334–5
Pringsheim, N. (1823–94), 117, 277, 317
Pritzel, G. A. (1815–74), 279, 318
Pryce, D. M., 219
Purkinje, J. E. (1787–1869), 121
Puttemans, A., 308, 330

Quélet, L. (1832–99), 226, 284, 318
Quevenne, T.-A. (1806–56), 212, 315
Quintanilha, A., 80, 128, 130, 330

Raab, J. (Hans) (1898–1971), 308, 335
Rabenhorst, G. L. (1806–81), 226, 273, 289, 320

Raciborski, M. (1863–1917), 123, 322
Raistrick, H. (1890–1971), 112–13, 328, 330
Ramsbottom, J. (1885–1974), 17, 24, 25, 299, 301, 308, 331, 333, 334
Raper, J. R. (1912–1974), 80, 139, 333
Raper, K. B., 80, 292, 304, 307, 331, 332, 335
Raspail, F. V. (1791–1878), 154–5
Raulin, J. (1836–96), 6, 106–10, 161, 277, 318
Ravn, M. K., 312
Ray, J. (1627–1705), 5, 14, 48, 225, 227, 232, 235, 243, 248, 270, 310
Raychaudhuri, S. P., 308, 336
Rayner, M. C. (?–1948), 102, 104, 327
Ré, F. (1763–1817), 143, 154
Rea, C. (1861–1946), 286, 326
Rea, E. A. (1864–1927), 299
Réaumur, R.-A. F. de (1683–1757), 25, 176, 310
Reay, M., 196
Redi, F. (1626–97), 18, 50
Reed, H. S., 308, 331
Rees, M. F. F. (1845–1901), 120, 213, 274, 283, 306, 318, 319
Reijnders, A. F. M., 80, 334–5
Reinke, J. (1849–1931), 98, 102, 318
Reissek, S. (1819–71), 102, 316
Remak, R. (1815–65), 121, 168–9, 315, 316
Rénon, L. (†), 176, 322
Retzius, A. J. (1742–1821), 259, 261
Richard, A. (1794–1852), 62, 313
Richard, L. C. M. (1754–1821), 62, 313
Richardson, A., 194
Richon, C. E. (1820–93), 120, 243, 320
Rivolta, S. (1832–93), 179
Rixford, E. (1865–1938), 178
Roach, W. A., 163
Robb, D. C. (†), 319
Robbins, W. J., 110, 330
Roberts, L. (1860–1949), 172
Roberts, W. (1830–99), 105, 211, 220, 319
Robin, C. P. (1821–85), 170, 237, 316
Robinson, W. (1838–1935), 209, 319
Rolfe, F. W. (1888–1947) & R. T. (1883–1958), 297, 308, 327
Rondelet, G. (1507–57), 40
Roper, J. A., 136, 332
Rose, A. H., 304
Rosen, F. (1863–1925), 122, 321
Rosenbach, A. J. F. (1842–1923), 172
Rossi, G. M., 232, 328
Rostafinski, J. T. von (1850–1928), 234, 283, 319
Rostrup, F. G. E. (1831–1907), 226
Rouget, A. (1818–86), 237
Roumeguère, C. (1828–92), 277
Rousseau, E. (†), 176
Rousseau, J. (1905–70), 299

Roux, E. (1853–1933), 214
Roze, E. (1833–1900), 120, 243, 297, 308, 320
Rylands, T. G. (1818–1900), 102, 315

Sabouraud, R. J. (1864–1938), 6, 172, 173, 174, 181, 308, 324
Saccardo, P. A. (1845–1920), 64, 226, 227, 228, 231, 233, 242, 268, 280–1, 282, 320, 322
Sachs, F. G. J. von (1832–97), 198, 283, 306, 308, 321
Sakamoto, K. (†), 231
Salmon, E. S. (1871–1959), 233, 323
Salter, C. T. C., 322
Sapin-Trouffy, F. P. S. (1865–†), 123–4, 322
Sargeant, K., 190, 334
Sarkisov, A. K., 190, 333
Sarton, J. (1884–1956), 13, 297, 304
Sass, J. E., 129, 328
Schacht, H. (1814–64), 74, 316
Schaeffer, J. C. (1718–90), 56, 68, 115, 226, 254, 273, 311
Schantz, H. L., 78, 325
Schatz, A., 304
Scheffer, R. P., 115, 192, 334, 335
Schenck, B. R. (?–1920), 177
Schilling, J. J. (†), 21, 22, 313
Schiønning, H. L. (1868–1942), 132, 133, 322
Schlechtendahl, D. F. L. (1794–1866), 274
Schleiden, M. J. (1804–81), 117, 121
Schmeideberg, O. (1838–1921), 185, 239, 318
Schmidel, C. C. (1718–92), 115, 311
Schmitz, C. J. F. (1850–95), 122, 123
Schmitz, J. (†), 197, 198, 316, 319
Schneiderhan, F. J., 320
Scholer, N. P., 152
Schönlein, J. L. (1793–1864), 168–9
Schopfer, W. H., 111, 138, 329
Schreber, J. C. D. von (1739–1810), 255
Schröder, H. G. F. (1810–85), 105
Schroeter, J. (1837–94), 157, 226, 319, 320
Schulthess, H. H. (†), 160
Schwann, T. (1810–82), 121, 211, 212, 304, 314
Schwarz, J., 308, 335
Schweinitz, L. D. von (1780–1834), 229, 255, 259, 263, 271, 282–3, 313, 314
Schwenckfelt, K. (1563–1609), 187
Schwendener, S. (1829–1919), 97–9, 318
Scopoli, J. A. (1723–88), 183, 234–5, 254, 255, 257, 311–12
Scott, G. D., 300
Scott, K. J., 112, 335
Secretan, L. (1758–1839), 226, 314
Sen, B., 336

Senebier, J. (1742–1809), 20, 312
Serrurier, (†), 176
Seth, L. N., 207, 330
Seyffert, C. (1719–†), 18, 311
Seymour, A. B. (1859–1933), 282, 328
de Seynes, J. (1833–1912), 120, 213, 285, 318
Shear, C. L. (1865–1956), 132, 229, 281, 327–8
Sherard, J. (1666–1737), 52
Sherard, W. (1659–1728), 52, 54, 246
Sheridan, A., 334
Shirai, M. (1863–1932), 231, 323
Sicard, G. (1829–86), 117
Simonart, P., 330
Singer, R., 207, 269, 334
Smith, A. L. (1854–1937), 96, 251, 308, 326
Smith, C. E., 178
Smith, D. T., 331
Smith, E. F. (1854–1927), 177
Smith, G. (1895–1967), 112, 221, 308, 330
Smith, J. E. (1759–1828), 289, 314
Smith, W. G. (1837–1917), 117, 285, 306
Sneath, P. H. A., 297
Sorauer, P. C. M. (1838–1916), 302, 308, 319, 325
Sowerby, J. (1757–1822), 8, 56, 91–2, 272, 312
Špaček, J., 308, 336
Spallanzani, L. (1729–99), 5, 18–20, 165, 312
Sparrow, F. K., 236, 237, 266, 331, 334, 335
Spegazzini, C. (1858–1926), 229
Splendore, A. (1871–1953), 31, 324
Stafleu, F. A., 305–6
Stahl, C.-E. (1848–1919), 98, 319
Stahl, G. E. (1660–1734), 210
Stakman, E. C., 157–8, 325, 326, 334
Stearn, W. T., 305–6
Steinberg, R. A., 110, 326, 329–30
Steiner, E., 335
van Sterbeeck, F. (1631–93), 42, 47, 48, 49, 206, 225, 232, 243, 267–8, 270–1, 309
Stevens, F. L. (1871–1934), 138, 328
Stevenson, J. A., 227, 307, 308, 319, 332, 333, 334
Stewart, D. M., 158, 334
Stockdale, P. M., 173, 334
Stoll, A., 188
Strasburger, E. (1844–1912), 121–2, 283, 320
Streinz, W. M. (1792–1876), 280, 317
Sturm, J. (1771–1865), 71, 116
Sturtevant, A. H. (1891–1970), 308, 335
Su, U. T., 207, 330
Suringar, W. F. R. (1832–98), 61
Sussman, A. S., 6, 79, 335
Sydow, H. (1879–1946), 230
Sydow, P. (1851–1925), 9, 280, 289, 308, 324
Syen (Sijen, Seyen), A. (1640–78), 49

Takamine, J. (†), 224, 322
Talon, J. (†), 207
Tanaka, S., 192, 329
Targioni-Tozzetti, A. (1785–1856), 55, 299
Targioni (Tozzetti), C. A. (1672–1748), 54
Targioni-Tozzetti, G. (1712–83), 54, 149–50, 299, 311
Targioni-Tozzetti, O. (1755–1829), 55
Tatum, E. L. (1910–75), 136, 330, 331
Tavel, R. F. von (1863–1941), 265, 275, 283, 321
Tehon, L. S. (1895–1954), 311
Ternetz, C., 104, 324
Teschmacher, J. E. (1790–1853), 155
Tessier, H.-A. (1740–1837), 147, 149, 160, 312
Thal, J. (1542/3–83), 186
Thaxter, R. (1858–1932), 237–8, 322
Theiler, A. (1867–†), 188, 327
Theophrastus (c. 300 B.C.), 35, 64, 140
Thévenard, P., 335
Thin, G. (?–1903), 172, 319, 321
Thistleton-Dyer, W. T. (1843–1928), 230
Thom, C. (1872–1956), 215, 219, 222, 291, 324, 327, 328, 332
Thomas, E. A., 98, 330
van Tieghem, P. E. L. (1839–1914), 89, 100, 106–7, 120, 130, 223, 277, 318, 319
Tillet, M. (1714–91), 145–7, 160, 311
Tode, H. J. (1733–97), 65, 226, 312
Tolaas, A. G. (1888–1972), 182, 325
Torrubia, J. (†), 25
Tournefort, J. P. de (1656–1708), 17, 49, 50, 65, 83, 85, 96, 141, 142, 144, 208, 244–5, 246, 250, 270, 276, 310
Tragus, H., see Bock, J.
Trattinick, L. (1764–1849), 65, 312, 313
Traube, M. (1826–94), 214, 317
Trog, J. G. (1781–1865), 226, 316
Tubaki, K., 308, 336
Tubeuf, K. von (1862–1941), 126, 283
Tucker, E. (?–1868), 160
Tulasne, C. (1816–84) & E. L.-R. (1815–85), 26–8, 61, 62, 96, 107, 117, 120, 123, 149, 187, 189, 191, 197, 232, 277, 316, 317, 318, 319
Tull, J. (1674–1741), 145, 311
Turian, G., 300
Turner, W. B., 113, 326
Turpin, P. J. F. (1775–1840), 154, 212, 315
Tyndall, J. (1820–93), 5, 18, 105, 198, 221, 320

Ubrizsy, G. (1919–73), 308, 336
Unger, F. J. A. N. (1800–70), 21, 154, 314
Ungerman, E., 292, 326
Unna, P. G. (1850–1929), 172

Vaillant, S. (1669–1722), 48, 50, 225, 250, 299, 310
Vanbreuseghem, R., 179, 333
Vandendries, R., 128, 326
van der Walt, J. P., 301
Vaucher, J. P. E. (1763–1841), 117
Venturi, A. (1805–64), 226, 316
Verma, J. P., 336
Viégas, A. P., 228
Villemet, P. R. (1735–1807), 24, 32, 312
Virchow, R. (1821–1902), 176
Virville, A. D. de, 308, 333
Vittadini, C. (1800–65), 71, 106, 226, 232, 301, 313, 315
Viviani, D. (1772–1840), 226, 314
Vuillemin, J. P. (1861–1932), 64, 242–3, 257, 264, 324, 325

Waddell, D. B., 128
Wager, H. (1862–1929), 122, 321
Waggoner, P. E., 191, 333
Wakefield, E. M. (1886–1972), 126, 283, 324
Wakker, J. H. (1859–1927), 283
Waksman, S. A., 233, 304, 325, 327
Waldeyer, W. (1836–1921), 121
Wallace, A. R. (1823–1913), 10
Wallroth, K. F. W. (1792–1857), 96, 313
Warcup, J. H., 232, 332
Ward, G. E., 223, 330
Ward, H. M. (1854–1906), 159, 284, 321
Wasson, R. G., 13, 193–5, 333, 335, 336
Wasson, V. P. (1901–58), 193, 333
Watson, J. D., 136
Watson, W. (1715–87), 25, 185, 311
Webster, J., 305
Wehmer, C. (1858–1935), 214, 222, 321
Wehnelt, B., 308, 330, 331
Weindling, R., 221, 330
Weinmann, J. A. (1782–1858), 227, 314
Weismann, A. (1834–1914), 301
Welsh, J. M. (†), 128
Went, F. A. F. C. (1863–1935), 289
Wernicke, R. (1852–1902), 178
Westerdijk, J. (1883–1961), 290–1
Weston Jr, W. H., 80, 287, 292, 331
Wheeler, H. 192, 335
Whetzel, H. H. (1877–1944), 287, 308, 326
White, C. J. (1833–1916), 172
Whitehouse, H. L. K., 133, 332
Whitfield, A. (1868–1947), 175, 302
Whittaker, R. H., 33, 34, 336
Wickerham, L. J., 133, 333
Wiegmann, A. F. (1771–1853), 154, 315
Wieland, H. (1877–1957), 185
Wieland, T., 185
Wildenow, C. L. (1765–1812), 65, 312, 313
Wildiers, E., 111, 323

Wilk, G. (†), 24
Williams, D. A., 204, 331
Williams, P. G., 112, 335
Willkomm, H. M. (1821–95), 93, 318
Wilson, I. M., 39
Wiltshire, S. P. (1891–1967), 270, 281
Windt, L. G. (†), 152, 313
Winge, Ø. (1886–1964), 133, 329
Winlock, H. E., 304
Winogradsky, S. N. (1856–1953), 283
Winslow, C.-E. A. (1877–1957), 291
Withering, W. (1741–99), 78, 300
Wöhler, F. (1800–82), 212
Wolffhügel, G. (1854–99), 105
Wood, R. K. S., 159, 192, 335

Wood, R. W. (1868–1955), 175
Woodham-Smith, C., 302
Woolman, H. M. (1853–1932), 145, 327
Woronin, M. S. (1833–1903), 62, 117, 283, 317

Young, T. W. K., 17, 336

Zahlbruckner, A. (1860–1938), 281
Zakon, S. J., 302
Zallinger, J. B. (1731–85), 143, 312
Zellner, J. (1870–1935), 112, 324
Zernike, F., 76, 299
Zopf, F. W. (1846–1909), 112, 182, 236, 321
Zosimos (3rd cent. A.D.), 210

Subject index

abstracting journals, 280
Académie des Sciences, Paris, 8, 31–2, 127, 276
Achlya, 29; *A. ambisexualis*, 139; *A. bisexualis*, 139; *A. prolifera*, 30
Achorion, 173; *A. (Trichophyton) schoenleinii*, 169, 171
Acrasiales, 80
acrasin, 80
Actinomycetes, 221
aecia, aeciospores, 123–4, 132, 153, 154
aflatoxin, 190
Africa, 177, 179, 230, 236
agar–agar, for solid media, 106
agaric (agarick, agaricum), 2, 37, 216, 243; 'male' and 'female', 214
agarics, 2, 4, 35, 43, 52, 69, 71, 80; classification of, 243, 247, 249, 269
Agaricum, 115, 252; *see also Fomes officinalis*
Agaricus, 23, 62, 246, 248, 250, 251, 252, 263; *A. aurantius*, 117; *A. bisporus*, 206; *A. campestris*, 23, 46, 120, 122, 206, 252; *A. metatus (Mycena metata)*, 120; *A. (Clitopilus) prunulus*, 71; *A. (Cortinarius) tabularis*, 78; *A. vaginatus (Amanita vaginata)*, 120; *A. (Crepidotus) variabilis*, 120
'airspora', 196, 198–202, 204
Albugo, 117, 236; *A. portulacae*, 62
Aleurodiscus amorphus, 122; *A. polygonius*, 129, 130
allergies, fungal, 196, 202–4
Allomyces, 80
Alternaria, 203, 204; *A. kikuchiana*, 192; *A. solani*, 192
alternaric acid, 192
alternation of generations, 124, 133
Amanita, 35, 247, 255; *A. caesarea*, 2, 46, 252; *A. muscaria* (fly agaric), 13, 185, 193, 205, 216, 299; *A. pantherina*, 185; *A. phalloides*, 185–6; *A. vaginata (Agaricus vaginatus)*, 120; *A. verna*, 185; *A. virosa*, 185
α-amanatine, 186
ambrosia beetles, 239
amphotericin B, antifungal antibiotic, 181
amylases, fungal, 224
Annellaria separata, 129

antamide, antitoxin of *Amanita phalloides*, 186
antheridia, 117, 120, 139
antibiotics: antifungal, 162, 181; from fungi, 8, 211, 217–22
Aphyllophorales, 75
apothecia, 96, 132
aquatic fungi, 236–7
arbuscules, of mycorrhiza, 104
Archimycetes, 266
Arctic regions, 231
Arcyria, 60
Argentina, 229
Armillaria mellea, 74
asci, 52, 66–8, 70; bitunicate and unitunicate, 266
Ascobolus magnificus, 132; *A. stercorarius*, 139
ascogonium, 120, 123
Ascomycetes, 6, 31, 74, 294; classification of, 265, 266; heterothallism in, 132–3; sexuality in, 117, 119, 120, 123; yeasts as, 213
ascospores, 66–8; 'free cell formation' of, 306
aspergillomas, 66
aspergillosis, 176
Aspergillus, 52, 60, 133, 136, 176, 182, 254; culture of, 85, 86, 87, 88; *A. flavus*, 190, 224; *A. fumigatus*, 66, 176, 203; *A. glaucus*, 21, 22, 28, 29; *A. itaconicus*, 223; *A. nidulans*, 136–8; *A. niger*, 6, 107–8, 110, 137, 161, 222, 223, 224, 302; *A. oryzae*, 214, 224; *A. terreus*, 223
Auricularia auricula, see Hirneola auricula-judae; *A. polytricha*, 206, 207; *A. pulverulenta*, 92
Auriscalpium vulgare, 251
Australia, 11, 75, 230–1, 308
Austria, 41, 225, 272, 308
autoclave, for sterilisation of media, 105

bacterial infections of fungi, 182
barberry, association of wheat rust with, 151–4; legislation against, 152
basidia, 70, 71; structure of, 71–4, 263
basidiocarps, *see* fruit-bodies
Basidiomycetes, 28, 74, 80, 294; classification of, 265, 266

basidiospores, 52, 68–70, 77, 128; discharge of, 197
Beauveria bassiana, muscardine disease of silkworms, 106, 166–7
Belgium, 207, 225, 226
Benomyl, systemic fungicide, 163
Biblical references to fungi, 90, 92, 140–1
Bibliography of systematic mycology, 281, 282
binomial nomenclature, 251–2, 254, 293
bios, 111
biotin, growth factor, 111, 138
biotypes, 158, 295
'blackfellow's bread', 206
Blastocladiella, 80
Blastomyces dermatitidis, blastomycosis, 31
boletes, 2, 43, 205
boletus, 2
Boletus, 23, 115, 196, 216; classification of, 246, 247, 250, 251; *B. edulis*, 2; *B. hybridus*, 92; *B. medulla-panis*, 92; *B. scaber*, 46
Bordeaux mixture, 161–2
Botrytis, 52, 60, 159, 254; culture of, 85, 86, 87, 88; *B. bassiana*, 166; *B. cinerea*, 159, 221; *B. (Phytophthora) infestans*, 156
Bovista, 38, 71, 246, 248; *B. nigrescens*, 43
brachymeiosis, 301
Brazil, 228, 308
Britain, 190, 204, 217, 292; fungi of, 56, 227, 234, 508; mycological societies in, 284–5; publications in, 277; timber decay in ships of, 90–3
Buller phenomenon, 130, 131, 132
bunt of wheat, *see Tilletia caries*
Burnettising, for preservation of timber, 93
Byssus, 200, 236, 251, 254

Canada, 203, 229, 287, 308
Candida, 180; *C. albicans*, 79, 170; *C. utilis*, 215
Cantharellus cibarius, 38, 40, 46, 48
carbon balance sheet of fungi, 113
carbon dioxide tension, and dimorphism, 31
carcithium, 65
Carlsberg Laboratory, Copenhagen, 8, 297
Carpobolus, 52
carpogonia, 120
cave, cellar, and mine fungi, 234–5
'cell theory', 7, 121
cell walls of fungi, 79; chemistry of, in classification, 269
cephalodium, 96
Cephalosporium, cephalosporins, 221
Ceratiomyxa fruticulosa, 60
Ceratostomella fimbriata, 192
Cetraria islandica, 216

Ceylon (Sri Lanka), 230
Chaetomium elatum, 202
Chaos Linnaeus, the genus, 23
cheese, fungi of, 215
chemotropism, 139
Chile, 229
China, 193, 206, 207, 214, 231; fungi in pharmacy of, 25, 216, 217, 304, 305
Chlorosplenium aeruginascens, 205
chromatin, 121
chromosomes, 121; maps of genes on, 135, 137–8
citric acid manufacture, 222
Cladosporium, 203; *C. fulvum* (*Fulvia fulva*), 204; *C. herbarum*, 204
clamp-connections, 116, 127
classical (Greek and Roman) references to fungi, 1, 2, 3, 12–13, 35, 64, 82; as food, 205; as medicine, 216; as plant pathogens, 140, 145; poisonous species, 38, 183–4; timber rots, 90, 92
classification of fungi, 4–5, 241–3, (before Linnaeus) 243–50, (Linnaeus) 250–2, (after Linnaeus) 252–5, (Persoon), 255–8, (Fries) 258–63, (after Fries) 263–6; biochemical and numerical, 7, 10, 269; speciation, 266–8
Clathroidastrum, 60
Clathroides, 60
Clathrus, 246, 251, 252; *C. cancellatus*, 43, 52, 53
Clavaria, 248, 250, 251; *C. fragilis*, 250; *C. (Cordyceps) sobolifera*, 25
Clavariaceae, 75
Claviceps paspali, 188; *C. purpurea*, 186–7, 189, 191, 216, 217 (*see also* ergot)
Clitopilus (Agaricus) prunulus, 71
Coccidioides immitis, coccidioidomycosis, 31, 177, 178
coccidioidin, 178
Code of Botanical Nomenclature, 293–4
Code de l'Écluse, 42, 267
Coelomycetes, 64
Collema, 96
colletotin, 192
Colletotrichum fuscum, 192
Collybia, 130; *C. fusipes*, 182; *C. velutipes*, 80, 94
Commonwealth Mycological Institute, Kew, 64, 230, 281, 282, 288, 292, 326
conida, 26, 64
Coniomycetes, 261, 262
Coniophora puteana, 92
Coniothecium, 30
conjugation in fungi, 117
conjugate nuclear division, 123, 127, 301
Conocybe, 195
copper fungicides, 160–2

Coprinus, 52, 68, 70, 205; *C. atramentarius*, 299; *C. cinereus*, 127; *C. comatus*, 70, 115, 129; *C. ephemerus*, 120; *C. ephemerus* f. *bisporus*, 129; *C. fimetarius*, 80, 127; *C. flocculosus*, 129; *C. lagopus*, 130; *C. micaceus*, 116; *C. narcoticus*, 129; *C. petasiformis*, 74; *C. radiatus*, 116, 120, 131; *C. rostrupianus*, 129; *C. stercorarius*, 120, 129; *C. sterquilinus*, 129
coprophilous fungi, 233
Coralloides, 246, 250
Cordyceps, 25, 250; *C. militaris*, 25; *C. sinensis*, 25, 216; *C. (Clavaria) sobolifera*, 25
Coriolus versicolor, 205
Corticium amorphum, 122; *C. caesium*, 70; *C. serum (Hypodontia sambuci)*, 127, 130; *C. varians*, 127
Cortinarius emodensis, 206; *C. (Agaricus) tabularis*, 78
Craterellus sinuosus, 48
creosote, for preservation of timber, 93
Crepidotus (Agaricus) variabilis, 120
Cryptococcus farciminosus, see Histoplasma farciminosum; C. neoformans, cryptococcosis, 180
Cryptogamia, 251
culture of fungi, commercial, 82-4, 206-9, 222-3
culture of fungi, experimental, 84-8; axenic, 11, 112; hanging-drop, 106; *see also* media *and* pure culture techniques
culture collections, 289-92
culture media *see* media for fungi
Cyathus, 14
cystidia, 52, 68, 69, 71, 116-17, 299
cytase, 159
cytochromes, 113; constitution of, in classification, 169
cytology, 121-5
cytoplasm, 121; continuity of, 76; inheritance through, 80
Cyttaria darwinii, 206
Czechoslovakia, 225, 226, 287, 289, 308

Daldinia concentrica, 216-17
dangeardium, 301
decay, fungi as agents of, 1, 89, 296
Dendrochium toxicum, 188
Denmark, 56, 213, 226, 233, 287, 308; *see also* Carlsberg Laboratory
dermatophytes, 10, 180; *see also Epidermophyton, Microsporum*, ringworm, *and Trichophyton*
diaporthin, 192
dikaryotic mycelium, 130
dimorphism, 30-1, 79
Diplodia zeae, 188
diploid chromosome number, 121, 123, 301

diploidisation, 130
Discomycetes, 28, 56, 66, 70; classification of, 257, 262; operculate and inoperculate, 266; spore discharge in, 197
distribution of fungi: ecological, 231-9; geographical, 225-31
dithiocarbamate fungicides, 162
DNA, 136
dolipores, 76

Egypt, 210
Elaphomyces, 102; *E. granulatus*, 216
Elvela, 251
Emericella nidulans, 137-8
Emmonsia, 178
Empusa muscae, 30
encarpium (fruit-body), 65
Endocarpon, 98
Endomyces decipiens, 74; *E. vernalis*, 224
Endomycetales, 239
Endothia parasitica, 192
entomogenous fungi, 163, 237-9
Entophyta (entophytes), 154, 168
enzymes of fungi, 112, 159, 211; commercial production of, 224
Ephebe, 96
Epidermophyton, 173
epizootic lymphangitis, of horses and mules, 179
ergot, ergotism, 65, 147, 186-8, 216, 217, 304
Ericaceae, mycorrhiza of, 64
Erinaceus, 247, 252
Erysiphaceae (powdery mildews), 112, 157
Erysiphe chicoracearum, 120
ethnomycology, 13, 140, 193-6, 210, 303
Eurotium herbariorum, 28, 29, 30
evolution, convergent, 75, 269
Exoascus, 123
exsiccati of fungi, 289

facial eczema of sheep and cattle, 190, 191
fairy rings, 2, 77-8, 300
favus, 168-70
fermentation industries, 8, 210-15
Finland, 226, 287
flagella, 76; flimmer or tinsel, and whiplash, 76, 77
fly agaric, *see Amanita muscaria*
Fomes fomentarius, 205, 215; *F. levigatus*, 75; *F. officinalis*, 2, 37, 114, 216, 243
food, fungi as, 6, 205-9; food yeast, 215
formae speciales, 157, 268, 294-5
fossil fungi, 281, 294
France, 8, 177, 201, 223, 284, 292; cultivation of fungi in, 82-3, 207, 209; fungi of, 56, 225, 226, 236, 308; publications in, 271, 272, 276, 277

freeze-drying of fungi, 292
fruit-bodies, 2, 65, 80, 120, 193, 205; in classification, 254, 261; monomitic, dimitic, and trimitic, 75
Fuligo, 60; *F. septica*, 14, 60
Fulvia fulva (*Cladosporium fulvum*), 204
fungi: in art, 4, 35, 297, 299; etymology of word, 1–2; in literature, 2–3, 297; pronunciation of word, 297; status of (animal, vegetable, or mineral), 21–5
fungi imperfecti, 64, 76, 81
fungicides, 92–3, 160–2, 302; systemic, 163
Fungoides, 246, 248, 250
Fungus, 114, 246, 249, 250, 255
fungus forays, 285
fungus stone (*Polyporus tuberaster*), 50, 299
fusaric acid, 192
Fusarium graminearum, 188; *F. sporotrichioides*, fusario-toxicosis, 190

Galera tenera, 197
Ganoderma applanatum, 205; *G. lucidum*, 193
Gasteromycetes, 71, 128, 294; classification of, 258, 261, 262, 265; conjugate division in, 123; heterothallism in, 128
Geastrum Pers. (*Geaster* Mich.), 49, 52, 246
genetics, 7, 133–5, 136–8; biochemical, 135–6; of plant resistance to fungi, 157, 158–9
Geoglossum, 250
geographical races, 129
geotropism, 198
Germany, 8, 100, 188, 204, 207, 287; fungi of, 226, 308; fungi in industry in, 223, 224, 225; publications in, 271, 272, 273, 277
Gibberella zeae, 188
gill hairs, seen as flowers by Micheli, 114–15
Gliocladium virens, 221; gliotoxin, 221
glycerol: commercial production of, by yeasts, 224
gonidia (algae in lichens), 26, 96–7
Greece, *see* classical references
Grete herball, The, 38
griseofulvin, 9–10, 175, 221
growth factors, 6, 111–12, 138, 301
Guatemala, 13, 194
Gymnosporangium, 52; *G. clavariiforme*, 124; *G. juniperi-virginianae*, 112

hair-baiting technique, for isolation of dermatophytes from soil, 10, 180
hallucinogenic fungi, 193–6
haploid chromosome number, 121, 123, 301
hay fever, 202

Helicobasidium mompa, 192
Helminthosporium victoriae, 192
Hemitrichia imperialis, 234
herbalists, 35, 38, 114, 145, 184, 293
herbaria, 288–9, 307
Heterobasidiomycetes, 266
heteroecism, 151, 154
heterothallism, 6, 121, 125–33; bipolar, 129, 131; morphological and physiological, 133; tetrapolar, 129
Hirneola auricula-judae, 38, 39, 216
Hirst trap for spores, 201–2
Histoplasma capsulatum, histoplasmosis, 31, 177, 179, 236; *H. duboisii*, 179; *H. farciminosum*, epizootic lymphangitis, 31, 179–80
histoplasmin, 179
Homobasidiomycetes, 266
Holland, *see* Netherlands
homothallism, 125; primary and secondary, 129
hormones, in control of sexuality, 80, 139
hosts of parasitic fungi: indexes of, 282; fungal relationships of, 156–9, 268
Hungary, 41, 225, 226, 308
hybridisation, experimental, 134, 135
Hydnum, 198, 247, 251; *H. auriscalpium*, 252; *H. coralloides*, 259
hymenium, 70, 71, 257
Hymenomycetes, 6, 49, 70, 71, 74, 76; in classification, 257, 261, 262, 263, 265, 269; conjugate division in, 123; heterothallism in, 126–31; sexuality in, 120–1; spore discharge in, 197
hyphae, 65; apical growth of, 79; binding, generative, and skeletal, 75
hyphal analysis, 75
hyperparasites (fungal parasites of fungi), 181–2
Hypholoma, 130
Hyphomycetes, 26, 64; aquatic, 236–7; classification of, 261, 262, 263, 264; predacious, 239; spore discharge in, 197
Hypodontia sambuci (*Corticium serum*), 127, 130
hypogeous fungi, 232–3

illustrations of fungi, 282, 299; chromolithograph, 56; copper-plate, 47–8, 272; cost of, 52; water and oil colour, 41–2, 43, 231; wood-block, 35–6, 46
immunology, 7; in classification, 7, 269
Index Kewensis, 281–2
Index of fungi, 268, 281–2
India, 8, 66, 140, 200, 230, 308
inositol, growth factor, 111
insects, fungi and, 163, 237–9

International Code of Botanical Nomenclature, 293–5
International Conference of Culture Collections, 292
International Mycological Congress, 288
International Society for Human and Animal Mycology, 181
Isaria, 30
isidium, 96
Italy, 13, 50, 84, 149, 163, 207; fungi of, 225–6, 308; *see also* classical references

Jamaica, 215
Japan, 8, 82, 179, 192, 292, 307; cultivation of fungi in, 206, 214; Emperor of, 234; fungi of, 231
Java, 231
journals, 156, 159, 276–9

koji fungus, 214
Kyanisation process, for preservation of timber, 93

Laboulbenia, 237–8; *L. rougetii*, 237
Laboulbeniomycetes, 133, 237–9, 265
Lactarius piperatus, 181
languages of mycological publications, 271–2
Lapland, 215
legislation against plant diseases, 152, 230
Lentinus edodes, 82, 206; *L. lepideus*, 235
Lenzites, 248
Lepiota procera, 43, 44
Lichen, 96, 251
Lichen-Agaricus, 66
Lichenoides, 66, 67
lichens, 5, 26, 96–100, 120; classification of, 32, 33, 143, 251, 265, 281; distribution of, 239–40; dyes from, 205; fungi parasitic on, 182; in medicine, 216; nomenclature of, 282, 294
lime sulphur, as fungicide, 160
ling chih (Chinese 'magic mushroom'), 193
literature, mycological, 270–82
lithophyte, 17
Lobaria pulmonaria, 216
Loculoascomycetes, 266
London County Council, 'ringworm school' of, 174
Lycogala, 60, 233
lycomarasmin, 192
Lycoperdon, 23, 244–5, 246, 250, 251; *L. bovista*, 2, 46
lyophilisation of fungi, 292
lysergic acid derivatives, in ergot, 188; *see also* ergot

macrocyst, 120
Madurella mycetomi, Madura foot, 66, 180
Marasmius oreades, 2, 78; *M. peronatus*, 55
marine fungi, 237
mating type factors, 78
media for fungi: for culture collections, 292; minimal, 136; solid, 106; sterile, 105–6; synthetic, 89, 108
medical mycology, *see* pathogenicity
medicine, fungi in, 215–17; *see also* antibiotics
Melampsora lini, 158–9
Melanospora destruens, 138
Mendelism, 7, 157
mercuric chloride, for preservation of timber, 93, 302
Merulius, 247
metabolism and metabolites of fungi, 112–13
Mexico, 13, 194
microfungi, elucidation of structure, 56–64
micron, 61
microscopy, 5, 56, 60–1, 297; electron, 76–7; phase contrast, 76
Microsphaera alni, 204
Microsporon (Trichophyton) mentagrophytes, 170
Microsporum, 173; *M. audouinii*, 170, 171, 174–5
mitosis, 121; conjugate, 123
monokaryotic mycelium, 130
Morchella, 1, 38, 39, 247; *M. esculenta* (morel), 43
morphogenesis of fungi, 77–81, 300
morphology of fungi: external, of larger, 35–56; micro-, of larger, 64–71; of microfungi, 56–64
moulds, 15
Mucedo, 60
Mucilago, 60, 233
Mucor, 23, 24, 29, 31, 58, 59, 60, 180; culture of, 85, 86, 87, 88; heterothallism in, 119; Linnaeus and, 251, 252; *M. fusiger*, 117; *M. mucedo*, 30, 126, 139; *M. racemosus*, 214; *M. rouxii*, 214
Mucorales (Mucorinae); homothallic and heterothallic, 125–6; phototropism in, 198
muscardine disease of silkworms, 165–8, 237
muscaridine, muscarine, 185
'musco-fungus' (lichens), 96
mushroom flies, 14
mushroom madness, 195
mushroom stones, 194–5
mushrooms, 2, 15; cultivation of, 82–4, 206–9, 223; diseases of, 182

mutants, of *Neurospora*, 80, 136
mycangia of ambrosia beetles, 239
mycelium, 17, 60, 64–5; diploid and haploid, 230; mycelial cords, 52, 65, 80
Mycena metata (*Agaricus metatus*), 120
mycetomas, 65–6
Mycetophilidae (mushroom flies), 14
Mycetozoa, 234; *see also* Myxomycetes
mycobionts, 98, 300
Mycoderma, 211
mycology, 1, 2, 4; factors in development of, 4–11; organisation for, 270–95
Mycophytes (fungi), Mycomycophytes (fungi proper), and Mycophycophytes (lichens), 266
mycorrhiza, 89, 100–4, 300
mycotoxicoses, 11, 186–91
myxamoebae, 80
Myxomycetes, 60, 233–4, 258, 265, 294

National Agricultural Advisory Service, 141
National Collection of Type Cultures, Britain, 292
National Fungus Collections, U.S.A., 288
Nepal, 215
Netherlands (Holland), 204, 208, 226–7, 287, 289
Neurospora, 132–5, 136; mutants of, 80, 136; *N. crassa*, 135; *N. tetrasperma*, 129, 132
New Guinea, 196
New Zealand, 75, 190, 231
nitrogen nutrition of fungi, 110
nomenclature, 293–5; binomial, 251–2, 254, 293
Nothofagus, 206
Nova plantarum genera, 4, 18, 50–2, 197, 225, 233; asci and ascospores, 66–7; basidial structure, 68–9; classification, 248–9; cultural experiments, 84–8
nucleoplasm, 121
nucleic acids, 136
nucleus of cell, 121, 122
nutrition of fungi, 6, 107–12
Nyctalis asterophora, 181; *N. parasitica*, 182
nystatin, antifungal antibiotic, 181

Octospora, 66
oidia, 130–1
Oidium (*Candida*) *albicans*, 170; *O. tuckeri*, 160
oleic acid, growth factor for *Pityrosporum*, 111
Olpidium brassicae, 76
oogonia, 117, 139
Oomycetes, 266
Orcheomyces, 104

orchids, mycorrhiza of, 100, 102–4, 300
organic acids, fungi in industrial production of, 222–3
Ovularia, 154

paddi straw mushroom, 206, 207
Panaeolus separatus, 129
Pannonia, 41, 225
pantothenic acid, growth factor, 111
Paracoccidioides brasiliensis, paracoccidioidomycosis, 31
paracyst, 120
paraphyses, 66, 70
parasexual cycle, parasexuality, 11, 136–7
parasites: facultative and obligate, 89; of fungi, 181–2; physiology of, 159–60; host relationships of, 156–9, 268
Paris inch, Paris line, 61
'paspalum staggers', 188
Pasteur flask, 105
pathogenicity of fungi: to animals and man, 5, 6, 21, 163–8; to fungi, 181–2; to plants, 6–7, 8, 11, 140–63, 230
pathotoxins, 192
patulin, 221
pectinase, 159
pelotons, of mycorrhiza, 104
Peltigera canina, 216
penicillin, 11, 217–20
Penicillium, 29, 60, 107, 182, 222, 224; *P. camemberti*, 215; *P. caseicola*, 215; *P. chrysogenum*, 137, 223; *P. glaucum*, 202, 220; *P. griseofulvum*, 221 (*see also* griseofulvin); *P. notatum*, 219; *P. patulum*, 221; *P. purpurogenum* var. *rubri-sclerotium*, 223; *P. roqueforti*, 215; *P. rubrum*, 219
Periconia circinata, 192
perithecium, 96
Peronospora, 117
Pertusaria, 66, 67
Petri dish, 106
Peziza, 123, 248, 251, 252, 306; spore discharge in, 304
pH: and acid production by *Aspergillus niger*, 222, 223; of medium for *Aspergillus*, 108
phalline, 185
phalloidin, 186
Phallus, 38, 41, 246, 251; *P. hadrianus*, 270, 299; *P. impudicus*, 216
Phellinus igniarius, 114
phototropism, 198
Phragmidium, 151; *P. mucronatum*, 58, 59, 144, 151; *P. violaceum*, 124
phycobionts, 98, 300
Phycomyces, 79; *P. blakesleeanus*, 111, 138; *P. nitens*, 126, 198

Phycomycetes, 265–6, 306; aquatic, 236
Phyllactinia guttata, 27
physiologic races, 158, 286, 295
Phytophthora (Botrytis) infestans, 156
phytotoxins, 191–2
α-picolinic acid, 192
Pietra fungaia, see Polyporus tuberaster
Pilobolus, 24, 150, 196, 198, 199
Piptoporus betulinus, 205, 267
Piricularia oryzae, 192
piriculin, 192
Pithomyces chartarum, 190
Pityrosporum ovale, 111
Plaincourault fresco, 299
plant pathology, *see* pathogenicity
plants: fungal diseases of, 6–7, 8, 11, 140–5, 230; fungal toxins affecting, 191–2
plasmodia, 60, 80
Plasmopara viticola, 161
pleomorphism, 26–30, 212, 294
Pleurotus ostreatus, 43, 44; *P. tuber-regnum*, 207
podetium, 96
poisonous fungi, 6, 38, 183–6; *see also* mycotoxicoses *and* phytotoxins
Poland, 225
polygalacturonase, polymethylgalacturonase, 159
polypores, 4, 35
Polyporus, 52; *P. brumalis*, 80; *P. hybridus*, 92; *P. mylittae*, 206; *P. nidulans*, 205; *P. rangiferinus*, 234; *P. squamosus*, 205, 234, 267, 268; *P. tuberaster*, 32, 50, 65, 207, 299
Polystictus xanthopus, 75
Pompeii, fresco of fungus at, 4, 35
Poria, 92; *P. cocos*, 206
potassium iodide, as systemic fungicide, 163, 177, 181
potato blight, 6, 155–6
predacious fungi, 239, 240
protein of cell wall, in yeast–mycelium dimorphism, 79–80
Protista, 33
Protomyces, 154, 265
protoplasm, 121
Pseudomonas tolaasi, 182
psilocin, psilocybin, 195
Psilocybe, 195; *P. cubensis*, 195
Puccinia, 52; *P. graminis*, 11, 132, 145, 149–54, 157, 203; *P. graminis* f. sp. *tritici*, 112, 157–8; *P. helianthi*, 132; *P. ramosa*, 60; *P. striiformis*, 157
pure culture techniques, 5, 28, 83, 89, 105–6, 213, 232
pycnia, pycnospores, 132
Pyrenomycetes, 28, 237, 262, 266
pyridoxine, growth factor, 111

Pyronema confluens, 120, 123, 138
Pythiaceae, 236
Pythium, 221

quinone fungicides, 162

Racodium cellare, 235
Ramaria botrytis, 43, 44
Ramularia, 154
Raulin's medium, 108
reproduction of fungi: early views on, 12–15; (1700–1850) 15–21
Reticularia, 60
Rhinocladiella cellaris, 235
Rhinosporidium seeberi, rhinosporidiosis, 180
Rhizoctonia, 103
rhizomorphs, 80
Rhizopus, 214; *R. arhizus*, 224; *R. oryzae*, 223; *R. stolonifer (nigricans)*, 19, 20–1, 30, 60, 117, 125–6, 198
Rhytisma salicinum, 197
ringworm, 6, 168–74; therapy for, 9–10, 174–5, 221
RNA, 136
Robigalia, Roman festival of the rust god, 140, 302
Rome, *see* classical references
Roumania, 205
Royal Society, 24, 25, 56, 58, 60, 234, 270, 276
Russia, *see* U.S.S.R.
Russula, 4, 196; *R. nigricans*, 182
rust fungi, *see* Uredinales

Saccharomyces, 30, 133, 136, 182, 212, 265; *S. cerevisiae*, 58, 133, 139, 212–13, 224, (mutants) 80
Saccharomycodes ludwigii, 132
sand dune fungi, 233
saprobes, 89
Saprolegnia, 29, 79, 138, 139, 175–6; *S. ferax*, 76, 77
Saprolegniaceae, 236
Saprolegniales, 117
saprophytes, 89
Schizophyllum commune, 126, 129, 292
Schizosaccharomyces octosporus, 132
sclerotia, 65, 186–7, 206, 207
Sclerotinia gladioli, 132; *S. sclerotiorum*, 159
Sclerotium clavus, 187
Scotland, 227, 308
Scutellinia scutellata, 68
Secale luxurians (ergoted rye), 186–7
self-inhibitors, in spores, 79
septa of hyphae, pores in, 76
Septobasidium, 239
serology, in classification, 7, 269

Serpula lacrimans (dry-rot fungus), 92, 94

sex chromosome, 135

sexuality in fungi, 5–6, 7, 114–21; cytology of, 121–5; heterothallism in, 125–33; hormonal control of, 80, 139; nutrition and, 138

shii-take (*Lentinus edoces*), cultivation of, 82, 205, 207

ships, fungal decay in, *see* timber decay

Siphonomycetes, 266

smut fungi, *see* Ustilaginales

societies: mycological, 181, 284–7; phyto-pathological, 156; *see also* Académie des Sciences *and* Royal Society

soil fungi, 232–3

Sordaria fimicola, 138

soredium, 96

Spain, 207, 217

speciation, 266–8

spermatia, 120, 123, 124, 132

spermogonia, 123

Sphacelia segetum, 187

Sphaeria, 306; *S. purpurea*, 187

Sphaerobolus, 52, 196

Sphaerotheca castagnei, 123

spontaneous generation, theory of, 18, 20–1, 212

sporangia, 62, 126

sporangioles, of mycorrhiza, 104

sporangiophores, 198

spores, 5, 13–14, 62; in air, 196, 198–202, 204; classification of, 64, 242, 254, 261, 263, 299; culture of fungi from, 52, 106; germination of, 18, 20–1, 79; of heterothallic fungi, 126, 129; liberation of, 196–8

sporidesmin, 190

sporidia, 26, 123, 153, 154, 299

Sporidiiferi, Sporiferi, 299

sporocarps, 80

Sporodinia grandis, see Syzygites megalo-carpus

Sporotrichum, 29, 177

Sporothrix schenckii, sporotrichosis, 31, 163, 177, 181, 236

sporulation, 80, 138–9

Sri Lanka (Ceylon), 230

Stachybotrys alternans, 188

Stemonitis, 60

Stereum hirsutum, 94, 197

sterigmata, 68

Sterigmatocysis nigra, 223

sterilisation of media, 105–6; discontinu-ous, 105

steroids, fungal transformations of, 224

stone fungus, *see Polyporus tuberaster*

Streptomyces, 181; *S. griseus*, 221

streptomycin, 221

Stropharia, 195; *S. semiglobata (stercoraria)*, 122

Suillus, 2, 115

sulphur, as fungicide, 160

Sweden, 226

Switzerland, 45, 217, 226, 287; fungi of, 273, 308

Sylloge fungorum, of Saccardo, 64, 233, 268, 280–1

symbiosis, 89, 98, 300

Synopsis methodica fungorum, of Persoon, 257–8, 268, 273, 294

Systema mycologicum, of Fries, 26, 65, 154, 261–2, 273, 294

systematics, defined, 241

Syzygites megalocarpus (Sporodinia grandis), 6, 117, 118, 126

takadiastase, 224

Taphrina, 31, 265; *T. deformans*, 160

taxon, 241, 295

taxonomy, 280; position of fungi in, 31–4; relation of, to classification, 241

Teliomycetes, 266

teliospores, 123, 124, 150, 151, 153, 154

temperature, and dimorphism, 31

teonanácatl, 'divine mushroom' of Mex-ico, 194

Terfezia, 2, 35

Termitomyces, termites, 239

textbooks of mycology, 181, 272–5

Thallophyta, 33

thallus, 96

theca, 66

Thecamycetes, 266

Thelephora sericea, 197

Thesaurus litteraturae mycologicae, 9, 280, 282

thiamin, growth factor, 111, 138

van Tieghem cell, 106

Tilletia caries (bunt of wheat), 145–9, 160–1, 189

timber: brown and white rots of, 94; decay, 6, 8, 90–4; dry rot, *see Serpula lacrimans*

tinder, fungal fruit-bodies as, 205, 214

toadstools, 2

Tokelau itch, 170–1

Torula, 30, 172, 212

Torulopsis, 30; *T. utilis*, 215

trace elements, in nutrition of fungi, 6, 110, 138

Tranzschelia anemones, 250

trees, mycorrhiza of, 100–4

Tremella, 251

Tricholoma georgii, 46

Trichomycetes, 239

Trichophyton, 173, 175; *T. concentricum*, 171; *T. (Microsporon) mentagrophytes*, 170; *T. (Achorion) schoenleinii*, 169, 171; *T. tonsurans*, 170, 172
truffles, 2, 12–13, 14, 35, 232; in classification, 4, 242, 243; cultivation of, 100, 207; *see also Tuber, etc.*
Tuber, 35, 66, 67, 246, 248; *T. melanospermum*, 207
Tubera, 246, 250
Tulostoma, 49, 246
Tyndallisation (discontinuous sterilisation), 105
Typhula erythropus, 130

ultraviolet and near-ultraviolet radiation, and sporulation, 138–9
Uncinula necator, 160
universities, mycology at, 7, 278, 282–4
Uredinales (rusts), 28, 112, 265, 294; axenic culture of, 11, 112; heterothallism in, 132
urediniospores, 150, 151, 153
Uredo antherarum (Ustilago violacea), 154
Uruguay, 229
U.S.A., 8, 173, 206, 217, 229, 308; fungal diseases in, 175, 177, 179, 181, 188, 203–4; fungi in industry in, 223, 224; herbaria and culture collections in, 286, 291–2; publications in, 271, 275, 277
Usnea, 95
U.S.S.R., 11, 188, 190, 193, 227
Ustilaginales (smuts), 23, 31, 265, 294; heterothallism in, 131; sexuality in, 123
Ustilago, 23, 24; *U. tritici*, 147, 149; *U. violacea*, 131, 154

'vegetable fly', 25
vesicles, of mycorrhiza, 104

veterinary mycology, *see* pathogenicity
victorin, 192
virus infections of fungi, 182
vitamins: produced by yeast, 215; required by fungi, 111, 136
vivotoxins, 191
Volvariella volvacea, 206, 207

water drop (bubble), associated with basidiospore discharge, 197, 198
Whitfield's ringworm ointment, 175
wheat: black (stem) rust of, *see Puccinia graminis*; bunt of, *see Tilletia caries*; loose smut of, *see Ustilago tritici*
Witwatersrand, sporotrichosis in gold mines of, 177, 236

Xanthoria, 98
X-ray epilation, for head ringworm, 174
Xylaria, 248; *X. hypoxylon*, 66; *X. polymorpha*, 16, 17, 32
Xyloma salicinum, 197
Xylostroma giganteum, 92

yeast, 29; cells first observed, 58, 60, 211–12; heterothallism in, 132; nutritional requirements of, 111; zymase of, 7, 211, 213–14
yeast–mycelial dimorphism, 30–1, 79

zinc chloride, for preservation of timber, 93
Zoopagales, 239
zoospores, 62, 77, 117, 236
Zygomycetes, 266
zygophores, 126, 139
zygospores, 117, 118, 125
zygotropism, 139
zymase, 7, 211, 213–14